Praise for *Paradise*

"In a remarkable feat of empathetic reporting, *Paradise* takes us inside the lives of people facing a rapidly unfolding horror. Lizzie Johnson has a masterful command of this rich and tragic story, and to read it is to experience a visceral sense of the fear, uncertainty, and ultimately resilience of those who fought their way through it. If you think you know the story of the Camp Fire, one of the deadliest in U.S. history, think again. Equal parts thriller, investigation, and deeply rendered portrait of an American town, *Paradise* will leave you breathless."

—Evan Ratliff, author of *The Mastermind*

"This book kept me up reading late into the night and then buried itself so deep in my head that I couldn't fall asleep. Lizzie Johnson has written the rare page-turner built on empathy and rigorous reporting. *Paradise* is both a definitive account of California's historic wildfire and a crucial warning of the disasters to come."

—Eli Saslow, author of *Rising Out of Hatred*

"In *Paradise,* Lizzie Johnson masterfully weaves together stories of improbable survival and immeasurable loss to create a compelling portrait of a community brought to its knees by a ferocious fire stoked by the forces of greed, mismanagement, and the worsening effects of a warming climate. This is a book about the strength of the human spirit, and also an urgent and necessary call for action."

—Fernanda Santos, author of *The Fire Line*

"This account of the deadliest wildfire in California history is a triumph of reportage and storytelling. Out of the ash, Lizzie Johnson has written a memorial to its victims, a tribute to its heroes and survivors, and a reckoning of its kindling and match. Among the culpable, we find ourselves."

—Mark Arax, author of *The Dreamt Land*

"*Paradise* is an extraordinary book. The enormity of nature, the humanity and dignity of individual people confronting it—Lizzie Johnson has woven it all together brilliantly."

—Jon Mooallem, author *This Is Chance!*

"[Johnson] balances the horror with compassion. . . . Crucial, comprehensive, and moving."

—*Publishers Weekly*

"Johnson had a firsthand view of [Paradise's] destruction (eighty-five people died) and has since put together this important minute-by-minute document of the consequences of climate change, based on frontline reporting, public records, and extensive interviews with survivors."

—*Literary Hub*

"A masterly account . . . Johnson does for California's deadliest wildfire what Sheri Fink did for Hurricane Katrina in *Five Days at Memorial*. With stellar reporting, she tells the moment-by-moment story of an unfolding disaster, showing its human dramas as well as the broader corporate and governmental missteps that fueled it. . . . The book is unmatched for the depth, breadth, and quality of its reporting on a major twenty-first-century wildfire, and it's likely to become the definitive account of the catastrophe in Paradise."

—*Kirkus Reviews* (starred review)

PARADISE

PARADISE

ONE TOWN'S STRUGGLE TO SURVIVE
AN AMERICAN WILDFIRE

LIZZIE JOHNSON

CROWN
NEW YORK

Library of Congress Cataloging-in-Publication Data
Names: Johnson, Lizzie, author.
Title: Paradise / Lizzie Johnson.
Description: First edition. | New York: Crown, [2021] |
Includes bibliographical references and index.
Identifiers: LCCN 2021012297 (print) | LCCN 2021012298 (ebook) |
ISBN 9780593136386 (hardcover) | ISBN 9780593136393 (ebook)
Subjects: LCSH: Pacific Gas and Electric Company—History—21st century. |
Camp Fire, Paradise, Calif., 2018. | Wildfires—California—Paradise. |
Paradise (Calif.)—History—21st century.
Classification: LCC SD421.32.C2 J64 2021 (print) |
LCC SD421.32.C2 (ebook) | DDC 363.37/909794—dc23
LC record available at lccn.loc.gov/2021012297
LC ebook record available at lccn.loc.gov/2021012298

PRINTED IN THE UNITED STATES OF AMERICA ON ACID-FREE PAPER

crownpublishing.com

1st Printing

First Edition

Maps by Jeffrey L. Ward

In memory of Phil John,
who believed in Paradise, even,
and especially,
when it was ruined.

Where we see the appearance of a chain of events, he sees one single catastrophe, which unceasingly piles rubble on top of rubble and hurls it before his feet. He would like to pause for a moment so fair, to awaken the dead and to piece together what has been smashed. But a storm is blowing in from Paradise, it has caught itself up in his wings and is so strong that the angel can no longer close them. The storm drives him irresistibly into the future, to which his back is turned, while the pile of debris before him grows sky-high.

—*Walter Benjamin, "On the Concept of History"*

CONTENTS

KONKOW LEGEND

first learned of the Konkow legend on a chilly spring day in March 2019, as I stood high on a plateau above the town of Concow. The director of the Butte County Fire Safe Council had invited me and two dozen others on a tour of the burn zone of the Camp Fire, the deadliest wildfire in California history, which had occurred four months before. By this point, she had given several of these tours to politicians and civilians alike; she hoped that people would leave better educated about the state's fire risk and what could be done about it.

From our vantage point, the reservoir was a dark bruise ringed by charred evergreen forest. As residents commiserated over their losses, remarking on how hot the Camp Fire had burned and how fast it had moved, a couple from the Konkow tribe, who happened to be on the tour, offered to share a story.

Their ancestors, they said, had once witnessed a wildfire similar to the one Butte County had just experienced. Two young boys had thrown pitch pine sticks onto a campfire, accidentally igniting a conflagration. The outcome was horrible. Most of the tribe died, and the few who survived had been forced to move north. This tale, they said, had been passed down through the generations and later translated into English. Hearing the story that afternoon, I was struck by how the Konkow legend offered a remarkable glimpse into the past—something never captured in modern statistics and rarely by history textbooks—of what tribal ancestors had witnessed long before white settlers arrived and displaced them, before housing developments were carved into sacred land.

Later, I asked the couple if they might be willing to send me a recording of the legend. They kindly agreed. Elements of their tale have been interspersed throughout this book.

To me, the legend illustrates the cyclical nature of wildfires and suggests how we can better adapt as the climate changes. Managing fire, as Native Americans have done for thousands of years, rather than fighting it at every turn, can prevent tragedy. The Konkows once cultivated low-intensity burns, scorching the forest floor as a vegetation management practice. The technique was widely used— until European settlers, who viewed fire as unnatural and evil, arrived and quashed it. Conflagrations could still prove deadly, as the Konkow tale shows, but the land always healed and regenerated, healthier for the burn. Even today, the tribe maintains a deep respect for fire.

We can all learn something from the Konkows' knowledge and stewardship of the environment, and their kinship with nature. Their legend serves as a call to protect these spaces for future generations— in part to honor those who lived on this land before us.

PARADISE

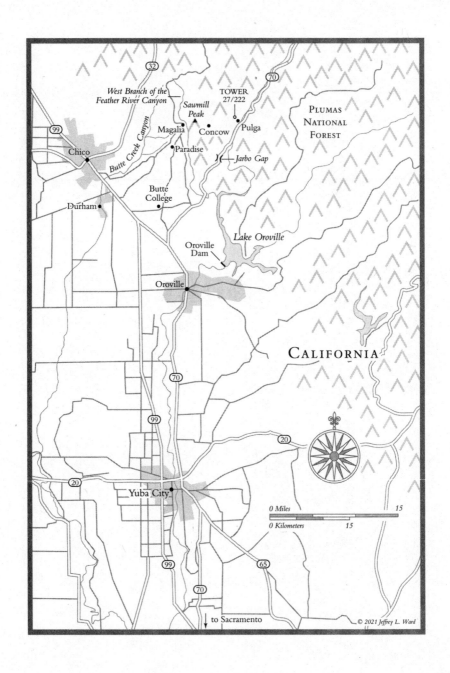

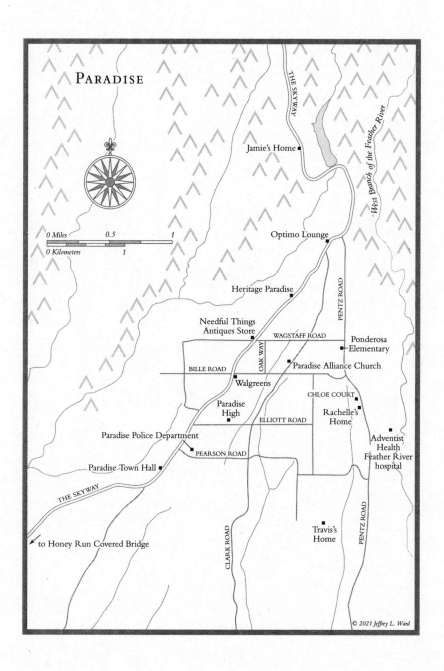

PART I

KINDLING

KONKOW LEGEND

In the beginning, Wahnonopem, the Great Spirit, made all things. Before he came, everything on the earth and in the skies was hidden in darkness and in gloom, but where he appeared, he was the light. From his essence, out of his breath, he made the sun, the moon, and the countless stars and pinned them in the blue vault of the heavens. And his Spirit came down upon the earth, and there was day; he departed, and the darkness of night closed again upon the place where he had stood. He returned, and the light shone upon the Konkows and all the other living creatures upon the earth, in the waters, and in the skies; the wildflowers bloomed in the valleys and on the mountainsides; the song of the birds was heard among the leaves of the madrone and on the boughs of the pines; and the hours of the day and of the night were permanently established.

This is the story of the Konkow tribe of the Meadow Valley Lands, as brought forward and told in the stillness of the nights, around the campfires, by the old men, the scholars, and the priests.

CHAPTER 1

DAWN AT JARBO GAP

For weeks, Captain Matt McKenzie had longed for rain. It would signal the end of wildfire season, which should have concluded by now, but November had brought only a parched wind. The jet stream was sluggish, failing to push rainclouds up and over the Sierra Nevada into Northern California. Since May 1, 2018, Butte County—150 miles northeast of San Francisco and 80 miles north of Sacramento—had received only 0.88 inches of precipitation. The low rainfall broke local records. It was now November 8, and with three weeks to go until Thanksgiving, the sky remained a stubborn, unbroken blue. Plants withered and died, their precious moisture sucked into the atmosphere. Oak and madrone shook off their brittle leaves.

Ponderosa pine needles fell like the raindrops that refused to come, pinging against the fire station's tin roof and waking McKenzie from a deep sleep around 5:30 A.M. A pinecone landed with a thud. He curled up on the twin bed in his station bedroom, feet poking from under the thin comforter, and oriented himself in the darkness. He didn't feel ready for the day to begin. Blackness edged the only window. Outside, gale force winds wailed through the hallway. He pulled aside the window blinds for confirmation: no rain. The sliver of a waxing moon and winking stars pricked the sky's endless dark. In an hour, the sun would rise.

After more than two decades of firefighting, McKenzie, forty-two, possessed a certain clairvoyance. He had dedicated half his life to the California Department of Forestry and Fire Protection, helping to battle conflagrations that sprouted in the vastness of California during its fire season. In such a huge state, urban departments could cover only so much ground; there had to be a larger force to stop fires before they burned too far or too fast in the wilderness bordering cities and towns. Known as Cal Fire, the state agency was one of the largest dedicated wildland firefighting forces in the world.

McKenzie had learned to read the agency's weather reports like tea leaves. When conditions were right, all it took was a spark to ignite an inferno. McKenzie and his crew were trained to anticipate and react aggressively, jumping into action while the fires were still small and easily contained. Nothing was left to chance. They did this the old-fashioned way, by digging dirt firebreaks and spraying water from their engines. The method was effective: Only 2 to 3 percent of the wildfires they tackled ever escaped their control. But fires broke out all over California every year, and members of his outpost, Station 36, were called upon to help quench the most destructive ones as part of the state's mutual aid agreement, by which jurisdictions pledged to help each other out during emergencies. The crew spent the year crisscrossing the state, from barren Siskiyou to coastal San Diego.

Innocuous mishaps—a golf club or lawn mower striking a rock, a malfunctioning electric livestock fence, a trailer dragging against the asphalt, a catalytic converter spewing hot carbon—could beget a blaze. More often, though, fires were started by downed electrical lines. They would snap and spark in high winds, showering embers and grief across entire communities. Lately, the Pacific Gas and Electric Company, the largest power provider in California, was experimenting with shutting off power when high fire risk was forecast.

In a remotely operated weather site near McKenzie's fire station, an anemometer was whirring, generating the next forecast. Surrounded by chain-link fencing, the instrument thrummed atop a slender tripod 20 feet tall, its three cupped hands circling faster and faster. It registered winds blowing at 32 mph, with gusts up to 52 mph. That November morning, wind wasn't the only problem. Relative humidity plummeted to 23 percent and continued dropping. It was forecast to hit 5 percent by noon—drier than the Sahara Desert.

MCKENZIE RAN A HAND through his silvered hair and swung his feet to the tile floor, trudging to the bathroom with a towel slung over one arm. Standing six foot one, he was tall and slim, with deep dimples and piercing blue-gray eyes. He had led Station 36 for four years and treasured its cowboy grit and strong camaraderie with the community, mostly retirees, loggers, off-the-gridders, and marijuana growers. McKenzie was now in the middle of a seventy-two-hour shift overseeing the station, one of the oldest and most fire-prone posts in Butte County. Covering 1,636 square miles in far Northern California, the county was nearer to the Oregon border than to Los Angeles, its small valley cities and hideaway mountain towns scattered along the leeward side of the Sierra Nevada. In the past twenty-five years, flames had ravaged the foothills 103 times. The worst of them—the Poe Fire, in 2001, and the Butte Lightning Complex and Humboldt fires, both in 2008—had devastated the county's rural communities, including those near McKenzie's station.

His outpost was perched on a knob of land off State Highway 70, the last stop before motorists entered U.S. Forest Service jurisdiction. At an elevation of 2,200 feet, the station overlooked the Feather River Canyon and abutted the western edge of Plumas National Forest. McKenzie joked that it was built on "the road to nowhere." A long driveway unspooled to a compound of squat tan buildings: a large garage for the fire engines, an office, and a twelve-bed barracks. Two captains—he was one of them—had rotating shifts and shared a private bedroom behind the kitchen. Everyone else slept in the dorm. At least six firefighters stayed on duty at all times, tasked with putting out house fires and responding to vehicle accidents or medical emergencies.

The men at Station 36 spent a lot of time together, much of it trying to impress McKenzie, whom they admired. They competed to hike the fastest or do the most push-ups, growing close through the friendly rivalry. On slow afternoons, they would pull weeds from the station's vegetable garden, tend its fruit trees, and play elaborate games of darts in the garage, storing their personalized game pieces in metal lockers labeled with tape. They would jam the living room armchairs against the wall and crank up the heater for floor exercises, sweating so profusely that the photos on the wall curled in their frames.

When they had a break, sometimes McKenzie and his crew would head to Scooters Café, a family-owned restaurant that—other than a hardware store, a stone lodge turned into a diner, and a market with two gas pumps—was the nearest business around. Motorcyclists choked its parking lot, waiting in a long line for Fatboy burgers—named after the Harley-Davidson motorcycle—or $2.00 beef tacos on Tuesdays. The owner of the red-walled café was a mild-mannered man who never called 911 or allowed his patrons to drive drunk. He served beer and "Scooteritas," but no wine, and he often dropped glazed doughnuts off for the firefighters. Sometimes he scheduled karaoke nights, hosted car shows, or booked concerts. McKenzie and his crew would sit on the station lawn and listen, the music echoing uphill in the summer air.

Station 36 was a quiet place, its stillness punctuated by the occasional grumble of highway traffic and the whoosh of wind. As one week in November turned to another, still with no rain, the crew hiked to a long-ago-burned home, its gardens lush with unkempt fig trees and wild blackberry thickets, and foraged for fruit to bake a cobbler. They responded to accidents at Sandy Beach, where swimmers like to launch themselves into the Feather River with a rope swing and sometimes got stuck in the currents. They scanned the canyon for smoke.

THE FEATHER RIVER CANYON had a long history of wind-driven wildfires. Station 36 existed in part because of its proximity to this yawning crack in the earth. The sixty-mile chasm snaked across Butte County, from Lassen National Forest to Lake Oroville; it trapped seasonal winds as they spun clockwise over the Sierra Nevada and pushed them toward the low-pressure coast. The winds blew day and night, billowing up the canyon walls as sunshine warmed the air and down them as temperatures cooled, clocking speeds upwards of 100 mph and blasting the towns of Magalia, Concow, and Paradise. They pelted homes and windshields with pine needles like obnoxious confetti. When there was a fire, the Feather River Canyon also funneled smoke south, directly to the hallway outside McKenzie's bedroom. The scent was always a swirling, ghostly harbinger of terrible things to come.

McKenzie, who had a laid-back nature and a dry sense of humor, considered himself a pessimist. This mentality served him well as a firefighter, but it had not served him so well in his personal life; always anticipating the worst was not good for relationships. He was divorced, and his daughter, Courtney, now eighteen, lived with her grandparents or his ex-wife during fire season. His senses, so finely attuned to potentially disastrous conditions on the landscape, failed him in more ordinary environments. He struggled to find a balance between his family and his job. He was rarely on time to Courtney's concerts and assemblies at Oroville Christian School, parking his red engine amid the rows of bumper-stickered sedans and minivans. With his handheld radio and bright yellow uniform, so unlike the other parents, he embarrassed her. Children often don't understand that kind of love. Now that she was older, he wished he'd shown up more often.

McKenzie had grown up near the county seat of Oroville, a city of nineteen thousand that was bisected by the Feather River about twenty miles south of Station 36. Teenagers raised in Butte County tend to fall into law enforcement, firefighting, construction, or methamphetamines; the latter account for 80 percent of the county's crime. (The district attorney had turned a glass laboratory vessel seized during a drug raid into an aquarium, which he proudly displayed in his office.) McKenzie's mother was a chiropractic assistant; his father managed a chain grocery store. They were strict, and

McKenzie and his older sister, Jennifer, were taught to work hard. His first job was bagging groceries to pay for the insurance on his Ford Ranger. He was eventually promoted to night shift janitor.

When McKenzie was seventeen, his father died of cancer. Afterward, McKenzie felt directionless. He debated signing up for the police academy and even completed courses in arrest methods and firearms, only to decide that it wasn't the right fit. At twenty, he enrolled in the firefighting academy at Butte College. Tuition was $2,500 plus the cost of uniforms. Coursework bored him, and his ornery streak and love of partying nearly got him thrown out. But he was excellent in the field, and thanks to an attentive mentor, his grades rose enough that he managed to graduate midpack. A few months later, Cal Fire hired him full time. McKenzie became a firefighter; his sister joined the local police force.

His father's cancer diagnosis had come quickly and unexpectedly, and in the decades that followed, McKenzie vowed he would not let the worst catch him by surprise again. In 2011, when he was on a dove hunting trip, a stranger accidentally shot him in the back with a .22-caliber handgun. McKenzie called dispatch and ordered his own medical helicopter instead of just phoning 911. The bullet permanently lodged in his left lung, robbing him of 25 percent of his breathing capacity. But he continued fighting fires, and whenever he struggled to keep pace with the younger men, he would blame his age, not his lungs.

When a blaze called the Wall Fire erupted in California in 2017, McKenzie was trawling for salmon off the sparsely populated Lost Coast. His phone lit up: The fire, it turned out, was headed toward Robinson Mill, where he lived on a ten-acre property, a factory-built unit at the end of a winding three-mile dirt road that stymied even ambulances. Over the course of the six-hour drive home, he convinced himself that his house had been destroyed, along with his menagerie of rescue animals, which included a goat mangled by coyotes, a boxer with damaged vocal cords, and an abandoned llama. Worst-case scenario.

He arrived early that afternoon, as an officer was barricading the dirt access road, and parked his truck, inflatable Zodiac boat in tow,

in the driveway. His neighbors' homes lay in rubble. Their marijuana patches smoldered, the air pungent with weed and smoke. But
he found his property mostly unscathed. The wildfire had scorched
a fence post, turned a small cabin to ash, and roasted half of the
mature oaks. His donkey—dubbed Ghost Donkey, or G.D., for his
shy nature—was still in his pen, as were the goat, boxer, and llama.

Perched on the roof, McKenzie poured himself a glass of tepid
scotch, no ice, and drank deeply. For once he was happy to be
proved wrong.

ON THIS NOVEMBER MORNING, the pine needles continued
to fall as McKenzie finished his shower and headed to the kitchen,
his hair damp and smelling of fruit-scented shampoo. The fridge
was stocked with fresh groceries, and he pulled vegetables from its
clear plastic drawers, lining up ingredients in a rainbow on the
counter. He sliced red potatoes for a corned beef hash—it was his
turn to cook breakfast. Normally the radios chattered and screeched,
reporting communications between stations across Butte County,
but he had muted the emergency radios overnight so he could
sleep. He kept them off, relishing the quiet. His cellphone rested
faceup on the counter. Wind thudded against the windows. His
knife clipped the butcher block.

Outside, the anemometer whirred.

THE FIRE: PREVAILING WINDS

Hundreds of miles from Butte County, a high-pressure system was brewing. The dry air mass settled above high deserts and sprawling salt flats, pooling in the Great Basin between the Rocky Mountains and the Sierra Nevada. To the west, California's sunbaked valleys beckoned. The air ached to reach them, pushing against the mountains that stud the Golden State's spine like an irregular dam, threatening to spill through their uneven peaks. The Sierra Nevada slants higher as the range runs south, with mountaintops rising 7,000 to 10,000 feet near Lake Tahoe and 14,000 feet near Mount Whitney—so it was over the leeward slope of the Northern Sierra that the air mass overflowed first, whistling through gaps and passes in the granite. Hours later, the tempest would crest even the immense peaks of the Southern Range, unleashing the winds on the lands below.

Southern Californians call the winds the Santa Anas; to Northern Californians, they're the Diablo Winds. Each locality has its own vernacular for these mountain gusts. Swiss Germans call them foehn winds. In the Rocky Mountains, they're the Chinook Winds. In Butte County, they're known as the Jarbo Winds. They pummeled Butte County's bluffs from the northeast—powerful enough to ground aircraft, set off car alarms, sandblast homes, and uproot trees.

At first, nothing made the weather on this particular day seem extraordinary. The Jarbo Winds were a well-known event, reversing California's prevailing airflow and depositing a ribbon of hot breath along the coast. The dry currents undulated above the white-capped Pacific Ocean, weaving their way across the water. Everything about the phenomenon was expected, except for the timing: The winds usually followed the seasonal rains, but this year, the rains had yet to come.

It had been more than seven months since even a half inch of water had tumbled from the sky. Tall grass that had thrived in the previous winter's rainstorms now cured in the sun. Brown husks from the state's historic drought, which had killed 150 million trees, matted the ground. Atop them, more needles and leaves fell. The ten-month period from January through October was the fourth warmest in California in more than 120 years, following five years of chart-topping heat. July was five degrees warmer than it should have been—the hottest in history. The whole world was warmer than it should have been.

Live fuel moisture—a measure of the water stored in a plant—was at 74 percent for a common evergreen shrub known as manzanita. The historical average during November was 93 percent. In the Northern Sierra Nevada, the National Fire Danger Rating System's energy release component—an estimate of how quickly a flaming front could consume a landscape—broke records all summer. Any of these signs would be troubling on its own: the curing vegetation, the parched landscape, the gales wailing like banshees.

Combined, they foretold an unprecedented peril.

CHAPTER 2

ALL ITS NAME IMPLIES

A warm breeze filtered through the town of Paradise on the afternoon of Wednesday, November 7, 2018. It ruffled Rachelle Sanders's blond hair as she made her way across the hospital parking lot, flip-flops thwapping against the pavement. It was 75 degrees outside, and sunshine bleached the sky. Rachelle paused to lock her white Suburban from a distance and then stepped through the facility's sliding glass doors. Her belly, full and round, strained the fabric of her athletic T-shirt.

The Adventist Health Feather River hospital sat on the town's eastern edge, overlooking the steep river canyon flecked with manzanita and gray pine. The hundred-bed facility served the largest community in the Sierra Nevada foothills, including its sizable elderly population. The average Paradise resident was fifty years old, and about a quarter of its roughly 26,500 residents had a disability of some kind, making the hospital indispensable. The Birth Day Place had become a respected labor and delivery unit that drew expectant mothers from across rural Northern California, where medical care was limited. Since it opened as a fifteen-bed sanatorium in 1950, only thirty-one thousand infants had been born within its walls, though these days there was more than one birth a day, marking a shifting demographic.

Rachelle, thirty-five, was due to give birth in three weeks and had scheduled a routine nonstress test, typical for older mothers in the last trimester. She had spent the morning with her personal trainer at the gym, doing Russian twists, lunging off a BOSU ball, and sipping the cold dregs of her morning coffee. She was still dressed in leggings, expecting the checkup to last less than thirty minutes. Her seven-year-old daughter and nine-year-old son would soon finish classes for the day; she would need to change clothes and pick them up at their charter school. "Full mom mode," she called it.

A nurse settled Rachelle in the pregnancy triage room, a sterile white-walled space with three beds partitioned by flimsy curtains. She squirted cold gel onto the globe of Rachelle's middle. Sensors measured her contractions and the baby's reaction. The prenatal heartbeat flooded the room, so thunderous that a patient resting in the far bed flipped over to scrutinize the scene. The volume had been turned up too high. The baby recoiled from the noise, and the rhythmic thudding of its heart sputtered—a sign of fetal distress. The nurse studied the grainy ultrasound screen and made a quick call to the obstetrician.

"Let's admit her," the doctor said.

DOWN PENTZ ROAD, the nursery in Rachelle's home was bare, and a stack of Amazon packages touched the ceiling in the entry-way. Neither she nor her husband, Chris, had wanted another child. Their home mirrored that apprehension. They had already given away their children's old clothing and toys and had been forced to rebuy it all. Refusing to unpack the looming tower of cardboard packages—a monument to the baby's expected arrival—was a way of making it feel less real. It was the sort of denial that Rachelle had learned from her parents as a child.

She had grown up three hundred miles south of Paradise in Fresno, a working-class city named after its ash trees. Her mother was a dental assistant; her father owned a contracting company. (He had somehow managed to charm her during a routine teeth cleaning.) As the oldest of three daughters, Rachelle was responsible for minding her sisters, two and six years her junior, while her parents worked. When her mother began to show signs of mental illness, it was Rachelle, an eight-year-old child, who managed the household. She used a two-step stool to scrub the kitchen counters clean and walked her younger sisters the half mile home after school. Her parents pretended nothing was amiss with this setup.

On weekends, Rachelle (rhymes with "Michelle") woke up early and accompanied her father to his work sites. She loved riding in his truck, listening to the local country radio station and watching the farmland whip past in a blur of tans and yellows. With him, Rachelle felt her responsibilities fade, and she enjoyed just being a kid. Even so, he could be stern, impatient with childlike behavior. If she promised to put away her art set and forgot, he wouldn't hesitate to throw it in the trash.

Some of her best memories took place during the winter holidays, when she and her family would drive to the mountains to visit her grandparents, who owned a retirement home on Pentz Road in Paradise. In the nineties, they had added a playroom off the garage that Rachelle and her sisters christened Motel 6 because it easily slept a half dozen people. Puzzles and game tables, including ping-pong and foosball, lined the walls. An exterior bulb with a frosted cover was the only source of light. It was ugly but couldn't be swapped for a new fixture because of updated code requirements. On those trips, Rachelle and her father would comb the many antiques stores for traditional piggy banks. In the evening, her grandfather, a retired cop, taught her to stargaze. He pointed out Orion's glittering belt, named the diamond point of the Big Dipper, and traced Cassiopeia's sharp peaks. To a city girl, the constellations were a revelation. So was the snow. Sometimes storms dusted Paradise in sugar-spun beauty—enough to marvel at, but never enough to shovel. Thickets of ponderosa pine cloaked entire blocks. Houses with long driveways were lined up as neatly as teeth. There were no streetlights and no sidewalks. In the spring, tens of thousands of butter-yellow daffodils bloomed along the roadsides.

By college, her trips to the mountains had grown more infrequent, though she still thought of Paradise often. Rachelle was studying to be a teacher at the state university in her hometown; she had always dreamed of teaching kindergarten. She was still an undergraduate when she met Mike Zuccolillo. She had finished job shadowing at a nearby high school and stopped at Chipotle for lunch, as did Mike, a recent divorcé with olive-hued skin and dark hair. They hit it off in line—complete strangers laughing over things neither now remembers—and began dating soon after. She took him to Paradise to visit her grandparents—it made for a cheap vacation. On that visit and the ones that followed, Mike came to love Paradise as much as Rachelle did.

They hatched a plan to settle in Paradise, moving first to Chico, twelve miles downhill. At 92,861 people, it was Butte County's largest city and sprawled across the valley floor. Mike applied for his broker's license and operated a real estate business from a home office in their backyard. They married and had their first child, Vincent, in June 2009. He was a cyclone of a baby, feisty and vocal, with his father's Italian coloring and slightly outturned ears. Six months later, they bought a house for $245,000 in Paradise and moved onto Castle Drive. Their neighborhood looked across Butte Creek Canyon, nicknamed the Little Grand Canyon for its rusty ridges and cavernous basin, on the western side of town. The house was an outdated foreclosure with mauve fixtures and appliances, but it was redeemed by its back patio. At night, the city lights below mirrored the starry sky. By Christmas, the family had settled in. Their neighbors welcomed them with a homemade gingerbread house, their new address iced above the candy door.

For a few years, they were happy. Mike was appointed to a seat on the town's planning commission. Rachelle quit her job teaching kindergarten to help with his real estate business. She looked up property listings; he attended the auctions. They had another child, Aubrey, a miniature of Rachelle with lake-blue eyes and pale hair. Mike nicknamed her Pop-Tart for her sweet, affectionate nature. They took their toddlers to play at Bille Park, a former olive orchard, which Rachelle had enjoyed on her childhood visits.

And then their marriage collapsed—an ordinary enough occurrence that in a small town like Paradise became big drama. They divorced in 2013. Mike kept the house overlooking Butte Creek Canyon and ran for Town Council, losing by 250 votes. Rachelle and the children moved in with her grandparents on Pentz Road, where that ugly frosted light fixture still shone from the wall. She got a job in the human resources department of an Oroville company.

As she drove through the outskirts of town day after day, she would pass Paradise's most famous icon, a wooden welcome sign. It had been there for forty-six years, topped by an actual bandsaw blade fashioned into a metal halo. It read: May you find Paradise to be all its name implies.

IN THE SUMMER OF 2014, as Rachelle settled into the routines of single parenting, she met Chris through a mutual friend. Chris was thirteen years older and managing a landscaping company in Chico, where he had been born. His mother had given birth when she was fifteen, a high school freshman unprepared for motherhood, and her parents had stepped in.

His grandparents, who had already raised three daughters, cared for Chris like the son they'd never had. His grandfather taught him to pitch a softball and root for the San Francisco Giants, to hook worms on a fishing line and grease a rifle for duck hunting. Chris called him Bompa. His grandfather worked long days delivering Wonder Bread, and his clothes always smelled of flour and yeast. After hours, he was one of the best baseball pitchers in Butte County—his photo even hung on the Wall of Fame at an Applebee's restaurant. Chris never saw him lose his temper, though he was known to be less patient with other family members. With the boy, he was gracious and giving, and Chris idolized him. For Chris's sixteenth birthday, his grandfather bought them tickets to see the San Francisco Giants play the New York Mets at Candlestick Park. Baseball legend Darryl Strawberry hit two home runs, the ball cracking across the stadium. Chris couldn't stop smiling.

He and Rachelle were bickering about the sport before they had even met in person. She had grown up in a Los Angeles Dodgers house; he remained faithful to the Giants. After a mutual friend put them in touch, they began messaging about their allegiances on Facebook. Their teams were scheduled to compete in a three-game series, so Chris and Rachelle made a bet: Whoever lost would have to snap a photo wearing the rival's hat. The Dodgers' Clayton Kershaw was pitching against the Giants' Tim Lincecum. Chris knew his chances weren't great, but he was willing to take the gamble.

He lost.

The next morning, a picture pinged on Rachelle's cellphone. In the snapshot, Chris wore a blue Dodgers cap and weakly smiled—it was more of a grimace than anything else. He was attractive, with light blue eyes and closely cropped blond hair, his arms muscled from running chainsaws and lawn mowers. They went on their first date soon after, meeting to watch a coed softball game in Chico. Chris brought his pitbull mix, Posey, who was named after a Giants catcher. Rachelle joked that she would refer to her only as "the dog that shall remain nameless." He laughed. They left the game early and walked in the park, stopping to rest on a bench. College students biked past, chains grinding as they shifted gears. Geese floated in the nearby pond. Rachelle felt she could talk about anything with him. Like her father, Chris was thoughtful and reserved, parsing what he wanted to say before speaking. And like her, he was direct. She had two children from a previous marriage, and so did he. They each had baggage. The obstacles seemed great.

"Do you still want to do this?" he asked.

"I'm a package deal," Rachelle said. "I have two little kids. I'm not going to waste my time or your time. If we are going to date, you need to know that this is what I come with."

"I'm in," he said.

Somehow it worked. Chris made her laugh, and Rachelle drew him out of his shell. She brought her children, Vincent and Aubrey, to his softball games. He introduced her to his adult daughters. Later, Chris moved into the house on Pentz Road. Rachelle's grandparents had relocated to Fresno. They hinted that they would sell the home to Rachelle soon, though on visits, her grandmother continued rearranging the cupboards and picture frames to her liking, unable to relinquish control.

Two years passed. Rachelle's ex-husband, Mike, ran again for a seat on the five-member Town Council. This time he won. He married a woman who had a daughter named Aubree, adding further complexity to their blended family. He nicknamed his step-daughter Peanut for her petite size, while his Aubrey was still Pop-Tart.

In December 2016, Rachelle remarried too. She and Chris drove three hours east to Reno and wed at the Chapel of the Bells. They didn't care about pageantry—no photo package, no flowers, no sheet cake—but they wanted their four children present. Chris's daughters were home from college for winter break. Rachelle wore a shiny silver dress with long sleeves, and Chris dressed in khaki slacks and a button-up shirt. The ceremony lasted less than an hour. Afterward, his daughters took their new stepsiblings to the circus, which also had arcade games. Aubrey won the hundred-ticket jackpot, shrieking as colored squares cranked out of the machine, enough to purchase a stuffed animal. Rachelle and Chris, meanwhile, sneaked back to the Sands Hotel, giddy with joy.

In early 2018, not long after their one-year wedding anniversary, as irises and crocuses stabbed the soil and the sun lingered on Sawmill Peak later into the evening, something had changed in the house on Pentz Road. Rachelle knew, intuiting the shifting and multiplying of the cells within her, long before letting herself acknowledge it. She had always been active, competing in adult softball leagues, training at the gym, and racing after two children. But she was increasingly exhausted. Her chest felt tender.

She was thirty-five; Chris was forty-eight. After a half dozen discussions, they had decided against conceiving a child. Instead, they invested in their future. Chris bought run-down yard equipment on sale—trucks, trailers, lawn mowers, weed-eaters—and opened his own landscaping company. He hired nine employees and signed lucrative contracts in Paradise, pruning homeowner association flowerbeds and edging the grass in manicured subdivisions. He and Rachelle reveled in the stable income and dreamed of travel: They wanted to lounge on white sand beaches and explore Mayan ruins. Chris had seen Nebraska's rippling prairies through a windshield and had whooshed by the Grand Canyon at 65 mph, both on road trips to visit his ex-wife's family in the Midwest—the extent of his previous adventures. He had worked full time since he was seventeen and hadn't been able to afford vacations. On the other hand, Rachelle had first left the United States when she was eleven years old, driving to Mexico in a tour bus with her church's youth group, and had visited every state except Alaska. She wanted Chris to see the world as she had. A baby wasn't part of their plan.

On that cool spring evening, Rachelle tucked her children into bed before fishing an expired pregnancy test out of the back of a bathroom cabinet. Chris rested in the living room, watching a true crime show in his broken-down easy chair. Canned dialogue pierced the stillness. Rachelle took the test, popped on the plastic cap, and walked outdoors to wait on the back patio. Chris followed. The sky bruised navy, then black. A breeze prickled her skin and agitated the dogwood tree, its limbs heavy with cream-colored buds. Hands shaking, she glanced down at the faintly intersecting blue lines.

PARADISE SPREAD ACROSS a wide ridge about 2,000 feet high, just beneath the snow-capped cathedrals of the Sierra Nevada. Here, in northeastern Butte County, the landscape hardened as it rose, the farmland, orchards, and rice paddies of the valleys giving way to volcanic benches and ravines. The town was shaped like a triangle and sculpted by two deep canyons: Butte Creek to the west and the West Branch Feather River to the east. Four main streets cut downhill. Only one connected to the north, looping past a reservoir and skirting Magalia, an unincorporated village of 11,500 people. Altogether, the area was nicknamed the Ridge.

The Konkow band of the Maidu tribe were the first to call this place home. The Indigenous group ignited small grass fires to clear brush and gathered acorns as their primary food staple. After the Gold Rush thrust California into statehood in 1850, stoking even greater dreams of easy wealth, the native population was slaughtered by the newcomers—through malaria and smallpox or outright genocide. Within two years of the Gold Rush, a hundred thousand had died. Much of the killing was state-sponsored, with Indigenous groups enslaved over petty offenses or forced into indentured servitude. Within twenty years, about 80 percent of California's total native population had vanished. In 1863, the 461 surviving Konkows were marched ninety miles west to the Round Valley Reservation. Only half of them survived the journey, called the Konkow Trail of Tears.

Then, in 1864, after a hot and dusty day of travel from the valley floor, one of the first prominent white settlers, a man named William Leonard, arrived in town. As oak land gave way to pine, he took a deep breath of crisp mountain air and said to his wagon crew, "Boys, this has got to be *paradise*." Or that's what some people say. Others maintain the town was named after an old saloon called Pair o' Dice. The true origin remains lost to history.

What *is* known is that the area delivered on its promise. Between 1848 and 1965, more than $150 million worth of gold was unearthed in Butte County. The most valuable find was a 54-pound nugget excavated in Magalia, then known as Dogtown, in April 1859. Naturally, this boom gave rise to others. The Diamond Match Company erected a lumber mill in nearby Stirling City, and Southern Pacific established railroad service from the northern Sacramento Valley. Leonard, the early settler responsible for the town's name, operated a sawmill in an area of town that became known as Old Paradise. The buzzing of blades and thumping of falling timber echoed across the mountain, and the scent of fresh-cut pine wafted on the breeze. "I felt as though I had been transported to another planet," remembered a visiting miner from Iowa. "There was nothing here I had ever seen or heard before. The great forests, the deep canyons with rivers of clear water dashing over the boulders, the azure sky with never a cloud was all new to me."

Agriculture soon topped timber as the main driver of the economy. Farmers seeded orchards in the red volcanic soil, cultivating Bartlett pears, kiwis, prunes, peaches, and apples. Attempts to grow olives largely failed. There were more than fifty orchards, the most famous belonging to the Noble family on Lot 26-G. Apple trees drooped with yellow and red fruit in the fall, the cool air sustaining them long after competitors' harvests had ripened on the hot valley floor. The fruit was popular, shipping as far east as Denver. To keep up with farmers' demand for water, in 1916 residents passed a $350,000 bond to establish Paradise Irrigation District. Crews dammed Little Butte Creek above Magalia—the former Dogtown, renamed for the Latin word for "cottages," for propriety's sake—to funnel water to farmers.

Despite its thriving economy, Paradise wasn't well known. The population was less than ten thousand. Only a handful of churches and bars operated on the town's main street, the Skyway. In 1934, a home sold for $205: $95 for lumber and another $110 for labor. That same year, the national average cost for a new home was $5,970. Developers hawked the town's affordability and location as part of a concerted effort to get more people to move to Paradise. "All roads in Paradise lead to some point of scenic beauty," a 1940s Chamber of Commerce brochure exclaimed. "Come lose yourself in Paradise awhile and learn what living really is."

The advertisement was a triumph. From 1962 to 1980, the population more than doubled, reaching twenty-two thousand. Wood-frame cabins with shake roofs dotted the heavy brush that had flourished in the wake of logging. Invasive Scotch broom and mustard coated the hillsides in flaxen splendor. The village expanded into a town with little oversight from the county. When a young woman applied for a permit to open the first ballet studio in the 1970s, there was no relevant building code, so officials tried to classify the establishment as a church, then a meeting hall, before creating a new category: dance hall. Meanwhile, logging roads and cow paths became residential streets, freezing the town's rambling and unplanned layout into place. Officials named the main streets after early settlers: Clark, Neal, Bille, Wagstaff. (Pence refused the honor, so that road became "Pentz.") More than 280 miles of additional private roads twisted along ridgelines to abrupt dead ends. Some of these remained entirely unpaved, ribbons of mud or dust depending on the season. They were so rough, local lore went, that a few miles of driving over their potholes could reduce a crate of berries to juice.

As the population of California skyrocketed from 3.3 million in 1919 to 39.4 million in 2018, retirees on fixed incomes and middle-class families increasingly began to feel the crunch of housing prices in urban areas. They pushed outward to "the wildland-urban inter-face," the name for the areas that reached into the mountains or deep into the woods. Their homes weren't always built to code, but they were cheap. Here there was room to dream. In November 2018, houses sold for an average of $304,000 in Sacramento, $671,000 in Los Angeles, $1.31 million in San Francisco, and $2.46 million in Palo Alto. The median property value in Paradise was $205,500.

People were drawn to the hamlet because it was affordable, or because it was a little out of the way, or because it had small-town charm. It was the kind of place that lined the Skyway with more than a thousand full-sized American flags on Memorial Day and allowed a ninety-nine-year-old to compete for the title of Choco-late Queen at the annual Chocolate Fest; the kind of place that celebrated holidays like Johnny Appleseed Day, when volunteers baked a thousand pies to sell for $15 each in Terry Ashe Park and an actor dressed up as Johnny, wearing a silver bucket atop his head and toting a wicker basket brimming with apples. In April, every-one looked forward to the weekend of Gold Nugget Days, when children competed in the costume contest and teenagers vied for the Miss Gold Nugget crown. Families cheered at the donkey derby and lined lawn chairs along the side of the road to view the parade, parents rubbing sunscreen onto their children's wriggling bodies as they waited for it to depart from the Holiday Market shopping center.

Retirees had always dominated the Paradise census, but the de-mographic was changing. In 2001, nearly 62 percent of the Ridge's population had been of retirement age; by 2018 the percentage had more than halved. Still, the population skewed conservative. Resi-dents could choose where to worship from among thirty churches, including the Center for Spiritual Living and the Paradise Church of Religious Science. Even the local hospital was faith-based, affiliated with the Seventh-day Adventist Church: no pork in the cafeteria, prayer-led meetings. But in town there were no synagogues, no mosques. In the 2016 presidential election, a majority of the county's vote went to the Republican candidate, and MAKE AMERICA GREAT AGAIN! stickers adorned many bumpers—causing tension downhill in liberal Chico. The older folk loved to gossip, chattering while volunteering at the Elks Lodge or getting their hair set at one of the many salons. They were a warm but skeptical bunch with a tendency toward nostalgia. Sometimes they posted on a Facebook page called "Paradise Rants and Raves!" They were mostly proud of their town.

For all of its wholesomeness, Paradise wasn't perfect. It was the largest community west of the Mississippi River that still relied on septic systems to dispose of wastewater. Without a central sewer system, the town struggled to draw businesses and bolster a strong commercial corridor. When Coffee's On opened a hut-sized joint across the street from the Holiday Market shopping center, it cost the owners $80,000 to install a septic system, which failed twice in only a few years. McDonald's paid $250,000 to update its system. The restriction scared away many larger corporations—though in 2018 Paradise finally attracted its first Starbucks, a cause for celebration. The mayor liked to say it was a sign they had "made it." The town did have a Cinema 7, an Ace Hardware, and a Kmart to anchor its sluggish economy. But there were no car washes, no major hotel chains.

For special occasions, residents drove downhill. Chico had nicer restaurants, as well as a Costco and a Walmart. Most locals spent their money outside Paradise, which was one reason the town had been known as Poverty Ridge as far back as the early 1900s— though old-timers always joked that it was a "darn nice place to starve if you have to." Social Security was the predominant source of income. Councilmembers earned a $300 monthly stipend, and their coffers were as limited as their constituents'. The Adventist

Health Feather River hospital was the largest private employer in town, providing a steady paycheck to thirteen hundred employees— many of them locals. The unified school district, which included Magalia, was the second-largest employer.

On Tuesday, November 6—one day before Rachelle was admitted to the hospital—voters approved the extension of a half-cent sales tax to cover public safety, road repair, and animal control, among other things. Without it, town leaders feared insolvency. It was bigger news than the new Starbucks or the fresh coat of paint on Town Hall: The measure would generate $1.4 million annually. The money would help the town's police force do its job. Paradise avoided some of the problems of an urban center—no gangs, little congestion. But outlaws moved to junglelike plots in the foothills to evade rules and dodge authority. Paradise Police's sixteen patrol officers and four sergeants stayed busy busting methamphetamine and opioid rings, cracking down on domestic violence, and responding to calls from the Ridge's thirty-seven mobile home parks. They rousted the homeless who camped in remote canyons.

The poverty affected Paradise's children the most. The Paradise Unified School District had among the highest number of students with Adverse Childhood Experiences, or ACEs, in the state. They were traumatized by abuse, poverty, addiction, or alcoholism in their homes. Nearly 70 percent of the thirty-four hundred students qualified for free or reduced-price lunch. Behavioral issues were so widespread that one academic year, the high school principal had handled thirty-one fights. The district had recently started teaching mindfulness, in the hope that it would help alleviate stress. Outside school, there wasn't much for teenagers to do aside from drinking beer, hiking the old mining flumes, four-wheeling, and shooting at street signs.

That changed on Friday evenings, when the townsfolk piled into the bleachers on Om Wraith Field, named after a former coach, to cheer on the high school football team. The boys converged on the field to the band's rendition of Johnny Cash's "God's Gonna Cut You Down." Students dressed in cutoffs and flannel, donning CMF headbands—an acronym for Crazy Mountain Folk, or something more crass, depending on one's interpretation. Parents wore the Bobcats' team colors: green and gold. The concessions stand sold hot dogs named after the head coach.

On Friday, November 9, the team was scheduled to play Red Bluff High in sectionals. Everyone was anticipating a win—a bright spot for administrators, who besides the daily challenge of fights in the school hallways were now contending with a dwindling faith in the local education system. Charter schools were springing up, siphoning 21 percent of the town's students. A new high school had opened in August on Nunneley Road, rattling the district, whose own buildings were run-down and couldn't compete with hallways that smelled of epoxy and fresh-milled lumber. At least voters had approved a $61 million bond on Election Day to rebuild Paradise Elementary and Ridgeview High, renovate Ponderosa Elementary, and replace the athletic fields at Paradise High. It was the town's biggest bond issue in thirty years—and a sign of hope.

WEDNESDAY BLED INTO THURSDAY. Green and blue bins lined Pentz Road for trash day. Campaign signs bristled in front yards. Parents dropped their children off at Ponderosa Elementary, then commuted to Chico, driving past the Ace Hardware, whose windows advertised that night's Christmas preview, and past the seasonal ice skating rink in Bille Park. Employees brushed ruddy leaves off the ice and pasted promotional posters on the bumper board, including for Adventist Health, whose $5,000 donation made it the biggest corporate sponsor. Less than a mile away, the Starbucks parking lot was full, though old-timers could still be seen at Dolly-"O" Donuts, whose sour brew they preferred.

Kevin McKay rolled his cherry-red Mustang through the Paradise Unified School District's gated parking lot. He clocked in at 6:45 A.M. and exchanged his keys for those to Bus 963. The vehicle was at the back of a long line of buses—at twenty-two years old, the most senior of the fleet. Kevin, forty-one, lifted himself into the driver's seat and began pulling levers and pushing buttons, making sure the brakes were lubricated and the engine purred. A tone pinged from the dashboard. The precheck list had twenty-five items, with another nine for the brakes. You could never be too careful when children were involved. Kevin dutifully checked off each box on the page, then pulled out of the lot and headed toward Magalia.

It had been a rough week. The night before, Kevin had euthanized his Bordeaux mastiff, Elvis. The dog—slobbery and loose-skinned—had been dying of cancer, though that hadn't made the veterinarian appointment any easier. Afterward, his twelve-year-old son, Shaun, had come down with the stomach flu. Kevin had been up with him all night. The boy was too sick to attend school, so Kevin's mother was taking care of him at home. Kevin planned to relieve her when he finished his route at 9 A.M.

He cranked open the small driver's window. Thick ponderosa pine and rows of squat homes decorated for Thanksgiving flashed past. Scuffed-up trucks with tall wheels zipped by in the other lane. Sometimes Kevin had to pause—to remember how he had wound up wearing a dark gray uniform behind the wheel of a two-decades-old school bus, to remind himself that the sacrifice was worth it. He worked part time and made $11 an hour shuttling raucous children from one end of the Ridge to the other. BE SAFE, RESPECTFUL, KIND, RESPONSIBLE, read the laminated yellow sign taped above the bus door. It fluttered in the breeze from the opened window.

The other drivers had warned him that these days, Magalia kids were particularly tough. Within the first month, a nine-year-old had punched his classmate in the nose, spraying blood across the seatback. But Kevin had a good way with children. He knew they often fought because they wanted to be seen, or because they hadn't eaten breakfast, or because they had learned aggressive behavior from their parents. He was firm, but sometimes all he had to do was listen. He made the nine-year-old sit in a seat near him, and after a week, the boy stuck around, confiding in Kevin about issues at home and showing off his homework assignments. His classmates followed suit. After school, they flooded the bus and clamored to give Kevin high-fives and pepper him with questions. Their mothers learned his name and waved hello during pickup and drop-off.

When the kids got too rowdy—throwing candy wrappers across the aisles or hitting each other with their backpacks or wedging gel pens in each other's ears—Kevin would whistle through his teeth. As his chocolate-brown eyes darted across the wide mirror nailed above the windshield, the bus would momentarily quiet.

Kevin had wanted to do more with his life. He'd moved to the Paradise area when he was twelve, relocating with his family from the historic logging town of Felton, near Henry Cowell Redwoods State Park, about seven miles inland from Santa Cruz's surfing scene. His parents were high school sweethearts who married the day after his mother's eighteenth birthday. They had ended up on the California coast after living in San Jose, where his father worked as a heating, ventilation, and air-conditioning technician, in an effort to give their sons a quality education. But the class sizes at the local school district in Felton still seemed too large, so they returned north to raise their children in their hometown.

The boys adored living in the foothills, where they had 11 acres to roam. Kevin and his brother fished for salmon in the Feather River, camped with their cousins in the woods, and shot BB guns at pine trees that oozed sap. When their parents traveled to Reno to play the slot machines—their father's idea of a vacation—they stayed home alone. Magalia was safe. Residents didn't bother to lock their doors.

Kevin grew into a barrel-chested and athletic teenager. He didn't take any flak—not even from his older brother's friends, who often stole his basketball to play keep-away. He was not one for mean-spirited jokes. To his parents' delight, he and his brother attended their alma mater. His senior year, Kevin's teammates voted him captain of Paradise High's football team. He dated a pretty brunette in his class and earned good grades, particularly in math and science, receiving an academic patch for his varsity jacket. He graduated in 1995 with the same aspirations his father had held dear: to purchase a house, get a good job, raise a family. He longed to attend the University of Oregon, where he had gone for summer football camp. But even with a partial scholarship to play on the team, his parents couldn't afford the tuition. He and his girlfriend enrolled at Chico State University.

Kevin was two years into a pre-med degree when they realized that she was pregnant. They married soon after, and their daughter was born in October 1997. Kevin named her Isabelle, choosing the French spelling, which seemed more proper. He nicknamed her Belle. He worked the graveyard shift at Safeway, stocking shelves while the rest of the world slept, then rushing straight to his 9 A.M. class, usually arriving late. He was resentful of his classmates, who enjoyed college in a way he couldn't. No one else was juggling the responsibilities of fatherhood. Within six months, Kevin dropped out, having failed zoology and psychology. It was difficult to study with a baby at home. Besides, school didn't put food on the table.

A few years later, he was hired at Walgreens and was soon promoted to store manager. The company deployed him across Northern California, to places like Red Bluff, Yuba City, Redding, and Chico, to fix problem stores. He drove twenty-five thousand miles annually for his new gig. Kevin didn't believe in firing troubled employees and relished helping them realize their potential. He knew how to inspire people. He was always on the road, and it didn't matter where he lived, so he bought his first home down the street from his parents' cabin in Magalia. The mortgage was $800 a month—comfortably affordable.

The travel was draining, but no matter how often he asked, the company refused to transfer him to its store in Paradise. It needed his skills elsewhere. To make the eighty-mile round trip drives less torturous, he listened to audiobooks. Historical nonfiction carried him across the Sacramento Valley, past almond orchards and olive groves, past rice paddies that reflected the clouds, past deadly car crashes on Highway 99. He listened to William Shirer's *The Rise and Fall of the Third Reich* and Timothy Snyder's *Bloodlands*. He was amazed by what human beings were capable of doing to one another, and for one another. Kevin had visited only one other country—on a road trip to Canada as a child—but in the car, it felt as if he was circling the globe.

He finished the *Game of Thrones* series and Jordan Peterson's *12 Rules for Life,* spellbound as the author explained that the most terrifying person to confront was the one in the mirror. Kevin wondered what he would see if he were brave enough to really examine his reflection. The audiobooks always clicked off too soon. Sometimes he took a few precious minutes to idle in the parking lot and finish a chapter. More often, he clocked in early to help the pharmacist fill prescriptions before the store opened at 8 A.M.

Kevin had studied to get his pharmacy technician license and encouraged others to do the same, as the title came with a bigger paycheck. In 2009, as an economic recession continued to shake the country, Walgreens reduced staff positions, and Kevin's workload grew. More was expected from every employee—particularly those who worked the pharmacy counter. The number of prescriptions was rising; the bulk of them were for opioids. Sometimes Kevin caught doctors refilling prescriptions before their end date. Sorting the white pills, his plastic knife clicking on the work tray, sickened him. His location in Red Bluff dispensed ten thousand narcotics prescriptions each week to an impoverished community of fifteen thousand people. He was witnessing the opioid epidemic unfold pill by pill.

The long commute and work responsibilities took a toll on his personal life. Kevin had by now married and divorced twice. He had two children by two different mothers and a new girlfriend, Melanie, who worked in the accounting department for the City of Chico. He had bought a foreclosed home in Paradise for $116,000 and spent nearly $80,000 fixing it up. The lot was overgrown with cedar, oak, and pine. Insulation tufted from the walls. Kevin finished the master bedroom, replaced the siding, and retiled the bathrooms. Once a year he took a week off from work to trim the trees, the scent of pine sap permeating the air.

Supporting his growing family had always been his priority. He wanted to be like his father, who for years had commuted to the Bay Area to work. Though he had missed huge chunks of Kevin's childhood, the family had never struggled to pay their bills. His father, who always aimed to achieve one goal a day, had built their home by hand, his biceps as thick as the tree trunks he split. Kevin took pride in what he could provide—he did make good money—but inside, he felt a gnawing emptiness. "Walgreens pays you just enough to give up on your dreams," a colleague had told him. The exchange haunted him.

In late 2016—as Rachelle and Chris were marrying in Reno—Kevin's father was diagnosed with a rare cancer that mutated his cartilage-producing cells. Only six other adults in the country had the same type of cancer. Soon after, his mother was diagnosed with full-blown melanoma. She had had a small tumor removed from her thigh two years before, with assurances that there was little chance the cancer would return. Now tumors budded in her lymph nodes. The McKays were fixers, and if it couldn't be fixed, they flew into a rage. Kevin could never remember what the stages of grief were, but he knew that anger was one of them—and his father was angry.

Kevin and his brother, a construction worker in Santa Clara, coordinated schedules to drive their father to appointments at the University of California, San Francisco Medical Center, a sprawling campus of steel and glass buildings near the city's ballpark. They met halfway, in Vacaville, transferring him between their cars. The sickness whittled their father to a shadow of his former self. Doctors nearly amputated his muscled right arm, the one that had once gripped an ax with resolve, then settled on excavating the disease-riddled bone and replacing it with a titanium rod. He would never use his arm again, they said.

The surgery didn't work. Neither did radiation. Kevin knew what was coming. The cancer's mortality rate was measured in years—not in percentages marking the likelihood of survival.

On one of their early morning drives to the hospital in San Francisco, his father rested in the passenger seat. The seatbelt pressed against his sunken chest. It was now 2017, and cancer had invaded his lungs and brain. The windshield wipers thrummed against the late January rain as they passed the toll plaza on the Richmond–San Rafael Bridge, the undulating roller coasters of Six Flags Discovery Kingdom in the distance. Kevin fiddled with the radio. "Hey, man," he said, as casually as he could, because no one ever told his father the way something was going to happen. "I don't think we should do this anymore. I think we're done."

"I don't think we should either," his father finally responded.

He died thirty-five days later, at sixty-six years of age. The man who had chopped firewood every weekend, who had inspired Kevin with his profound determination, was gone. He had taught his son to stay out of debt, to love and provide for his family, to dream. Kevin felt his absence everywhere. His mother, alone after forty-seven years of marriage, moved into Kevin's guest bedroom, where she kept her husband's ashes in a polished oak urn. Kevin reflected on his bad job and his good paycheck. A year later, he quit Walgreens.

By early 2018, Kevin had finalized plans to enroll in Butte Community College to "reactivate his brain," then transfer to Chico State for his diploma. He wanted to teach history at his alma mater, Paradise High. But first he needed to find a way to support his family.

He noticed the advertisement on a highway exit one morning as he drove to Red Bluff. The blue banner was plastered on the side of a school bus. Paradise Unified School District was looking for drivers.

Kevin had driven nearly four hours each day for years. Doing it in a bus couldn't be that hard. Plus, if he was employed by the district, he thought, the administrators might be more likely to hire him full time as a teacher after graduation. He invested $500 in training and paperwork for his commercial driver's license. The California Highway Patrol ran background checks. An instructor taught him to make seven-point turns in a 35,000-pound vehicle and listen for oncoming trains where the road intersected with railroad tracks. He completed first aid training. That summer, his new license landed in the mailbox.

The job interview was at the district office in Paradise. Kevin wore slacks and a button-up shirt. No tie, because he thought that was a bit much. The panel asked him why he wanted to be a bus driver. He told them about Karen, who had driven Bus 3 when he and his brother were in middle school. She always had a smile on her face when she picked the boys up. They lived a half mile from the last stop, and rather than make the boys walk, she asked the superintendent to list their driveway as the final stop on her route. Karen was the sweetest woman, Kevin said, but strong enough to throw a wheezing ten-speed bus into gear. She even handed out bags of chocolate candies at Christmas.

The hiring committee had wet eyes. Unbeknownst to Kevin, the administrators and bus drivers had plans to join Karen, now retired, that day for lunch.

Kevin got the job—it paid far less than Walgreens, but it gave him the time he needed. He and his girlfriend, Melanie, celebrated at the Panama Bar Café in Chico, which boasted thirty-three varieties of $4.00 Long Island iced teas. He ordered a Mai Tai; she ordered an orange cream. They split an order of potato skins.

As the school year began in August, the district assigned a small bus of special needs students to Kevin. They wanted to ease him into the gig. He triple-checked the route every morning, rereading the stops and familiarizing himself with the roads. The bus rattled as he drove. "Good morning, y'all," he shouted to the students, the doors squeaking as they opened. "Time to get some kiddos to class." He was a large man, still built like a quarterback but soft with middle age. He slicked his dark hair back with gel and had a goatee. Blue-gray knots were inked around his right forearm, next to a tattoo of his last name, visible beneath his shirtsleeve. He greeted the students with a smile, just as Karen had done. The children revered him. In September, his boss reassigned him to Magalia. She knew Kevin could handle the tough kids.

Some days were encouraging. He finished his morning route early and made it to his macroeconomics class on time, even nabbing a parking spot in Butte Community College's crowded lot. On other days—when traffic caused delays, or the kids were particularly rowdy, or he was too exhausted to complete his own homework—his goals felt out of reach. This morning felt like such a day. He was tired. He ached over how much he already missed Elvis's wet nose.

A breeze rippled the water in Magalia Reservoir. The sky was an endless blue plane against the autumn foliage. Kevin twisted down backroads through the Sierra Nevada foothills, the wind pushing through his window. He had sixty children to pick up. He didn't want to be late.

ACROSS TOWN, as dawn blushed on the horizon, Rachelle nursed her newborn in a private birthing suite. The lights were off, and the cabinets lining the wall were dark outlines. Her son's bassinet sat next to her. The baby's face puckered as he nursed, one fist clenching and unclenching. She and Chris had named him Lincoln. He was warm in her arms, all of six pounds and five ounces. Nearby, Chris lay slumped in a teal armchair, exhausted but awake. He had passed a sleepless night at their home on Pentz Road, where his mother was also staying.

Their son had arrived at 8:33 the previous evening, sticky with vernix and blood. The minute that Chris and Rachelle laid eyes on him, all their trepidation over a new baby dissolved. Lincoln completed them. Instead of referring to his children or hers, Rachelle and Chris could use new words: us, ours. After surgery, when Rachelle cradled the baby against her chest, Chris had snapped a photo on his cellphone, overcome. He had dropped his stepchildren, Vincent and Aubrey, off at their father's house after school, anxious not to miss his son's birth. Rachelle was clear-headed enough to insist that the surgeon remove her fallopian tubes, demanding to see them, pink and bulging, in a glass jar. This was their family now.

Chris and Rachelle made plans to finally slice open the packages stacked in the entryway, coat the nursery in fresh paint, and assemble the high chair. They would decorate the wall with baseball decals and fold the onesies—blue for the Dodgers, orange for the Giants. Lincoln would pick his own favorite team one day, they promised each other, though they were already making bets on which one he would favor. Rachelle stroked the soft curve of his head. The baby was a marvel, his fingertips as delicate as pearls and his skin faintly sweet-smelling. He was premature, but he was healthy.

Nurses changed shifts at 6:45 A.M., and Rachelle and Chris overheard the murmur of hallway conversations. A woman from Redding, ninety miles to the northwest, had arrived and was scheduled to be induced. The fluorescent ceiling light flicked on. A nurse's aide tapped on Rachelle's door.

CHAPTER 3

RED FLAG OVER PARADISE

O fficials had predicted strong northeast winds beginning late Wednesday evening. The National Weather Service issued a Red Flag Warning—the highest level of fire risk, determined by a combination of dryness and strong winds—in the Sacramento Valley and parts of the Bay Area, including the Oakland Hills. Across Northern California, Cal Fire held additional firefighters on duty in case a wildfire sparked. The crews were stationed in regions where a blaze was most likely to ignite: Lake County and Sonoma County, Marin County and Alameda County. Butte County, too, which staffed three additional engines.

Jamie Mansanares, thirty-five, didn't think anything of the yellow pine needles collecting on his roof. It was standard for November. He usually woke before dawn and drove straight to the gym, but this morning he let himself sleep in. The smell of hot oil permeated his Magalia house and roused him from bed. His wife, Erin, was frying sausage. As it sizzled and popped in the pan, she whipped glaze for homemade doughnuts. He could hear the clink of dishes and the laughter of their daughters in the other room. He crawled out of bed, then helped Erin feed and dress their three girls, shoving small limbs into leggings and long-sleeved shirts. Erin packed lunch for Tezzrah, their lanky seven-year-old, a second grader at Achieve Charter School—in the same class as Rachelle's daughter,

Aubrey. Mariah, four, and Arrianah, two, were recovering from bronchitis. They curled up on the new leather couch, which smelled of plastic wrap and was so big it made the living room feel like a movie theater. Jamie hadn't snipped off the tags yet.

He had grown up above a pizza parlor in San Francisco's Richmond neighborhood, without the luxury of new furniture. His parents had rented a two-bedroom apartment in a six-story building. Their unit always smelled of melting cheese and rising dough from the pizzeria below. Jamie had shared a bedroom with his older sister and whichever member of his extended family happened to be visiting. His father washed airplanes at San Francisco International Airport, and his mother worked as a technician in an X-ray laboratory. They did the best they could with what they had. Jamie and his sister had entertained themselves by cruising down the city's steep hills on their skateboards, jamming their heels against the pavement to stop.

When Jamie was six, the 1989 Loma Prieta earthquake struck just before the World Series game at Candlestick Park. The Oakland Athletics were playing the San Francisco Giants—rival teams across the bay from each other. The earthquake registered a 6.9 magnitude in the twenty seconds that it shook the ground and managed to collapse a section of the double-deck Nimitz Freeway, killing forty-two motorists. Across the Bay Area, sixty-three people died.

Jamie's family moved soon after, settling in Oroville, near where his aunt lived. San Francisco was getting pricier and more populated, and they didn't want to face another natural disaster. His parents bought a house this time, with enough space for Jamie and his sister to have their own bedrooms. His father got a job with the city's recreation and parks department while his mother stayed home with the children. Their new neighborhood had bumpy dirt roads and no sidewalks. The snick of cicadas and keening of red-tailed hawks replaced the hum of trash trucks and the hydraulic screech of city buses. Jamie and his sister encountered their first hillbillies. They had mullets and dirt-lined nails and were tweaked out on something—probably methamphetamine—but they were surprisingly nice, Jamie thought. His family still didn't have much, but what they had, they shared. His mother never turned anyone away from her dinner table, cooking elaborate Mexican dishes. Jamie's new classmates were poor and white, but none of them treated him differently because of his brown skin. He could make friends anywhere, even in Butte County.

Jamie finished high school in Oroville and returned to the Bay Area to work for the California Conservation Corps, a state-run program that provided job training to young people. They advertised their programs with a tongue-in-cheek campaign: "Hard work, low pay, miserable conditions and more! Learn skills, earn scholarships, make friends—the best year of your life." In the following years, he bounced between Oroville, Paradise, Las Vegas, and Alaska doing handyman and seasonal jobs. He liked to work with his hands and earned money plastering walls, tiling stucco, processing salmon, and, once, laying thousand-year-old stones for a masonry job. His life didn't gain direction until he met Erin at a friend's backyard barbecue.

The friend had promised that she was worth meeting. Erin was tough but warm, hardheaded but gracious, with enough drive for both of them. She was petite, with a heart-shaped face and dark, fluffy hair that brushed her waist. Erin thought Jamie looked like a stereotypical bad boy. He had faux diamond studs in each earlobe and spoke with a drawl. Every thought was a run-on sentence. But she loved his booming laugh and shy smile. She discovered that Jamie was in fact a teddy bear. They became inseparable. After their rental on Almond Street in Paradise nearly burned down in the Humboldt Fire of 2008, they moved into a travel trailer on her parents' property in Magalia to save money for their own place. Though they had been fortunate that time, and their rental survived, Erin knew she wanted her own home.

She planned to purchase a house whether or not she and Jamie stayed together. Erin had lived with an older brother and three younger siblings in a trailer until she started elementary school, when her family moved into their two-bedroom home in Magalia. For the family of seven, space had always been tight, and at fourteen, Erin decided to put aside her babysitting money for a house. She had been saving ever since. She longed for the freedom to nail whatever holes in the wall she wanted and to decorate a chef-themed kitchen, complete with wallpaper emblazoned with cartoon men in tall white hats. That spring, Erin paid $126,000 for a three-bedroom, two-bathroom house built in 1984. The driveway was matted with dead leaves and pine needles. She had to drag her feet to uncover the asphalt. But the closets: The hallway one was four feet deep, and the master bedroom had a walk-in. She had never had a closet of her own. She made the down payment in cash before Jamie ever saw it. She was twenty years old.

Erin's younger brother moved in with her and Jamie. They insulated the garage, laid a new roof, and installed gutters. Jamie repaved the asphalt driveway with concrete. A wooden fence bordered the backyard, and they stacked cords of firewood along its length. After two years of dating, they had a baby, Tezzrah. She was dark-complexioned, with thick black hair like Jamie's, though his had mostly receded by that point. Erin's family liked to tease that she didn't look much like her mother, though they loved Jamie and didn't mind having a mixed-race family.

When Tezzrah was two, Jamie and Erin got married. They held the ceremony at Merlo Park in Stirling City, a logging community higher up in the foothills. His family cooked a Mexican buffet. Hers carved an archway as a backdrop for their vows. They lit a hand-dipped unity candle. Erin baked a lemon cake with cream cheese frosting, adding two extra layers so the three tiers would stand straight. She placed a glass heart engraved with their names on top. The cake topper later disappeared—stolen or thrown away, they never figured it out. Tezzrah danced until she crumpled into a family friend's lap. It was the happiest day of Jamie's life.

Mariah and Arrianah soon followed. Jamie was outnumbered, but he treasured his girls, and he told them so all the time. They took their first steps down the hallway with the four-foot-deep closet and liked to do gymnastics in the living room. On Easter, the family's favorite holiday, they hunted for hundreds of plastic eggs in the backyard, ducking under the massive oak tree with the tire swing hitched to a branch. One year, Erin's brother surprised the girls with a crate of baby rabbits. Tezzrah named the tiniest one Cinnamon and adopted it as a pet. The rest eventually became dinner.

Jamie worked as the maintenance man at a long-term care facility called Heritage Paradise, where Erin was a certified nurse's assistant. They had both been hired recently. Erin's grandparents hadn't been around when she was a child, and she loved braiding the women's long hair and sneaking in donuts from Dolly-"O" to selected residents. Everyone looked forward to Jamie's visits, and he gave his phone number to anyone who asked. "Just call me," he would say, always willing to fit in a quick chat. He was responsible for the building's upkeep and access, making sure the hallways were cleared of medical carts and locking the emergency exits so patients with memory issues couldn't wander off.

Even more beloved at Heritage Paradise were Jamie and Erin's three girls. Recently, on Halloween, the couple had brought Tezzrah and her younger sisters to the facility for trick-or-treating. They went from room to room asking for candy to plonk into their plastic pumpkin buckets. The residents kissed and hugged the girls, who were bright-eyed and happy, who never shied away from their wheelchairs or oxygen machines, who made them laugh with their absurd questions. Given their parents' jobs, the girls were used to being around the sick and infirm. Death didn't scare them.

THAT THURSDAY MORNING, Jamie finished breakfast and changed into his uniform, packing his work bag. The facility, located just off the Skyway, had opened that summer under new management. Jamie believed in their mission, and he liked the staff. Administrators had poached the best nurses and aides from nearby facilities, promising a good paycheck and a lighter patient load. They had also restained the oak furniture and hung hummingbird feeders on the back patio. After opening on June 1, 2018, Heritage Paradise was finding its rhythm—as was Jamie. He had been promised a promotion and was being trained in the kitchen by Jill Fassler, the facility's chef. Jamie's life had been defined by quiet striving—now for his daughters—and this was no different.

He and Erin had heard rumors that the electricity might be cut because of high fire danger, so they had pulled out their cache of oil lamps, just in case. When Tezzrah was a baby, an oak tree had snagged a distribution line and PG&E had stranded them without power for more than ten days. For now, at least, everything was still working. The refrigerator hummed. The television blasted cartoons. "I'll take Tezzrah to school," he announced, heading outside to warm up his Subaru Outback. Mornings in the mountains were chilly. Erin was stuffing a load of damp laundry into the dryer when he returned to give her a kiss.

Whenever he left the house, without fail, Jamie told his family that he loved them. He kissed Mariah and Arrianah, who squirmed on the black leather couch, and grabbed Tezzrah's backpack. He walked out the front door, hand in hand with his oldest daughter. She chattered about her classmates, and how she wanted to stay up late on Friday to build a blanket fort and make buttered popcorn in the microwave. She was working on a word search at school and rehearsing a play about dogs. Jamie listened absentmindedly. "I love you," he called to his wife over his shoulder.

MORE THAN 125,000 MILES of Pacific Gas & Electric Company electrical lines stretched across Central and Northern California—a distance five times the circumference of the earth. PG&E was the largest power provider in the state, delivering electricity and natural gas to more than 16 million people. Its service area covered seventy thousand square miles, and most of California—equivalent to one in twenty Americans—received power from the private utility.

Nearly half of its grid crossed land that the state considered high fire risk. The network consisted of two types of power lines. Transmission lines were high-voltage electrical freeways that ran vast distances to cities; distribution lines were low-voltage cables strung along wooden poles that crisscrossed roadways and neighborhoods. These caused PG&E the most grief because they were closer to the ground and toppled when tree branches snapped, as had happened when Jamie and Erin lost power. For the previous decade, PG&E—criticized for its failures at maintaining both types of lines—had been struggling to invest in its infrastructure and regain public confidence. As millions of trees across the Sierra Nevada died and the landscape withered, it became a losing battle.

The company's problems had begun in 1993 in Hinkley, an unincorporated community in the high Mojave Desert. A file clerk named Erin Brockovich discovered that PG&E had dumped 370 million gallons of tainted waste from a natural gas pumping station into adjacent unlined ponds. From there, the carcinogenic compounds—used to prevent rust—leached into the groundwater, contaminating the town's tap water. The utility tried to conceal its role, but the confirmation of hexavalent chromium in the drinking water led to one of the largest direct action lawsuit settlements in American history. More than six hundred residents of the rural San Bernardino County town were paid $333 million, and the case inspired an award-winning film, *Erin Brockovich,* starring Julia Roberts.

One year later, a valley oak brushed a 21,000-volt distribution line in a hamlet east of Sacramento. The ensuing Trauner Fire blackened 500 acres in Rough and Ready, a town of 950 people, destroying twelve homes and a historic schoolhouse constructed in 1868. The district attorney sued PG&E. She found that the company had diverted more than $77.6 million from its tree trimming budget and put most of it toward corporate profits. Investigators uncovered two hundred safety violations. PG&E was convicted of 739 counts of criminal negligence and forced to pay $24 million in fines. "*People v. PG&E* affords us a rare (and brief) opportunity to peer at the inner workings of a corporation that claims to be our benefactor," wrote a local reporter who covered the case. "However, when the curtains part, we discover a cast of characters and a plot that would have suited Dickens or Runyon."

The company's safety record only got worse. More than a decade after the Trauner Fire, a natural gas pipeline owned by PG&E ruptured around dinnertime on September 9, 2010, in the San

Francisco suburb of San Bruno. The explosion in the Crestmoor subdivision was so loud that first responders initially thought a jet-liner had crashed. The 3,000-pound steel pipe registered as a 1.1 magnitude earthquake as it burst, shifting homes from their foundations and gouging a 72-foot-long crater in the concrete. A thousand-foot wall of fire engulfed thirty-eight homes and burned eight people to death, including a mother and her thirteen-year-old daughter. Another fifty-eight were injured. PG&E struggled for more than an hour and a half to shut off the gas.

"It was beyond words," one resident said of losing her teenage son, husband, mother-in-law, and dog in the explosion. "The loss of my loved ones, my personal belongings, my neighborhood, and my life happened due to the negligence of a greedy company."

When Crestmoor had been built in 1956, crews had welded the thirty-inch natural gas pipeline incorrectly. The company hadn't caught that the pipeline was slowly cracking; it didn't even know where many of its pipes were located. In 2016, a federal jury convicted PG&E on six felony counts of violating safety standards and obstructing the investigation into the explosion. There was subterfuge on a grand scale: Documents were found to contain fabricated data or be printed in erasable ink. State regulators had already fined PG&E $1.6 billion for the incident in 2015—at the time, the largest penalty ever imposed on a utility in the United States. After the federal court sentencing, PG&E was slapped with a second, $3 million fine. Its executives were ordered to perform ten thousand hours of community service. The company was put on court-ordered federal monitoring, like a felon.

The California Public Utilities Commission, known as the CPUC, was tasked with supervising the state's investor-owned electric and gas utilities, as well as railroads, rail transit, water, telephone, and ride hailing companies like Lyft and Uber. The commission was an economic regulator, not a safety enforcer, and it struggled to oversee all of these entities, particularly PG&E, which was essentially a monopoly: too big, too many resources. And the wall between them was perforated with revolving doors. Employees left the CPUC to work for PG&E or as lobbyists on the company's behalf, helping to write the laws that governed them. PG&E was a top political donor in Sacramento, shelling out $5.3 million to campaigns in 2017 and 2018. The relationship was so inappropriate that the utility paid nearly $100 million in fines following the San Bruno explosion for communications that violated state law. In one of tens of thousands of emails made public, it was revealed that then-CPUC President Michael Peevey had shared a couple of bottles of "good Pinot" over dinner with a PG&E executive and his wife.

PG&E failed to change. People continued to die because of its negligence—in fact, the disasters only increased in scale. On a 102-degree day in September 2015, a 44-foot gray pine fell into a 12,000-volt power line east of Sacramento in Calaveras County. The Butte Fire (nowhere near Butte County, despite the name) destroyed 549 houses. Two people died. Calaveras County sued, on the grounds that PG&E had known the tree was a hazard and had chosen not to trim it. The CPUC imposed an $8 million fine.

Then, on a blustery evening in October 2017, the worst wildfires in modern state history ignited. They ripped across Northern California, pushed by the Diablo Winds. The infernos killed 44 people and hospitalized another 192. They incinerated fabled vineyards and the working-class Santa Rosa neighborhood of Coffey Park. People died in swimming pools, in mobile home parks, in their bedrooms and their cars. A fourteen-year-old perished at the end of his family's driveway, unable to outrun the flames. PG&E was held responsible for seventeen of the twenty-one wildfires—which burned an area eight times the size of San Francisco—though the company escaped blame for the worst of the bunch. The Tubbs Fire killed twenty-two people in Santa Rosa and the unincorporated neighborhoods on its outskirts. (In 1964, the Hanly Fire had burned the same footprint, destroying only orchard land and a hundred homes.) The culprit, Cal Fire investigators found, was a private power pole owned by a ninety-one-year-old woman near Tubbs Lane that had been damaged by woodpeckers.

This fall, PG&E was trying something new. As the National Weather Service forecast a Red Flag event in Northern California, the utility cut electricity to distribution lines in seven counties, hoping to prevent a wildfire. The CPUC approved the de-energization. San Diego Gas & Electric had pioneered this approach after sparking two huge blazes in 2007, which had merged with a third wildfire to kill two people and level more than thirteen hundred homes. Cutting electricity was controversial, and PG&E had resisted imposing such outages for years. But management felt it had few options. As the planet warmed, nighttime temperatures had risen twice as fast as daytime temperatures, lowering relative humidity during the early morning hours, when firefighters had heretofore been able to contain a wildfire more easily. In the previous four years alone, more than 1,550 wildfires had been linked to PG&E equipment failures. There was no room for error.

The electricity cutoff, which took place in October, left nearly sixty thousand customers without power. San Diego Gas & Electric, or SDG&E, followed PG&E's lead and blocked electricity to an additional 360 customers in the foothills communities sixty miles northeast of San Diego, near the Cleveland National Forest. In the last half decade, SDG&E had used shutoffs as many as twelve times. They seemed to work. The smaller utility's power lines had not sparked any significant wildfires since then.

In Northern California, gas stations closed. Convenience stores sold out of batteries and ice. Residents brushed their teeth by flashlights and cellphone screens. Tourists with out-of-state license plates wandered through the empty streets of Calistoga—a tony hot springs town in Sonoma County—in search of a cup of coffee. With traffic signals out, the police propped up temporary stop signs. Local school districts canceled classes, and children, enjoying the rare day off, pedaled their bikes in the hushed streets.

ON THE EVENING OF Wednesday, November 7, PG&E was considering cutting electricity off again, this time to parts of eight counties. Paradise Town Manager Lauren Gill sent out an email update to the Town Council and staff. The chance of deactivation in

Paradise was 35 percent, Gill wrote at 3:29 P.M. If PG&E chose to cut electricity to the town, she added, it would happen between 4 A.M. and 8 A.M. the next morning.

"PG&E will have 6-8 helicopters and 135 personnel inspecting the lines to get the power back on as soon as possible," she continued. "Although this is not our emergency and we are not in charge of it, this may be the new normal. All residents of California need to be prepared for these planned outages. We have windy days every year—it is actually more dangerous when power lines are charged. Let's hope that these outages save lives and property."

PG&E managers had been hesitant to order this outage. Each mile that was deactivated would have to be carefully inspected before electricity could start flowing again. With more than 3,100 miles of lines—equivalent to the distance between San Francisco and Boston—this would take time, and PG&E knew there would be pushback. More than seventy thousand customers stood to be impacted—though the actual number of people affected was far higher. A single "customer" could be a household of four people, an office building, or an entire mobile home park. No electricity meant freezers of spoiled food and run-down cellphone batteries. Vital medical equipment, like oxygen and dialysis machines, wouldn't work.

Already, residents were emailing the Butte County Board of Supervisors to complain. At 7:18 P.M., one of these notes landed in Supervisor Doug Teeter's inbox. Teeter lived with his wife and two young daughters on Rockford Lane in Paradise. His grandfather had built the house by hand and named the street after his hometown in North Dakota. Teeter, who had switched to the Republican Party in recent years, represented Paradise and the unincorporated village of Magalia. He was a man of principle—though the gas station up the street was closer, he preferred the one just outside town, driving the extra two miles so the sales tax would go to the county. He had no control over the shutoffs, but his constituents liked to grouse. "We have just been informed by PG&E that power to our community may be turned off potentially for several days," the email read. "I believe it is counterproductive and punitive to residents of the affected areas. PG&E has been providing power for decades and has only now engaged in this ridiculous policy!!"

PG&E had plans to avoid such outages in the future by increasing surveillance of its grid, with a goal of adding another thirteen hundred weather stations and six hundred high-definition cameras by 2022, but those measures weren't in place yet. On Tuesday, November 6, the utility had activated its new Wildfire Safety Operations Center in San Francisco, which had only just opened eight months before. Its meteorology team was working twenty-four hours a day, monitoring wind speeds, measuring humidity levels, and calculating fuel moisture to determine how quickly a wildfire might move. Workers on the ground in Butte County were assigned to keep an eye on the lines. Some of them were positioned near Camp Creek Road, one mile from the neighboring town of Pulga. A woman from the Bay Area had bought the ghost town—named after the Spanish word for "flea"—for $499,000 in 2015 and dubbed it Ladytown. Her 64 acres were a few miles down the highway from Station 36 in Jarbo Gap.

The fifty-six-mile Caribou-Palermo Line delivered power from Poe Dam, a popular kayaking spot. The dam was part of a chain of seven hydroelectric powerhouses that plugged the Feather River. Known as the Stairway of Power, they were a technological wonder, churning out enough clean electricity to power San Francisco. The transmission towers that connected them, snaking past Pulga along the Feather River Canyon, were so old they had been considered for inclusion on the National Register of Historic Places. The average tower near Jarbo Gap was sixty-eight years old. The oldest was 108. Workers struggled to maintain them without injuring themselves. In October 2016, a contractor painting a tower had grasped a piece of cross bracing to shift positions. The metal snapped.

PG&E management knew the towers were at risk of collapse, but they assumed any damage would occur during a wet winter storm—they hadn't factored in what would happen if those storms arrived late. In December 2012, a rainstorm had pummeled the

Caribou-Palermo Line and knocked down five steel towers, an event the utility termed a "catastrophic failure." The next year, PG&E replaced them—along with one additional tower that had been damaged—with fifteen temporary wooden poles. Workers didn't swap them out for permanent steel poles until 2016—four years after they'd fallen.

One of the oldest pieces of infrastructure was Tower 27/222. Captain McKenzie drove past the tower, visible from Highway 70, every week on his way to Station 36. It had been designed in 1917 and put into service on May 6, 1921, shortly after World War I ended. At the time, Warren G. Harding was the country's twenty-ninth president, and the first Miss America Pageant was being held in Atlantic City. As crews finished its construction, Prohibition had just taken effect and women had recently gained the right to vote.

Between November 6 and November 7, PG&E posted fifteen tweets warning of an impending shutoff. These outages didn't include powerful transmission lines like the Caribou-Palermo—only local distribution lines. The company also sent email and text message notifications to customers. At 6:13 P.M. on November 7, Gill dispatched a final email to her colleagues: "We will all know tomorrow morning. The Town is ready."

FOR DAYS, TRAVIS WRIGHT had been bombarded with text notifications from PG&E. Now his phone was quiet. No new updates. His wife, Carole, had left for work, pausing to snap a photo of the morning sky for its eerie beauty. Travis headed outdoors to clean up the yard. He paid attention to every Red Flag Warning. Danger had struck in the foothills before—it was the price of living in the forest. At fifty-three, he tried to be responsible. And by this point, he knew how to prepare for the possibility of fire.

Travis cleared out the pine needles spilling from the gutters and dragged sprinklers in a semicircle around his backyard. He returned to the front, where ornamental grass whispered by the front door. He headed inside to round up their possessions. The 14-by-17 wedding portrait in the living room went into the back of his

Subaru Crosstrek first. He added rifles, computers, hard drives, photo albums, and emergency cash, then a bright yellow folder of important documents—Social Security cards, passports, birth certificates—and, finally, a stack of jeans from his wife's dresser drawer.

His work boots crunched on the gravel. He squirted a garden hose at the roof. Across the street, he saw his close friends and neighbors Paul and Suzie Ernest swing open the front door of their house. Paul often wandered over to chat in the afternoons, barefoot and bare-chested, swigging from a bottle of pale ale from the local brewery. Travis watched them climb into Paul's old blue truck. He waved hello, then continued working.

Their homes were clustered along a narrow ridge near the southern edge of Paradise, where pavement turned to gravel before dead-ending into wilderness. One of his neighbors had bought a tractor to grade and maintain the road. Everyone chipped in some cash to help. Most people in town didn't even know that Edgewood Lane existed, but it reminded Carole of her longtime family home. Her relatives had bought a 40-acre property in Paradise with a few gold bars in the early 1900s. The land had since been parceled off and sold to developers, but she treasured her memories of her childhood visits. It was her tales of soaring ponderosa pines and the sweet taste of wild blackberries that had convinced Travis to relocate to Butte County.

Born in Southern California, he had spent much of his childhood with his grandparents in Bakersfield before moving to Napa to attend high school. He met Carole there, in the school gymnasium. Their first date was after church. They shared homemade manicotti that his mother had prepared. After their graduation, Travis proposed with a rectangular Seiko wristwatch. He couldn't afford a diamond ring. The proposal was nothing special—they were sitting in the front seat of his Toyota Celica—but he felt it was the right thing to do. He was twenty; she was nineteen. Carole always said "almost twenty" when she recounted the story, because it sounded better. Looking back, Travis realized how young they had been. Maybe they were wise beyond their years or maybe they were just stupid, they liked to joke.

They moved to Pasadena, where Carole trained as a dental hygienist and Travis studied radiology. He liked seeing the proof of a patient's pain materialize on the X-ray screen, the small crack in a bone crystallizing to black. It was a small way he could help others. They had a son, Jefferson, who slept in the home office of their historic bungalow. They considered having a second child, but there wasn't enough space or time. Travis read his son books, played Legos and Pokémon with him, and taught him to ride his tricycle on city sidewalks. He didn't mind staying up past midnight with the boy when he was sick. He included him in everything. Jefferson grew, and so did Pasadena. The Town Council voted to knock down the historic homes on Colorado Boulevard for condominiums, and parking meters sprang up along the streets. Traffic snarled. Their family's story of coming to Paradise was like so many others in that way. It was safer and more affordable to raise a child in the foothills.

Travis had agreed to check out the mountain town on the condition that they not live on a gravel road pitted with potholes. But he fell in love with Edgewood Lane, ruts and all. That was the charm of Paradise. After a failed bid to purchase a historic home that everyone called the Dr. Mac house, for the prominent local doctor who had built it, he and Carole designed a Craftsman-style home for themselves. Rabbits and deer nibbled at their garden, and invasive brush crept up to their patio. Travis struggled to keep it at bay, plowing two hundred feet of defensible space. In 2008, when the Humboldt Fire nearly engulfed the town, fire officials had parked an engine on the lawn, preparing to save his home if the flames crept closer. The structure was built with fire-resistant materials and ringed with cleared land, while the neighboring lots were too overgrown to safely accommodate an engine. The officials assured Travis that had a fire encroached, his house would have survived.

Travis was proud to hear it. He had spent a lot of time doing yardwork since retiring from his job at Feather River hospital. He had taken CAT scans and X-rays until recently, when rheumatoid arthritis made his hands swell and ache, causing him to drop expensive film and struggle to lift patients from their gurneys and wheelchairs. Retirement was a tough transition, but Travis soon realized that he enjoyed the solitude. He spent many afternoons at the antiques shops on the Skyway. The closets in his home overflowed with purchases, which sometimes frustrated Carole, though she was mostly just happy that he had found something to do with his free time. On weekends, Travis rode his four-wheeler along the Ridge's lava-capped trails with Paul, his neighbor. They had recently discovered some caves and dilapidated gold mines. They snacked on tomatoes from Paul's garden, warm from the afternoon sun.

On this Thursday morning, Travis shuffled through pine needles a foot deep, as dense as snowdrifts. They sucked at his leather boots and crowded the crannies of his home. The wind was relentless. His cellphone pinged with an alert. He ignored it and continued working.

THE FIRE: THE CARIBOU-PALERMO LINE

The wind slammed against the Harding-era transmission tower, ripping a heavy electrical line from its brittle iron hook. It was 6:15 A.M. The 143-pound, 115-kilovolt braided aluminum wire—known as a jumper cable—fell through the air. A piece of the rusted hook fell with it. The energized line produced a huge bolt of electricity, reaching temperatures up to 10,000 degrees Fahrenheit and zapping the steel tower like lightning as it charred the pillar black. Droplets of molten metal sprayed into the dry grass.

That's all it took. Where the hot metal droplets landed, fire ignited. The wind stoked the flames, which glittered in the darkness. For ten minutes, or maybe fifteen, the fire feasted. Tiny flames licked along tiny stalks and burrowed into vegetation beaten back by herbicides. Wisps of white smoke filled the air, eddying in swirls and curlicues, as delicate as milk dissipating in coffee, before disappearing. The fire glowered. Near the fringes of the cleared land, dead timber was stacked several feet high. Manzanita, pine saplings, and grass bunched around the dead logs. The fire blasted toward the tangled and overgrown brush.

The flames stretched higher, grasping for oxygen, then hunched and bent horizontally with the wind, pushing southwest. The heat broiled the flat volcanic benches ahead, where national forestland touched private logging fields. The fire didn't care about man-made divisions. It was ravenous. The smoke thickened, impossible to ignore. As the hot air rose, cooler air rushed in to take its place, pushing the flames up the slope. The canyon lay in the distance—a ready racetrack. There was nothing, and nobody, ahead to halt the fire's advance.

7 A.M., near Pulga

Paradise

PART II

SPARK

KONKOW LEGEND

As the days and nights interchanged in the countless moons of the past, the Konkows and all the other people on the face of the earth became very wicked and bad, until one day the spirit of Wahnonopem, borne upon the beams of the rising sun, came through the pines and appeared unto some very wise old men, and said to them: "My children, whom I have made out of my breath, shall not bow down and worship the mountains, the waters, the rocks, or the trees, or anything which I have made upon the earth, or in the waters, or in all the skies; but go to all my people and say that they shall bow down to me, and me alone; and all who do not believe in and worship me shall be devoured by the wild beasts and the demon birds of the forests, or destroyed by the great fire, Sahm."

This the Great Spirit said to the teachers of our tribe, and then he passed away into Hepeningkoy, the blue land of the stars. But his words were not heard, and wickedness increased and went wild and rampant about the whole land, and Wahnonopem caused Yanekanumkala, the White Spirit, to appear in the flesh unto the people, that he might enlighten and turn them from their evil ways. This good man began his teachings and for many years he lived among our people, teaching the young men and the maidens many lessons of love and wisdom, many songs and games and gentle pastimes; and all these years they loved him more and more. But he died, and the lessons were forgotten. The songs died away in the forests, and in their stead came the war whoop, the shrieks of struggling women, and the groans of the wounded and the dying; and the name of Yanekanumkala became a jibe and a mockery all over the land.

As time went on, the Great Spirit sent two more good men, white spirits from the Yudicna, the unreachable frozen regions at the end of the earth, to explain once more the teachings of wisdom and of love, and the worship of Wahnonopem; and to show them that they came from the Great Spirit. He made the streams issue forth from solid rock, the mountains dissolve into lakes and the waters of the sea. They healed the sick and gave back the spirit of life to the dead, who, as they quickened into life again, bowed down for a time before the Great Spirit and worshiped him. But these good men accomplished no lasting good. Wickedness went about roaring as fiercely as before; and they passed away, carried by the wind to their homes in the frozen seas, amid the floating ice mountains, and the golden auroras of the far-off Yudicna.

Wahnonopem, after the good men had departed, became wrathful against his children and sent a great drought upon the land. The gentle rain fell no more upon the earth, and it baked and cracked and yielded no more food. The sweet summer grasses and the white clover shrank away and became as wisps; the pine tree bore no more of its nutty cone; the brown balls of the buckeye and the red grape of the manzanita were nowhere to be found; and the flesh of the roebuck, the black bear, and the wild game in the woods was a frothy poison. The people worked hard digging for the socomme, the sweet roots of the swamps, which had become as rocks, and when found they were molded away or wasted into strings. Suffering and hunger were all over the land, and the old men, the young men, the women and the maidens cried in their anguish for the Black Spirit of Death to come to their relief.

THE FIRE: DESCENDING INTO CONCOW

How easily this parched land burned, incinerating with a soft crackle that deepened to a howl. The wildfire lapped Flea Valley Creek and climbed upward. Embers hurtled one to two miles ahead of the flaming front, carried on the wind. Every second, flames raced a distance greater than a football field. Every eight minutes, a swath of land the size of Central Park—or downtown Chicago or the country of Monaco—burned. The fire overtook every living being. Flame darted from brush and vine into the canopy of old-growth timber, where it burned brighter and hotter. Its hunger was endless, but there was plenty of fuel. The flaming head of the wind-driven fire sprinted faster than its flanks, carving a trail like a shooting star.

Pine needles baked in place on their branches, seemingly frozen in time. At 212 degrees Fahrenheit, the water and sap stored in tree trunks began to boil. The trees sweated until—their cell walls bursting—they combusted. The pine forest had settled into dormancy for the winter and awoke just in time to die. Leaf and limb turned to carbon in seconds. Dead timber and rotten stumps smoldered. Their shriveled roots carried fire laterally along the oxygen-rich subsurface, threatening to ignite vegetation up to forty feet away.

As the wildfire crested the ridgeline above Concow, it slowed and trundled downhill. But not for long—the Jarbo Winds were spilling over the Sierra Nevada, juddering from high to low pressure. The descending air stoked the flames, forcing them faster down the incline. The smoke column, which had been building upward in place, sheared with the wind. It coated the ground in a white miasma that caramelized to yellow, then russet and gray. In places, temperatures exceeded more than 3,000 degrees Fahrenheit. A single breath could blister lungs. The minerals in the volcanic soil—iron, lime, and silica—melted and hardened to slivers.

The smoke heaved, churning with ash as it punched through the lower levels of the atmosphere, now visible to two satellites that floated twenty-two thousand miles above the earth.

CHAPTER 4

CODE RED

A t Station 36 in Jarbo Gap, the crock pot hissed, releasing a thin waft of steam. The smell of corned beef hung in the air. Captain McKenzie diced the last of the red potatoes and dumped them into a bowl of water so they wouldn't turn brown. His higher rank meant that he didn't have to cook breakfast for the other firefighters, but he enjoyed it. He took pride in his reputation as one of the best station chefs in Butte County.

The ceiling lights remained off, and a television brightened the kitchen. The local news channel was playing clips from Southern California, where a gunman in Thousand Oaks had shot up a Western-style bar popular among college students. At least a dozen people had died, the newscaster announced, in the deadliest shooting for California that year. The screen flashed to a reporter in a rumpled suit, interviewing a young man who had been inside the bar. "That's what's really blowing my mind," the student said, shaking his head in disbelief. "It's a really safe area." The program flipped to a commercial. McKenzie ground coffee beans and spooned the granules of French roast into a paper filter. Less was needed these days: The seasonal employees at Station 36 had recently left their posts. Their departure marked another end to fire season.

The crew normally woke up by 6:15 A.M., lining up barefoot to shower, then tucking their navy T-shirts into their navy cargo pants and lacing up their leather work boots. Breakfast was at 7 A.M. sharp. But it was November, and the men of Station 36 moved more slowly now, savoring the calm after a ruthless fire season that had killed six wildland firefighters. Only an afternoon training hike was on the schedule. McKenzie could see the barracks from the kitchen window, his own movements distorted in the glass. The building was still dark. He chopped bell peppers and a yellow onion. The news program returned to the Thousand Oaks shooting. At 6:29 A.M., his phone blinked with a notification from Cal Fire's Emergency Command Center, or ECC, in Oroville. A worker at Poe Dam had called dispatch to report a vegetation fire under the transmission lines off Highway 70, possibly in the Camp Creek Road area. Jarbo Gap was the closest fire station.

"Bullshit," McKenzie said, setting down the knife.

He didn't smell smoke—it had to be a false alarm. He walked across the kitchen and flung open the back door. The wind ripped the knob from his grip, slamming the door against the metal porch railing. The air was crisp, no trace of ash. The ponderosa pines moaned in the wind.

When a wildfire was reported, the ECC played emergency tones over its radio frequency, making sure to catch the attention of first responders who might be exercising, sleeping, or otherwise distracted. In the barracks, the firefighters awoke to a high-pitched shrieking that lasted ten to fifteen seconds. McKenzie saw the lights in the building flick on. He jogged to his bedroom, shedding his sandals and athletic shorts for a fire-resistant suit.

He headed quickly to the garage and climbed up into Engine 2161. Sweaty gear was piled in the backseat. McKenzie registered the stink with annoyance. A new recruit slid into the passenger seat, and a veteran with twelve seasons of firefighting experience crawled into the back. Their seatbelts clicked. McKenzie started the engine with a rumble, a bobble-head Yoda bowing atop its radio mount. "Everybody in?" he said. His passengers nodded. At 6:35 A.M., McKenzie steered down the driveway and hooked right onto Highway 70, his headlights sweeping across the empty parking lot at Scooters Café and reflecting off the fog lines. A second engine with another three firefighters followed, its dashboard sporting a rubber duck.

"I didn't smell anything. Did you?" asked the older firefighter.

"No," McKenzie said. "Nothing."

The next six minutes were silent.

Highway 70 paralleled the Feather River Canyon as it wound through the foothills. The new firefighter flipped through a book of maps, trying to match the latitude and longitude on the dispatch with local landmarks. He held a flashlight to illuminate the pages, swaying with each curve of the road.

Wildfires were reliant on heat, oxygen, and fuel—a combination known as the "ignition triangle"—for sustained combustion. When the three factors aligned, flames grew in size. When a variable was removed, as with damp grass, a sealed room, or mineral soil, flames starved and sputtered out. Fire was a science; it obeyed the laws of physics and could therefore be studied. But like the weather, it couldn't always be predicted. A good firefighter relied on experience and deep knowledge of fire behavior—and also gut instinct.

McKenzie careened around another bend in the highway, passing the sign for Plumas National Forest, its letters inscribed in looping yellow cursive. Near the town of Pulga, the hummocks dropped to reveal PG&E's Caribou-Palermo Line crisscrossing the valley. The system was a feat of engineering, some of its towers drilled into granite rock outcroppings, others situated atop steep drainages. Had McKenzie been closer, he might have seen the line break near Tower 27/222, the snapped cable flapping in the wind. The hook had fallen forty-seven feet to the ground. But from a distance, he couldn't at first make out the reason for the emergency dispatch.

"Oh, shit," McKenzie heard from the backseat. "I see the fire."

Flames rippled atop the canyon wall. They were short and squat, indicating a low-intensity burn, and they were billowing white smoke. McKenzie knew the blaze wouldn't stay small for long. It seemed to be creeping toward Flea Valley Creek, where the air would eddy before funneling up the sheer draw and blasting over the mountains, coaxed along by the Jarbo Winds. He still couldn't tell where, exactly, the fire had sparked.

Fourteen minutes had passed since he received the alert. He continued northeast.

At 6:43 A.M., he parked on Pulga Bridge, upriver from Poe Dam. Suspended by a web of steel girders, the bridge sat high above the Union Pacific Railroad tracks, which snaked below, following the gentle curve of the Feather River until they crossed the canyon on their own three-span bridge. The two bridges appeared to be stacked one atop the other, extending over the same water at different angles, the pine-speckled gorge slanting upward on either side. The view was striking, an iconic landscape that graced postcards, calendars, and coffee mugs—but it did not register with McKenzie. He rolled down his window and peered west. The tops of the mountains, 3,000 to 5,000 feet high, were soft with orange light. Through the smoke he made out a single transmission tower rising out of a singed hole in the canopy. The tower's proximity to the flames made McKenzie wonder if it had caused the wildfire.

The second engine shuddered to a stop behind him. The flames appeared much closer now, lighting up Douglas fir, ponderosa pine, and incense cedar like birthday candles and clawing through the manzanita. Embers winged westward on the wind. It looked like the wildfire was 10 acres in size, maybe more, as it blasted along the boulder-strewn slope toward Pulga. The town, which a geologist had built in 1904 as a base for railroad workers, had emptied out after the heyday of gold mining and steam passenger trains diminished. When Betsy Ann Cowley bought Pulga, in March 2015, she and her friends became the only residents, though she hoped to turn her "Ladytown" into an events space.

Camp Creek Road, a rugged dirt track named after a nearby stream, heaved across the southwestern corner of her property. PG&E workers known as linemen used the route to access and check on the utility's grid, including the Caribou-Palermo Line. The day before, a right-of-way consultant working for PG&E had sent Cowley an email alerting her that the company would be completing maintenance on one of its electrical circuits that week. The work involved a swarm of ground vehicles and the maddening clap of helicopters. On a cruise in the Dominican Republic, Cowley didn't respond.

McKenzie knew that the only way to reach the fire burning near Pulga was to wend up Camp Creek Road. His crew could try to chase the wildfire on foot, lugging hoses connected to their engines and spraying water at the flames—but it would be a death wish. The dirt track was so treacherous that the blaze would no doubt barbecue them first. The previous autumn, McKenzie had driven the road to try to reach a smoldering pine that had been struck by lightning. But the track hugged the cliffside so tightly that he had nearly scraped the engine's side mirrors off. Firefighters had walked ahead of him, craning their necks to check for erosion under each tire lest he topple into the adjacent canyon. It took an hour to drive one mile before McKenzie was able to turn around, spooked by the risk.

There was nothing to do but radio for extra help. He studied the blaze as it scuttled along the escarpment and thought about how many fire engines he should request. Normally, for a new wildfire, he might ask for five. But watching the flames, he decided to request fifteen, along with four bulldozers, two water tenders, and some hand crews. He knew that the size of his request—nearly half of the unit's thirty-nine engines—would strike fear in anyone listening to the scanner at that moment. He steadied his voice.

"Engine 2161 responding," McKenzie said over the radio at 6:44 A.M. "Possible power lines down. We have eyes on the vegetation fire. It's going to be very difficult to access because Camp Creek Road is nearly inaccessible. It's on the west side of the river, underneath the transmission lines, with a sustained wind on it. This has the potential [to be a] major incident."

The radio crackled and cut out.

SEVERAL MILES SOUTH of Pulga Bridge, Curtis Lawrie parked his Town of Paradise–issued Ford Expedition outside Scooters Café.

A battalion chief for Cal Fire, Lawrie was on call that day, responsible for handling any major crises that popped up along Highway 70. He was one of many leaders in Cal Fire's strict hierarchy. The agency oversaw units in thirty-five of California's counties; Butte County's was organized into several tiers, the better to oversee the twenty-three municipal fire stations and sixteen volunteer stations scattered across its seven regions. The unit was led by Chief Darren Read. Beneath him were division and battalion chiefs, then captains, engineers, full-time firefighters, and seasonal workers. As a captain, McKenzie supervised a single fire station; as a battalion chief, Lawrie oversaw several. Lawrie's boss, John Messina, who served as a division chief, managed three or more battalions at once.

When a battalion chief went off duty, a peer filled in for him, which is what Lawrie was doing that morning. Jarbo Gap wasn't his usual assignment—his home station was in Paradise. When Lawrie's pager had beeped at 6:29 A.M. with an automated dispatch—the same ECC alert McKenzie had received—he was already awake. He kissed his wife, Tessa, goodbye, asking her to keep an eye on the local news as she got their children ready for school. Then he made the twenty-five-minute drive east to Jarbo Gap, the rolling foothills turning to arid mountains.

Wind thudded against his windshield. Lawrie looked up, taking in the DON'T TREAD ON ME sign nailed to the front of the restaurant, next to a painting of a skeleton lounging on a white sand beach smoking a joint. In one of the windows, the neon Budweiser sign had gone dark. Lawrie scanned the sky—still impossibly clear. He rubbed his temples, where his brown hair was graying, and slid out of the SUV. He was fifty but usually looked a decade younger, with a round, youthful face. But on this morning, his eyes looked tired, his face aged by shadows. He hadn't slept much the past few nights, worried that PG&E was going to shut off power.

On Tuesday, Paradise Fire chief David Hawks, who was also a Cal Fire division chief, had forwarded Lawrie an email from a PG&E public safety specialist. The specialist warned that Butte County, of all the areas at risk of blackouts, stood to receive the strongest of the forecast winds. The email had nagged at Lawrie—

and now it bothered him even more. He hadn't gotten around to purchasing a backup generator for his home on Pentz Road. His daughter would be upset if the tropical fish in their heated 30-gallon tank died.

Lawrie was now the incident commander for this new wildfire, tasked with coordinating the efforts to contain it. Wildfires are named after the road or landmark nearest to their origin, and his colleagues at the ECC had already dubbed this one the Camp Fire. Dispatchers had considered designating the blaze the Seventy Fire, after Highway 70, or the Poe Fire, after Poe Dam, but those names had already been taken—and much like the National Hurricane Center, which retired the monikers of destructive storms, Cal Fire tried not to recycle names. Even so, the name Camp Fire had also been used once before, in 2008, for one of the twenty-seven fires that merged to form the Butte Lightning Complex. That Camp Fire had burned fifty homes to the ground.

Just over the ridge from Pulga was the unincorporated town of Concow. Lawrie knew he needed to get a handle on this new inferno, and fast. The Butte Lightning Complex had been devastating to Concow, and he didn't want it to suffer another hit. The community's population numbered 710—mostly retirees and those who preferred seclusion. The median household income was about $25,000. Mobile homes and travel trailers sat along the edges of Concow Reservoir, in the shadow of Flea Mountain. There was one elementary school with six teachers. No grocery stores, no medical facilities, no public transportation.

Lawrie's radio crackled. It was an update from McKenzie. The captain reported that the Camp Fire was already thirty times bigger than when he and his crew had arrived on scene just half an hour earlier. "My best guess would be about two hundred to three hundred acres, possibly, with a rapid rate of spread," said McKenzie. "It's just above Pulga now. Heading in a direction toward Concow."

MORE THAN TWENTY MILES to the southwest, the smoke column was already visible on the Lake Oroville Wildfire Camera, part of a statewide network of cameras used for fire surveillance. In Cal Fire's Emergency Command Center, dispatchers monitored the footage on a row of televisions, which broadcast feeds from every fire camera in the county. The column had begun to lean west with the wind—this is why McKenzie hadn't smelled it from his fire station in Jarbo Gap.

The ECC sat within the headquarters of Cal Fire's Butte County unit—known colloquially as BTU—which took up four and a half acres in Oroville, across the street from a Home Depot and a Dollar General. Dispatchers answered more than twenty thousand 911 calls annually and directed firefighters and equipment across the seven battalions. When an incident commander requested an evacuation, ECC dispatchers called his orders in to the Sheriff's Office—unless, of course, a sergeant was already on scene to do it. Law enforcement officers then spread the message to evacuate using CodeRed, an emergency platform to which they, and some local city managers, had exclusive access. The Butte County Sheriff had already issued the first alert: "Due to a fire in the area, an evacuation order has been issued for the town of Pulga. If assistance is needed in evacuating, call 911."

Beth Bowersox, thirty-three, was one of more than a half dozen ECC dispatchers on shift that morning. With straight, shoulder-length brown hair and a dimpled chin, Bowersox was tough and assertive, traits that came in handy as a woman in a male-dominated workplace. She wasn't always as fast or as strong as the men she worked with, but she knew she made up for it by being smarter. She had gotten her start as a seasonal firefighter in Tehama County, northwest of Butte, defending homes and scraping firebreaks in the dirt during the state's blistering summer months, before becoming a dispatcher. She had worked the overnight shift in the ECC and was the one to suggest "Camp Fire" as the name for the blaze—a name that would stick throughout all the chaos and disaster to follow. In the dispatch pods adjacent to hers, Bowersox could hear her colleagues fielding a growing volume of phone calls. As they spoke with residents who were reporting fire sightings, she communicated over the radio with firefighters in the field. That day, Bowersox would be Lawrie's primary contact for receiving and passing on information.

Thanks to the camera feeds near her dispatch pod and the civilian reports, Bowersox had a better idea of where the Camp Fire was burning than Lawrie did. But despite the observations she could provide him, he didn't have enough to go on yet to create a plan of action. The standard approach was to rely on the Incident Command System, a protocol for communicating and coordinating response efforts during an emergency. The U.S. Forest Service, along with several Southern California fire agencies, had developed it in 1974 after a devastating series of wildfires killed sixteen people. The system—a means of imposing order and creating a common language in the midst of chaos—was so effective that the Federal Emergency Management Agency, or FEMA, had adopted it in the 1980s, along with smaller governing bodies like the Paradise Town Council and the Butte County Board of Supervisors. As incident commander, Lawrie needed to divide the wildfire into sections and subsections—called branches and divisions—and establish a chain of command. Firefighters would then manage each section like a fiefdom, funneling information up to him. Lawrie needed their on-the-ground reports in order to make important, fateful decisions, such as when to evacuate people and where to assign his limited resources.

His colleagues, though, hadn't reached their posts yet, and McKenzie's last dispatch was several minutes old. It was clear that the wildfire was well past Pulga, though no one yet knew whether it had traveled the three miles to Concow. Lawrie was left with questions: Where was the fire burning? How big was it? And how close was it to homes and businesses? Over the radio, he heard flames chasing McKenzie's crew as they inched down Pulga Road. The route was strewn with gravel and boulders and spray-painted with warnings not to speed. McKenzie tossed lawn furniture and firewood from the patios of small cabins and yoga huts. The cushions and chopped logs were easy kindling. He had protected all kinds of structures over the course of his career: barns and outbuildings, a wooden bridge and a sweat lodge, and, once, a carved statue of an eagle clutching a salmon in its talons. But he had never defended yoga huts, he thought wryly.

Lawrie stood behind his SUV, studying the map he had unfurled on his tailgate. The autumn morning was chilly, the temperature 42 degrees. Born in Camden, Indiana, Lawrie had moved with his family to San Andreas, California, as a young child. He earned a degree in industrial technology before catching the "fire bug" while working for his college's fire department. Lawrie had devoted thirty years to the fire service—twenty-six of them in Paradise. He had even met Tessa, his wife, through the department. She and Lawrie had clicked after her stepfather—a Cal Fire engineer—invited Lawrie to her college graduation party.

Now he scanned the paper topography before him, squinting in concentration. The vehicle was custom-built to serve as a mobile command post, its hatchback outfitted with a radio system and drawers that were filled with paperwork and evacuation plans. He listened as Bowersox relayed more civilian reports. The two worked with each other often, and he was comforted by the familiar cadence of her voice. Bowersox's parents had both been firefighters like Lawrie—her mother was Cal Fire's very first female hire—and her brother worked as a battalion chief in the neighboring county. Naturally, Bowersox had joined the family profession. At fourteen, she enrolled in the Butte County Fire Explorers program. After she was hired full time, Lawrie had become a trusted colleague. Bowersox appreciated his inherent kindness. She was often teased by her colleagues for having the hobbies of an "eighty-year-old woman and a thirteen-year-old boy." She liked snowmobiling and cross-stitching, playing video games and drinking tea. But she was best known for her unique blend of compassion and professionalism while working the radio, as she was this morning with Lawrie.

Bowersox advised him that a PG&E distribution line above Concow had just toppled—downed by a falling ponderosa pine—and sparked a second wildfire, causing blackouts for four customers: the Yankee Hill Fire Safe Council, the Internet provider Digital Path, the cell service provider AT&T, and the local PG&E office.

Lawrie named the fire Camp B and called for a crew to check it out using wildland engines. The firefighters nicknamed these engines "weed wagons" because they were taller than city engines, with the four-wheel drive needed to chase down grass fires. At 7:20 A.M., Lawrie asked Bowersox to request that the county Sheriff's Office issue a warning—but not a mandatory evacuation—for Concow. He didn't realize that the winds had already heaved firebrands over Concow Reservoir and dumped them on a block of homes well within town limits. The spot fires quickly engulfed them all.

As the morning brightened and people awoke to the sight of flames out their windows, 911 calls began to inundate police departments across Butte County. Civilians couldn't call Cal Fire for information directly, so local police dispatchers had to transfer them to the ECC. At the Paradise Police Department, on Black Olive Drive, only one dispatcher was on duty. Carol Ladrini, fifty-eight, had arrived for her shift at 6 A.M., her uniform clean and ironed, her makeup immaculate. She always woke up early, allowing time to curl her short blond hair before work. Some of the other "early morning shift girls," as she called her fellow dispatchers, didn't bother, but Ladrini took pride in her appearance. She knew that she represented Paradise Police and brought a no-nonsense attitude to everything she did. Before joining the force eleven years earlier, Ladrini had worked as a digital technician for a company that manufactured picture plaques for headstones. Looking for a change of pace, she had applied for the dispatch job along with a friend. She got the gig; her friend didn't. Ladrini had been sworn in the day before the first Camp Fire—part of the Butte Lightning Complex—broke out in 2008.

Now, in a matter-of-fact tone, Ladrini told the frantic residents calling 911 not to worry. Other reports had come in, she reassured them, and the wildfire was still several miles away. From her pod, she could see only her circuit board flicking with incoming calls, not the smoke drifting over Paradise. If Ladrini didn't answer immediately, the call was redirected to another law enforcement dispatch center in the county.

One caller, a woman who lived on Dean Road, on the lip of the Feather River Canyon, wanted to know where the wildfire was burning.

"As far as I know, it's north of Concow," Ladrini replied, offering to transfer her to Cal Fire.

"So we're not in danger?" the woman asked.

"Not so far."

IN THE PARKING LOT at Scooters Café, men in yellow fire suits and tan uniforms paced back and forth. They pressed cellphones to their ears or spoke into handheld radios. Among them were representatives for PG&E, the Highway Patrol, State Parks, and the Sheriff's Office. They had descended on the incident command post to receive information and orders from Lawrie, relaying what he told them to their respective agencies. Meeting in person was supposed to prevent the spread of misinformation. A police sergeant from Paradise would soon be on his way too.

A fellow battalion chief parked next to Lawrie's SUV. He had been assigned to oversee firefighting efforts in Concow. He lumbered out of his Ford F-250 and began studying the map on Lawrie's tailgate. The battalion chief was edgy. His two sons, twelve and seventeen, were home alone on the family's three acres in Paradise, and his wife was at a work conference in Sacramento. He sensed that this fire was galloping toward his town. On the drive to Scooters Café, he had watched the smoke swell and darken with every curve of the highway. It was as wild and runny as an inkblot. Normally, plumes rose straight up, billowing tens of thousands of feet in the atmosphere—striking in their awful, purling beauty—but this column pointed toward Paradise like an arrow.

The man's panic manifested as aggression. He didn't think Lawrie was moving fast enough. A PG&E representative tried to cut in as they talked, asking about de-energizing the distribution system, but the battalion chief interrupted him to address Lawrie: "You need to start issuing evacuation warnings for the eastern edge of Paradise. Order forty-five engines, too."

"Are you serious?" Lawrie replied. He tried to process the demand—the first useful request he had received since McKenzie's last dispatch. There were only thirty-five available engines in their unit, excluding the four set aside for daily medical emergencies and automobile accidents. The battalion chief was asking for more engines than the county owned.

"Yes!" he said, practically shouting. "I'm serious!"

Lawrie stared at him, taking a moment to think. The news coming in on his radio was not good. A power line had tumbled onto the Skyway and trapped a school bus. A sheriff's deputy was trying to rescue residents from the Concow subdivision of Camelot Park, which was advertised as a gated mobile home community—though it actually lacked a gate, which residents had always thought was fitting for Butte County. (What sounded glamorous was in reality often far from it.) A half hour had passed since Lawrie assumed the role of incident commander. At the ECC, Bowersox was struggling to determine the fire's perimeter. She eyed footage from the high-definition cameras on local mountaintops. The lenses bruised with smoke, then went dark as the infrastructure connecting them melted. Lawrie still hadn't requested a mandatory evacuation for Concow, so Bowersox did it for him.

Meanwhile, a Cal Fire captain was corkscrewing along a ridgeline above Concow, tailed by a second engine. They had been dispatched to "Camp B" in order to catch the wildfire and douse it, fast. Firefighting depended on drawing a perimeter and keeping flames within it to "box in" a conflagration. But that method didn't work in an ember-driven wildfire, which spewed spot fires and defied every attempt at containment. The Camp Fire had already coughed a flurry of orange coals past Concow. Now the fire's leading edge was racing downhill toward the town, about to overtake the smaller wildfire on Rim Road. "It's just above the town of Pulga right now," the captain said over the radio. "It's headed toward Concow Lake."

"We need to move now!" the battalion chief said to Lawrie, his voice rising.

The restaurant parking lot would not be able to accommodate

the number of engines now bound for Jarbo Gap. Bowersox had already anticipated the need for help and requested engines from nearby counties. Calling for mutual aid was like throwing a rock into a pond. The appeal rippled outward, with neighboring cities and counties responding first, followed by departments from the Bay Area, the Central Valley, and Southern California, then from border states like Oregon and Arizona, and sometimes from countries as far away as New Zealand and Australia, which had similar Mediterranean climates and firefighting strategies. Their fire season was flip-flopped, meaning the region could send people during its own rainy off-season—though recently global wildfires had begun to overlap, exhausting firefighting resources worldwide.

Lawrie knew his fellow battalion chief was right. At 7:32 A.M., he requested a warning for the western side of Pentz Road in Paradise. Afterward, he sent the battalion chief down to Concow. Then, to make space for all the equipment heading their way, Lawrie instructed the command post to follow him a few minutes south to Pines Yankee Hill Hardware. The business, owned by a Tea Party supporter, had a massive gravel parking lot and served a community of 330 people. Posters for Recipe 420 potting soil and Valspar paint were nailed above the front door, next to the green-and-yellow flag of the State of Jefferson, an underground movement that called for conservative Northern California and Southern Oregon to secede from their respective states and form a new one. Signs clipped to the shop's barbed wire fence announced it as a wildfire safety zone—a place for people to gather when their homes were threatened. It was the perfect spot for a command post.

Lawrie parked in the unpaved lot at 7:40 A.M. On average, wildfires moved 1.4 to 1.9 miles per hour. In 2008, it had taken five weeks for the Butte Lightning Complex to reach the Feather River Canyon before sputtering out. Somehow, within an hour, the Camp Fire had already hit Concow. Like the battalion chief, Lawrie was increasingly anxious about his family's safety. He struggled to focus on the task at hand. His sixteen-year-old daughter was at Paradise High for a before-school meeting. His wife was getting breakfast on the table for their eighteen-year-old son and twelve-year-old daughter and planning a dinner to celebrate Lawrie's birthday. He was turning fifty-one the next day. "Get the animals and get out," he texted her, his hands trembling. "This is real."

Lawrie's radio crackled with another update from Bowersox. At the ECC, 911 calls continued to roll in. Extra dispatchers had reported to work in order to help handle the influx. At 7:41 A.M., a captain answered a call from the county's emergency services officer, who wanted to know whether there was going to be an evacuation. "I'd say there is going to be a large-scale evacuation order," the captain said. He couldn't say when the order would come—that depended on the incident commander, who still didn't have a coherent picture of what was happening.

Lawrie was distracted. He juggled the radio and his cellphone, trying to find out if his family was okay. His calls weren't going through. Beneath the pall of smoke, important infrastructure was melting. By the day's end, fifty-one cell sites and seventeen towers would go offline, either damaged in key sections or entirely incinerated. Made of glass strands as thin as human hair, fiber optic cables had revolutionized long-distance communications—but they were also extremely vulnerable to fire damage. Already the shoddy coverage in Jarbo Gap had become even more limited.

Lawrie had never witnessed a blaze like the Camp Fire in all his years of firefighting. Wildfires burned aggressively in the heat of the afternoon—rarely in the morning, when the hillsides were typically damp with dew. No amount of training or experience could have prepared him for this moment. There was no precedent. Now he stood to lose more than colleagues who lived in other places. His entire family and community lay in the path of the wildfire.

John Messina, forty-six, a Cal Fire division chief who had just arrived, saw the worry deepen on Lawrie's brow and offered to take over as incident commander. He preferred to let lower-ranking colleagues like Lawrie serve in that role, because it gave them valuable experience, but he could tell that Lawrie wasn't mentally present. Lawrie accepted Messina's offer with relief and dropped down to lead operations, gathering intelligence from other firefighters over the radio.

Messina wore a navy Cal Fire baseball cap over his close-cropped hair and a clean yellow fire suit. His wide smile revealed a gap in his front teeth. He was personable and well liked, with twenty-eight years of firefighting experience and an earnest demeanor. These days he spent a lot of time doing paperwork, but he loved his job. Colleagues described Messina as their Rain Man, after the film character, for his photographic memory. Raised by small-town grocers, he had grown up in Alturas in northeastern California and attended college in San Diego. It wasn't happenstance—he had purposefully moved as far away from his hometown as possible. He took off entire semesters to work as a firefighter, banking the salary to pay tuition. After graduation, a job with Cal Fire had kept him in Butte County. His wife and two children, eleven and fourteen, were safe in their orchard-screened home in the farming community of Durham, where tractors often slowed down traffic. He couldn't imagine how worried Lawrie must be, not knowing whether his family was in harm's way.

Messina asked for all aircraft capable of dousing wildfires in Northern California—ten air tankers—to be dispatched as soon as possible. At the same time, he also prodded the ECC to order a regional incident management team, essentially a group of experts drawn from across the state. Such teams were usually not requested until twenty-four hours after a wildfire had ignited. And only when local authorities were completely overwhelmed.

OVER THE RADIO, Messina and Bowersox heard their colleagues descending into the belly of the Camp Fire. The flaming front blasted through Concow, past gravel roads and marijuana patches, past NO TRESPASSING and WELCOME, JOE DIRT signs, past scalloped barbed wire that laced the wooded hillsides. Residents scrambled to evacuate in their cars, still in pajamas, hair uncombed and teeth unbrushed. An ominous fog descended, forcing them to crack their vehicle doors as they drove just to make sure they were still on the road. The landscape blurred, the spot fires on the other side of

Concow Reservoir ballooning in size to 30 or 40 acres each, puffing with gunmetal-colored smoke. The wildfire stoked its own weather, superheating the air and causing it to rise as cool air rushed into the vacuum, spawning delicate fire whirls. Wind speeds topped 90 mph, ramming through Camelot Park and funneling up Cirby Creek, stripping the trees bare.

Firefighters treated several victims with severe burn injuries. One woman—a mother of five—had been found beneath a truck, where she had depressed the tire's valve stem with a nail for air to breathe. She was laid in the back of a fire engine, then transferred to an ambulance. On the engine's bench seat, she left a delicate sheet of molten skin. By now, Messina knew that the situation in Concow had become dire—but he still didn't expect flames to reach Paradise. A few minutes later, though, flames were spotted near the Feather River Canyon.

Just before 7:50 A.M., a woman who lived on Drayer Drive near Feather River hospital reported spot fires all over Sawmill Peak. She could see them from her bedroom window. "The wind is blowing like hell, and nobody notified us or anything," the woman said. "That's 'cause there's no evacuations at this point," the police dispatcher replied.

Bowersox called Messina and Lawrie on the radio, informing the men that fires were lapping the mountain northeast of Paradise. Messina had already sketched a box around the Camp Fire on his whiteboard, hoping to keep it within the marker-drawn square. Pulga was the western boundary, and Pentz Road was the eastern boundary. Bowersox's news stunned him—Sawmill Peak was outside his box. The blaze had blitzed seven miles. Realizing it was about to reach the eastern edge of Paradise, Messina asked the Butte County Sheriff's sergeant to call in a mandatory evacuation for everything north of Highway 70, as well as three zones along Pentz Road. He reiterated the same request to Bowersox, so that everyone was on the same page. In the fear and chaos, civilians needed evacuation orders for a clear understanding of when to leave their homes and which roads they should use to safely get out of danger.

He limited the order to a portion of Paradise, hoping to leave roads accessible for the four thousand people in those zones urgently needing to escape.

In the Sheriff's Office, there were five people who were trained to send emergency messages using CodeRed. Of them, only one person—an information systems analyst—had shown up for work. The rest were out of town or tending to their families. The analyst now struggled to field all the incoming phone calls, with two people—a Cal Fire dispatcher at the ECC and the sergeant in Yankee Hill—calling in with identical requests, including this one for Pentz Road.

The CodeRed system was owned by the California-based software company OnSolve, which maintained twelve data centers around the world and billed itself as the "Nation's #1 Public Safety Alerting Solution." In Massachusetts, the Newton Police Department had used CodeRed to alert residents of the Boston Marathon bombings in 2013; in Florida, Sarasota County had relied on the system during Hurricane Irma in 2017. The platform was capable of sending a single message to hundreds of thousands of phone numbers and email addresses every minute. It worked well—as long as people had the foresight to sign up. As of early 2018, only 11 percent of Butte County's population had registered, but dispatchers knew that word spread quickly in small communities, and even reaching those people could save many lives.

At 7:57 A.M., the analyst keyed Messina's first evacuation order for Pentz Road into CodeRed. He thought about how to communicate the emergency in ninety characters or less, then typed: "Due to a fire in the area, an evacuation order has been issued for all of Pentz Road in Paradise East to Highway 70." In the hubbub, he never received the order to evacuate Concow. It was buried under other requests. A CodeRed alert wouldn't be sent to that community for another two hours. Aside from firefighters pounding on front doors, residents received no warning. After pressing Send on the Pentz Road order, he dispatched an Amber alert–style message through a system run by FEMA. Called the Integrated Public Alert and Warning System, or IPAWS, it was capable of reaching every cellphone in a geographic area regardless of whether the user had signed up, making it even more effective than CodeRed. However, Butte County had never tested the system. Unbeknownst to the analyst, a system error prevented the alert from sending.

Paradise Police dispatcher Carol Ladrini hadn't received the order for Pentz Road when she answered a call from a woman in the evacuation zone. "Um, are we supposed to be evacuated, or what?" the woman asked.

"No, you'll be notified. There's a fire north of Concow up off of Highway 70. No danger to Paradise, okay?"

"Well, it said evacuation west side of Pentz Road."

"What says that?" Ladrini said, flustered.

"Butte, um, Fire."

"Okay, we haven't been advised of that."

AT CHICO MUNICIPAL AIRPORT, it was calm. Not a breath of air whooshed over the two-lane runway as David Kelly, forty-two, took off at 7:48 A.M. One thousand feet, then two thousand feet, then three thousand feet between his aircraft and the earth. The Cal Fire contract pilot hit a patch of turbulence so jarring it rattled his teeth, the 30,000-pound tanker jolting like it was in the spin cycle of a washing machine. The wings convulsed as if they were going to snap off. Kelly climbed until he hit a smooth patch of air at five thousand feet. He steered east toward the foothills.

More than a thousand gallons of bubblegum-pink fire retardant were stowed in the hull of his Grumman S-2 Tracker. The propeller plane had been designed to hunt Soviet submarines during the Cold War but had been gutted and repurposed. Cal Fire operated twenty-three of these air tankers. They were used for dumping a chemical gel—made of a salt compound, water, coloring, and a thickening agent made of clay or gum—atop anything combustible, slowing a wildfire's burn by lowering the temperature and reducing flammability. The retardant, also called "slurry," was stored in the Grumman's old bomb bay. Kelly had been flying aircraft like this for more than twenty years. His father had flown air tankers for the

Forest Service for decades, but Kelly hadn't followed in his footsteps until taking a break from college. He got a gig pumping gas at the local airport one winter and began chatting with the aviation folks. He was immediately hooked—and never returned for his music education degree.

With fifty fixed-wing and rotary-wing planes, Cal Fire had the largest fleet of firefighting aircraft in the world. They were housed at thirteen air bases and nine helicopter bases across the state. Tankers, like the Grummans, were the most effective at containing a wildfire early. Pilots in them could navigate steep and rocky areas before ground forces gained access, helping douse flames within the first two hours of ignition. The Grumman was svelte and fast, able to operate with precision in terrain that larger aircraft couldn't navigate. Kelly's air tanker had two red stripes and the identification number 93 painted on the tail, as well as red stripes on each wing.

Kelly loved the uncertainty of his job; he enjoyed weighing the probabilities and unknowns and working out a solution. A wildfire was a giant puzzle. His favorite days were when he knocked down a blaze within a half hour of ignition, allowing firefighters on the ground to extinguish it before anyone ever learned its name. But as he careened toward the mouth of the Feather River Canyon, it didn't seem as if it was going to be that kind of day. The fire was already at 1,000 acres and roaring. He circled over Flea Mountain, where he could see a radio repeater topped by a towering antenna. The Sheriff's Office and ambulance crews relied on the repeater to communicate. Kelly knew it was about to burn—but the way the winds were blowing, he couldn't hit his target with retardant without endangering himself.

Better to protect lives and homes, he thought.

Kelly abandoned the repeater and flew down the mountain's northernmost flank, tracking toward the flaming front, where he might be more effective. As he began to descend, debating where to drop his retardant, the turbulence suddenly caused his Grumman to bottom out. For a moment, he lost control of the aircraft. He pulled on the throttle. Nothing happened. The downdraft sucked the plane toward the forest floor—only to reverse seconds later, when an updraft popped him back up. This happened twice in a few minutes. He climbed back to the safety of 5,000 feet to regroup.

His airspeed hovered around 140 knots, standard for circling. His ground speed, on the other hand, registered at 210 knots. Kelly did a quick mental calculation. The wind was thrusting his tanker 70 knots faster than he wanted. To drop the retardant, he would need to flick his control levers to "flight idle" until he slowed to 125 knots, providing just enough fuel to keep the propellers spinning; then he could descend, dropping like a rock and releasing his cargo. The goal was to spray the pink slurry like rain, evenly coating a house or a clump of trees from above. But with an added tailwind, he couldn't be effective. When the retardant fell toward the ground, it would shadow his target, coagulating on one side and leaving the other side bare. Even worse, it might just evaporate.

In any case, the wind was pummeling him too hard from every side for this maneuver to work. His tanker's small size and the variability of its weight made it dangerously vulnerable in these unpredictable gales. He couldn't crawl any slower, to feel out safe pockets of air. If Kelly was stalled when he hit a downdraft, he would smash into the mountain.

Still, he was determined. He would at least try to douse a neighborhood or an evacuation route near the wildfire's edge. He banked to the right, toward Paradise, struggling against the wind. Houses spread downhill below him, barely visible through the curling black smoke. Square and humble, arranged in orderly rows, the town's homes looked like Monopoly game pieces. With a sinking feeling, Kelly realized that the flames were already licking at them. The wildfire was now 2,000 to 3,000 acres in size. He pictured the people below—retired folks in their robes and slippers, still asleep or getting ready for the day. People were good at evacuating when a fire arrived in the afternoon, he knew, but it was barely 8 A.M.

He lowered the flaps on his tanker's wings and quieted the engine as he approached the town, planning to descend so he could unleash the retardant. But instead of gliding downward, his tanker shot up 1,000 feet, cresting a massive updraft like an invisible ocean wave. Kelly could fight fire—but he couldn't fight weather. It was over. He turned around, landing in Chico at 8:29 A.M. without having released a single drop of retardant. His air attack supervisor immediately grounded all fixed-wing aircraft until the winds subsided. Even so, desperate firefighters would continue to call for retardant drops for the next six hours.

"You're going to have significant structure-threat issues," Kelly's supervisor said over the radio at 8:01 A.M. "The fire is now in Paradise."

THE REPORT FROM air attack seemed inconceivable to the command post. Wildfires rarely traveled eight miles in just over an hour, or threatened such a large community. But it was true: The blaze had breached the town limits. Just after 8 A.M., Messina asked both Bowersox and the sergeant at Pines Yankee Hill Hardware to request a second round of CodeRed evacuations in another four zones in Paradise. The sergeant processed Messina's request, calling the message in to the Sheriff's Office. In Oroville, however, Bowersox hesitated. *They think they're going to stop this fire, but they can't,* she thought. On the radio, she could hear firefighters debating how to put out the flames.

What neither she nor the firefighters knew yet was that nine people in Concow had already been charred to death. Dozens more had been trapped in a plowed field—or submerged themselves in the cold reservoir—as flames rose around them. On Pentz Road, in Paradise, Lawrie's home was about to ignite. Bowersox couldn't see the carnage from her post at the ECC, but she did have a dawning sense of just how sweeping the devastation was becoming. *They aren't seeing this for what it is,* she thought. She agonized over Messina's request to evacuate only four more zones. Bowersox's own bungalow was in Zone 8, on a cul-de-sac near Pentz Road. She lived across the street from a seventy-five-year-old widow named Sara Magnuson. When Bowersox had moved in a few years earlier, Sara had left a bag of apples on her doorstep. For Christmas, she brought Hershey's Kisses. The elderly woman suffered from dementia and lived alone.

Bowersox had overheard enough 911 calls to realize that all fourteen zones in Paradise were being threatened by fire—not just the additional four that Messina wanted to evacuate. The air report from Kelly had been the final confirmation. Her head felt woolen. She needed to do something. The answer came in a flash. *Get them out,* she thought. *Just get them all out.*

As someone who had grown up in a household of firefighters, Bowersox was acutely aware that it wasn't her place to adjust evacuation orders. She was lower-ranking—and she hadn't run the idea past the captain on duty in the ECC or received approval from the command post. She knew she stood to suffer the consequences if something went awry: a pay cut, a suspension, or worse. But time was of the essence. The evacuation might save the lives of people like Sara. For the first time in her almost-fourteen-year career, Bowersox flouted orders.

She turned to a colleague in a nearby pod.

"Tell the Sheriff's Office to evacuate all of Paradise," Bowersox said.

"That's not what Messina said," the dispatcher replied, confused.

"I don't care. Who's it going to hurt? Do it."

Bowersox listened as her colleague called the law enforcement office and requested that Paradise be emptied. Then Bowersox ran to her car and drove the eighteen miles to Drendel Circle, in Paradise, so she could evacuate her own bungalow. She caged her three rescue cats, Gus, Bailey, and Peter, and collected some paperwork. She saw no sign of Sara.

AT PINES YANKEE HILL Hardware, incident commander Messina called a fellow division chief in the ECC to inquire about extra help. They needed more fire engines. Lawrie stood by Messina's side, still trying to reach his family. He hadn't heard from Tessa in more than an hour. Before Messina could ask his question, his colleague on the line spoke first: "We have fire in Paradise, and we have fire in Concow. I want to order one hundred strike teams of fire engines."

Messina paused.

A single strike team was made of five engines led by an SUV. Their unit would be requesting a total of five hundred engines—a state record for the number of resources ordered within the first hour of a wildfire.

Does Messina think I'm nuts? the division chief thought.

"That's completely accurate," Messina replied. "Let's go with it."

The division chief hung up, then turned to his phone once again, scrolling through his contacts. He clicked on the name of Cal Fire unit chief Darren Read. Read, forty-nine, oversaw every firefighter, every fire engine, every fire station in Butte County. He was formidable—in both experience and physique. He had worked as a firefighter for thirty-one years, across five different units, and had also trained as a peace officer. In a photo taken with former governor Arnold Schwarzenegger, Read made even the Terminator look tiny. He towered a whole head above Paradise Fire chief David Hawks, who many anticipated would take his job someday. Read was away for the day, attending a meeting of the Cal Fire Northern Region Leadership Team in Marin County, three hours south of Butte County and just across the Golden Gate Bridge from San Francisco. Every Cal Fire unit chief—the "boss" of each North State county—had gathered at Skywalker Ranch to bid farewell to Cal Fire director Ken Pimlott, who was soon to retire. A group photo of all the leaders—navy uniforms, gold bugles pinned on collars, smiles—was planned for the end of the day.

Now, a few minutes before the meeting was set to begin, Read picked up the phone. "If you stick around for that picture," the division chief told him, "you won't have a goddamn unit left by the time you get back here." Read raced to his car and headed for Oroville. Within a few minutes, the meeting would be canceled. Director Pimlott would leave to help open the state's Emergency Operations Center in Sacramento. The Cal Fire chiefs would delay for only a few minutes, snapping a quick photo together in front of a historic covered bridge before scattering to their posts. Only the tallest among their ranks was missing from the picture.

MESSINA GAZED ACROSS the parking lot at his colleagues. Their slack faces betrayed a sense of anxiety and discouragement that he was starting to feel too. For the past hour, he had tried to remain poised as every plan fell apart. At 8:49 A.M., with no options left, he had asked the sergeant to call for a final townwide evacuation in Paradise. This confused the county analyst, who had already dispatched an alert based on Bowersox's request. (Messina wouldn't realize she had defied his authority until four days later.)

The Camp Fire roared through the foothills and gullies, unstoppable, turning the sky a livid purple. As soon as Messina drew a new containment border on his whiteboard, he had to erase it and redefine the boundaries. What he could really use, to be honest, was a piece of goddamn Scotch tape so the wind would stop snatching the papers from his tailgate. He grabbed a rock the size of a soup can and anchored his map. It held down the chaos for a bit, but he knew even it wouldn't hold. Messina rubbed the stone with his hand, turning to speak to Lawrie. "If you need to take care of your family, then go."

Lawrie's home had been in the first evacuation zone. Neither man realized that it had likely already burned down. "Where am I going to go?" Lawrie said, shrugging, trying to hold it together. "I don't know where my family is. The best thing I can do is stay here. This is my job. I have to stay."

"Are you sure?" Messina asked.

"Yes."

The Paradise Police sergeant had not made it to the command post at Yankee Hill. The Highway 70 corridor had become too treacherous. With a sinking feeling, Messina realized that he and his colleagues could soon be stranded themselves, their communication cut off. After nearly three decades in the fire service, he had seen plenty of wildfires—and the toll from this one, he knew, was going to be record-breaking. The horizon roiled darker with every passing minute.

At 9:30 A.M., he directed his colleagues into a line of white trucks, SUVs, and sheriff's squad cars. They would take Highway 70 toward Oroville, where the 928-acre campus of Butte Community College would provide plenty of space to set up a new command post. The Incident Command Team could meet them there and take over. But as the cars headed down the highway, they were immediately caught in the crush of residents evacuating from communities near Plumas National Forest, part of Messina's first evacuation order. The two-lane highway was swollen with gridlock traffic.

The brains of the firefighting operation were trapped. Messina considered snipping a nearby farm's barbed-wire fence and driving through the harvested hayfields. Instead, he jerked onto the highway shoulder and into the ditch, his wheels ripping through the palomino grass. The sheriff's deputies flicked on their sirens and followed. Over the radio, Messina heard the air attack supervisor announce that the Camp Fire had progressed halfway through Paradise. The town was now threatened by fire on three sides. Embers rained down on Butte Creek Canyon and Magalia.

As he pulled in to Oroville, Messina knew that no one had had a good read on the fire to begin with. They couldn't fight it. The inferno was too big, too fast. His final command over the radio carried all the resignation of that knowledge. "I don't want to see anybody laying hose or going after a spot fire," he said. "There will be zero firefighting. Your only mission is to protect evacuation routes and the lives of civilians. Get people moving, now!"

CHAPTER 5

THE IRON MAIDEN

As Messina and his team were packing up to leave for Oroville, Paradise town manager Lauren Gill was rushing from her home to Town Hall. She grabbed her coffee, still steaming in its mug, and dashed out to the bright red Volvo in her garage. She was in a hurry to set up the local Emergency Operations Center. Similar to Cal Fire's ECC, the EOC would serve as the town's hub for emergency management. As its director, Gill was responsible for gathering intelligence and distributing information to Paradise residents. As she was backing out of the driveway, a neighbor waved her down.

"Did you see the smoke?" the woman asked.

"It's in Pulga, far away," Gill answered tersely. "I have to get to work."

As Gill navigated toward the lower Skyway, slaloming through the labyrinth of residential streets, she realized she had forgotten her mascara on the countertop in her kitchen. She turned onto Elliott Road and sighed. She really could have used some mascara this morning, to make her tired eyes appear brighter.

Already the day hadn't gone as she'd planned. Gill, sixty, had set her alarm for 6 A.M., leaving ample time to get ready for a Rotary Club board meeting at the Cozy Diner. But she had missed the alarm completely, oversleeping by more than an hour. The election earlier that week had worn her out. As she was powering up the coffee maker, she had glanced out the kitchen window. The sky above Butte Creek Canyon was golden, misted with smog, and didn't look quite right. When Gill stepped onto her front porch to get a better view, a sharp, acrid note hit her nose. She ducked back inside to call Paradise fire chief David Hawks. He was off duty that morning and had just finished working out with a friend in Chico. As far as Hawks knew, the wildfire was still miles away, near Pulga. At 7:28 A.M., Gill sent a text message to the Town Council chat group: "Fire in Concow. Fire chief says it's a long way from paradise. I'll keep you posted. Only retweet or share info from cal fire."

A few minutes later, she joined a brief conference call with several agencies in town, passing on those same details. By the time councilman Greg Bolin had texted the chat group back—"Are there evacuations in paradise yet?"—Gill was leaving for Town Hall. "Checking," she replied. She didn't know that Cal Fire had just requested the evacuation of Pentz Road. High above Paradise, contract pilot David Kelly was flying his air tanker over town for the first time. At 8 A.M., Gill's phone pinged again, this time with a message from councilman Mike Zuccolillo, Rachelle's ex-husband: "This was a text to me by friend on dean rd. . . . not verified. (Fwd:) Fire on dean rd. Apple view off of Pentz. Sawmill peak, 15 spot fires." A second text followed from Bolin, whose family owned a local contracting business: "I just received a call from my business partners that there is brush burning at Mountain View off Stark. Could someone get that word out."

Gill parked at Town Hall, then tapped out a reply: "Call fire or 911 with that." She looked up, noting the assistant town manager's Chevy truck and the clerk's Honda CRV already in the lot. Ash was wafting from the sky, dusting their windshields in white. Gill hurried indoors. Normally, she dressed in elegant skirt-suit sets and spindly high heels that *tick-tick-tick*ed down the hallway, announcing her arrival. But today, with no time to get ready, she had gone with black jeans and shiny leather loafers. Gill ducked her head into the clerk's office and asked her to send an internal CodeRed alert to the town staff. The clerk nodded and turned to her computer.

"The EOC is activated. Please report to Town Hall," she typed into the system at 8:07 A.M.

Down the carpeted corridor, the EOC was coming together in the council chambers—just as staff had practiced in simulations. As Gill was speaking with the clerk, a few more employees had trickled in behind her. They were now taping maps to the walls, flipping through binders of evacuation plans, unloading plastic tubs of supplies. They scoured social media posts on their laptops, trying to distinguish between rumor and truth. The assistant manager had already posted an update to the town's Facebook page: "Cal fire reports a fire in the concow area. Approx. 50 acres. We will update as we get more info." Now the staff's cellphones rang furiously as people called for more information.

Gill continued down the hallway and opened the door to her office, logging on to the computer and loading her email. She scrolled through the new messages. She was hearing that a blaze had sparked near Noble Orchards and Feather River hospital. Was it a separate fire?

Gill had come to know Paradise and its local landmarks well over the past few decades. She had moved to the town from Orange County in 1981 to be closer to her then husband's family. His parents had retired to the Ridge because they loved the ponderosa pines and the hushed surroundings, and she—then twenty-three years old—had followed because she loved him. Paradise had incorporated two years earlier, and the town manager hired Gill as a secretary. It was a temporary role that involved answering phones and residents' questions. As the new administrative clerk, Gill hadn't even known what culverts and septic tanks were, because they didn't have those in her hometown of Garden Grove, but she learned. She also began to understand the impact one could have in a small community, where local politicians actually made things happen.

After the demise of her marriage—which produced no children, though she had always longed for a family—Gill focused on her career. Her father had always told her, "If you work hard and you're good at what you do, you will be rewarded for it." It wasn't her intention to stay in Paradise, and she might have enjoyed a different career path—teaching literature or working as a librarian, perhaps, because she loved books—but department heads continued to tap her for openings, and Gill jumped at the opportunities. By 2013, she had been appointed to the most powerful position in Town Hall. With a degree in public administration—which she had gone back to school for—and nearly three decades of experience working for the town, she was a natural choice to become Paradise's first female chief executive officer. Gill knew the position wasn't going to be a "fun" job. More than 20 percent of Town Hall's workforce had been laid off during the economic recession, and the budget was shot. But she was glad to have a hardworking staff and backing from the council. She also had the support of her family: two sisters and a brother, who lived across town, and later her fiancé, Don, who had five grown children and twelve grandchildren, whom Gill adored and treated as her own.

Not everyone in Paradise felt as warmly about her. Slender and petite, with a reliably perfect pouf of blow-dried brown hair, Gill sang beautifully at the staff Christmas party every December and served as the president of the local Rotary Club. But some people found her cold and calculating, preoccupied with appearance and control. She cared a lot—maybe too much—about everything. A smattering of locals had taken to calling her the Iron Maiden. Whether or not constituents liked her, however, Gill couldn't be recalled, because she was not an elected official. Unless the majority of the five-person Town Council voted to terminate her contract, the job was hers to keep. It was a lucrative gig, too. In 2017, Jody Jones, the sixty-two-year-old ceremonial mayor, earned less than $5,000. Gill, who guided the town with a firm hand, brought home nearly $180,000 in salary and benefits.

Gill wanted Paradise to be a good place to live. She was responsible for drafting the policy reports that councilmembers used for making decisions, making the case, for instance, for a half-cent sales tax or a new sewer system in the commercial district in order to attract new businesses.

In 2013, Gill had hired Marc Mattox as town engineer. He was thirty-three but he looked even younger, with flyaway blond hair that he combed to the right. Laid-back but diligent, he had a civil engineering degree. He was always the first to volunteer to help out, whether by explaining a technical policy decision or refilling the paper towel dispenser in the men's bathroom. Gill told him his job was to "green and black"—get money for road improvements and lay asphalt. Mattox cobbled together transportation grants meant to lessen car dependency and improve safety. The town would agree to install a bike lane or a sidewalk—even though few people used such things—just so the road could be repaved. Despite his efforts, Mattox never found funding to widen the town's narrowest streets, like Roe Road. Gill promoted him to assistant town manager, then assistant EOC director. Gill hoped he might take her job when she retired. It was difficult to persuade educated candidates to move to a community as remote as Paradise, and the town's staff had traditionally created succession hierarchies. Gill and Mattox were both hands-on people who didn't like to be told what did or didn't work. She liked this about him.

Gill, Mattox, and the rest of the Town Council were preoccupied with development and beautification, keeping residents happy. But they were also tasked with keeping them safe—and there was danger on every side. On hazard maps, the community was an island in a splotchy sea of colors marking past wildfires. Since 1999, thirteen monster blazes had barely missed the town limits. Until now, luck had always been on their side.

But fire was unpredictable, and in recent decades, Paradise leaders had watched as wildfires swept across California with an unprecedented ferocity. In 1991, the Oakland-Berkeley Hills Fire killed twenty-five people across the bay from San Francisco, hopping both a four-lane and an eight-lane freeway and leveling nearly thirty-five hundred homes. The tragedy showed just how vulnerable a community could be if it lacked clear evacuation routes.

The truth was, California had always burned—as much as 5.5 to 19 million acres annually in prehistoric times, or up to 19 percent of the state's land. Flames were as typical of the changing seasons as rainstorms and blizzards. The same swath of hillside blackened by the Oakland–Berkeley Hills Fire of 1991 had also burned in 1923, 1970, and 1980. But residents were quick to forget the past, and amnesia was part of California's identity. The state legislature left it up to local governments to protect their constituents, and thus development continued unfettered, and more and more homes became kindling.

In Paradise, the threat of an ember-driven wildfire, akin to the blaze in the East Bay of San Francisco, was very real. State documents cautioned that a fire could cause "catastrophic life and property loss." The town was a tinderbox nestled between two geological chimneys. As in so many other Sierra Nevada towns, miners and developers had carelessly pitched homes on fire-prone terrain, never grasping the consequences of taming their Eden. The risk compounded as dry brush accumulated and trees died. Though the Town Council tried to enact stronger evacuation policies, it didn't succeed until after wildfires torched Paradise's outskirts in 2001 and 2008.

The Poe Fire erupted just before the terrorist attacks of 2001, after a windstorm knocked a ponderosa pine onto a PG&E power line. It blackened 8,333 acres, destroyed twenty-six homes, and caused $6 million in damage. It nearly killed Captain McKenzie and his crew as they tried to save an elderly couple and their dogs, the fire burning so hot that it melted the hose off the back of their engine.

Then, seven years after the Poe Fire, at the start of a dangerously dry summer, an arsonist lit the Humboldt Fire beneath some power lines. The blaze slunk out of Butte Creek Canyon and thundered toward the Skyway, taking out eighty-seven homes at the edge of town. The heavy smoke and flame forced three of the town's four main roads to close, and more than ten thousand people were left stuck on the one remaining route: Pentz Road. Traffic crawled at 4 mph. The radio station that broadcast emergency alerts transmitted outdated information. As a strange orange light overtook the sky, residents began to panic. Then, at the last moment, the wind shifted, saving Paradise.

One week after the Humboldt Fire, dry lightning strikes sparked a dozen more wildfires near the opposite edge of town. When the slow-moving Butte Lightning Complex hit the Feather River Canyon more than a month later, residents were forced to evacuate again. The hospital was emptied of patients. Altogether, that fire scorched 60,000 acres and leveled two hundred homes—the worst season in local history.

"The greatest fear is fire on the Ridge," a thirty-seven-year-old woman told a reporter of the ordeal. "There's no way out. You're trapped."

The Town Council was determined to heed these hard-learned lessons. In 2009, it carved Paradise and Magalia into fourteen zones and created a staggered evacuation plan to prevent gridlock. It contracted with CodeRed to issue warnings to cellphones and landlines. Crews paved the nine-mile upper stretch of the Skyway—at the time a gravel road. That same year, a local Civil Grand Jury issued a report cautioning against further population growth until Paradise did more to address its challenges. "The unpredictability, intensity, and locations of the 2008 wildfires . . . emphasized the critical shortcomings of the area's readiness for extreme fire situations," the report read.

Frankie Rutledge, the mayor at the time, responded with a scathing public letter. She thought Town Hall had done enough. "None of the findings concerning the lack of emergency evacuation plans, emergency communications, or notifications are applicable to the town of Paradise," she wrote.

Instead, in 2014, with Gill firmly in charge, the Town Council decided to explore the idea of narrowing a portion of the Skyway, the town's major artery, which over the years had been widened to a four-lane highway. That year, thirteen people had been injured in crashes on that corridor. The goal, then, was to lessen the "expressway" feel of that stretch and reduce pedestrian accidents by means of a "road diet." Gill was in support. Like the rest of the Town Council, she liked that it would create additional street parking, hopefully drawing business to the shopping district, and only downtown would be affected. During public comment, though, an eighty-eight-year-old named Mildred Eselin called the project "insane."

"I can't believe it's being seriously considered," said Eselin, one of the few dissenting voices. "Imagine if there were a fire, the traffic coming down the Skyway. If the council is searching for a way to diminish the population of Paradise, this would do it."

The proposal passed unanimously.

The Bay Area engineering firm that consulted on the downtown project urged the council to remember that the roadway was the town's main evacuation route. It discouraged raised medians and other "improvements." But the council went ahead with its plan, and in late 2014, road crews slashed the Skyway from four lanes of traffic to two. Several other roads were also narrowed that year, reducing the number of vehicles that could pass through Paradise. On Pearson Road, bike lanes were added. On Clark Road, travel lanes became turn lanes. And where six concrete curbs— called bulb-outs—jutted into traffic on the Skyway, volunteers took a simple beautification approach. Instead of raised medians, they planted flowers and firmly bolted donated benches to the concrete.

In 2015, the Town Council devised a plan to reverse flow on the lower section of the Skyway in an emergency, sending four lanes of traffic downhill to Chico. Deciding on the logistics had been a particular challenge, because the plan called for the police to first block the arterial roads that fed onto the Skyway and Pentz Road to prevent collisions. Anticipating human reactions, too, complicated things. Gill never knew how residents would react: panicking, abandoning their vehicles to walk, refusing to leave their pets behind.

The following year, a few days before the seventh anniversary of the Humboldt Fire, Town Hall led an evacuation drill during morning rush hour. Though up to twenty-four thousand vehicles traveled the road every day, only about seventy people participated in the drill, inching past volunteers in reflective orange vests before heading out of town through the tapered portion of the Skyway. Most everyone else, informed of the drill beforehand, simply avoided downtown. Meanwhile, firefighters staged their own drill, a wildland-urban interface training exercise in which captains roped off dozens of trees and twenty-one homes with toilet paper—which disintegrated when hit with water—to represent flames. Firefighters evacuated bedridden patients, dealt with low-pressure hydrants, hopped "downed" electrical lines, rounded up horses. The drill was so intense that one firefighter was taken to the hospital with symptoms of cardiac arrest. The fictitious scenario burned three hundred homes—a hundred more than the Butte Lightning Complex.

In the end, for all its efforts, the Town Council never designed a system for emptying the entire ridge at once. Evacuating every resident would take eight hours under perfect conditions—or five hours with the one-way traffic plan in effect—but the estimates didn't factor in congestion or the extra time required to help vulnerable residents leave their homes. There wasn't even an accurate local accounting of such people, since the town's roster hadn't been updated in years.

FROM AN OFFICE at Feather River hospital, Chief Financial Officer Ryan Ashlock jumped off a conference call with a handful of agencies in Paradise. He tipped back in his chair, thinking. Gill had originally scheduled the call to discuss PG&E's planned outage, but since the blackout hadn't occurred, there wasn't much to discuss. Gill explained that PG&E had not gone forward with its plan because it "didn't seem appropriate": Weather reports had not registered sufficiently strong winds. "Are there any questions?" Gill had asked.

"We're concerned about fire in the area of Paradise," Ashlock replied.

"We know there's a fire in Pulga," Gill said. "We aren't sure what's going on, but we don't feel it's a threat to our town."

As Ashlock finally left the office and walked down the hallway, he noticed hospital staff congregating near the windows facing the Feather River Canyon. The clouds were a violent churn of red and orange. Singed pine needles and blackened twigs pinged against the double-pane windows. The sight startled him into activating the hospital's Emergency Command Center, just in case. Ashlock, thirty-three, was the only executive on campus that day; the rest were at regional meetings. He called a Code Triage External, the signal for an emergency occurring outside the hospital's walls that could potentially threaten patients. A medical administrator's first instinct was never to evacuate: A full evacuation was not in the best interest of the hospital's most vulnerable patients, and besides, the facility had fire-rated walls and doors that could withstand flames for two to four hours. Ashlock simply wanted the staff to be aware of the situation unfolding in Pulga.

The morning had begun with the typical lineup of general surgeries. Orthopedic cases were reserved for earlier in the week, so patients could be discharged before the weekend. All other operations were slated for Thursday or Friday, though PG&E's plans to de-energize the town meant that fewer elective operations had been scheduled. The first patient was readied for rotator cuff surgery, expected to last three hours. Another patient prepared to have his gallbladder removed. A child was slated for a tonsillectomy. In the Birth Day Place, a nurse greeted a woman from Redding who had an appointment to be induced.

Nurse supervisor Bev Roberson walked through the hospital's main doors from the parking lot, a fall-themed coffee mug in her hand. Unlike most medical facilities, the hospital was spread out across several acres on a slope near its namesake, the Feather River Canyon, and to move between departments, patients and medical professionals often had to duck outdoors. Roberson, fifty, had just finished an hourlong meeting in a conference room in the lower wing. She had watched a PowerPoint about the hospital's annual accreditation process as other staff members nibbled from a continental breakfast buffet.

Why do I even take the time to get myself ready for work? she thought now, the wind disheveling her neatly brushed hair. She ducked into the climate-controlled interior, the hospital doors *swish-swish*ing behind her. After twenty-two years as an emergency room nurse, Roberson had only recently been tapped as the new in-patient nurse manager, overseeing sixty-seven hospital employees. They were a tight-knit group. ("Don't talk crap about anyone, because someone on staff is always related or married to that person," the ICU director liked to say.) Her promotion meant more meetings, and Roberson was headed to another one, this time led by Ashlock.

She glanced down as she walked and recoiled. A handful of embers, almond-sized and steaming, bobbed on the surface of her coffee.

Down the hallway, Ed Beltran, the charge nurse, examined his hard plastic name badge. It had partially melted as he'd stood in the parking lot to make a phone call. A flaming stick had smacked him in the chest, scorching a small hole in his scrubs. He couldn't believe it. Management had already instructed Beltran, forty-four, and the rest of the staff to pull down the window shades so patients wouldn't see the blood-orange horizon and charcoal showers. "Close the doors, too," Ashlock had added.

Ashlock was now weighing whether or not the hospital needed to evacuate. Boyish, with bright blue eyes and ginger hair, Ashlock had worked in Paradise for only a year. He and his wife, who had grown up on the Ridge, had relocated from Pasadena in 2017 for his job. He wasn't familiar with wildfires, and he wanted to be cautious. He hadn't heard anything more from Gill about the fire's spread. It was Gill's responsibility, as the director of Paradise's EOC, to communicate relevant information to cooperating agencies, like the school district, the hospital, and the parks department. She received training for this role four times a year. A retired fire chief schooled her on the Incident Command System; the town paid him $1 annually for his help. He trained Gill on how to handle a wildfire and a wet winter storm. She also learned to manage an active shooter and a pandemic.

Ashlock asked the nurses and physicians to discharge as many able-bodied patients as possible so there would be fewer people to move in case of an emergency. There were eight patients in the intensive care unit; one was intubated. The ICU director began printing their medical charts and tracking down gurneys and wheelchairs. Ashlock thanked him and walked into the wood-paneled boardroom for the 8 A.M. meeting. When everyone had arrived, they linked hands to pray. "Keep us safe as we go throughout this day and give us the wisdom to make the right decisions," Ashlock said. "Amen."

He started the meeting four minutes early, at 7:56 A.M., and was interrupted soon after by a technician seated in the back of the room. The technician had been listening to the first responders' radio traffic on his cellphone. "Ryan, Cal Fire just issued a mandatory evacuation for everything east of Pentz Road," he said. The order included the hospital—information that normally should have funneled to Ashlock through Gill. But she wasn't yet aware of the evacuation order.

A hospital director held out a chunk of burned bark he had found in the parking lot for everyone to see. It was black and narrow, as long as a carrot. Ashlock knew he needed to act.

Four minutes later, his phone rang. His brother, who worked in the Human Resources Department on the northern end of campus, had spotted flames in the Feather River Canyon. As soon as Ashlock hung up, the hospital's information technology director called. From the southern end of the campus, he could see fire, too. The flames stretched taller than the one-story hospital, reaching toward the sky as if they might torch the clouds. Administrators had called 911, but there were only two fire engines left in Paradise, and no one was available to respond to the campus. Most of the force was in Concow or Jarbo Gap or on their way there. The remaining firefighters were snuffing out a grass fire near Ponderosa Elementary.

The technician announced that he had more news. On the radio, the police had called for a townwide evacuation. Bowersox's order was making its way through the proper channels. Ashlock called a Code Black—everyone needed to leave immediately. "Bring all patients to the ambulance bay," he instructed the employees. He dialed Enloe Medical Center in Chico, making sure it had room for their patients, and requested eight ambulances and two helicopters. At 8:10 A.M., as Kelly was realizing he'd be unable to drop his load of fire retardant above Paradise, the hospital administration posted on its Facebook page: "We are evacuating. Don't come here."

In the Birth Day Place—on the opposite end of the same hospital building, away from the growing pandemonium—a nurse's aide knocked on Rachelle's door and stuck her head inside to greet Rachelle and Chris. She checked Rachelle's heartbeat and blood pressure, inputting the numbers into a computer. For a few minutes they chatted and laughed, the three adults focused utterly on the baby as if he were the entire world. Then the nurse's aide left to finish her morning rounds. Chris smiled at Rachelle, then stepped outside to take a phone call in the parking lot.

JUST AFTER 8 A.M., Paradise Police dispatcher Carol Ladrini received a call from Cal Fire's Emergency Command Center. "Hey, we've just issued mandatory evacuations for the entire town of Paradise," the dispatcher said.

"Are you serious?" Ladrini asked.

The last large-scale evacuation had taken place a decade ago during the Butte Lightning Complex. Ladrini passed word of the townwide evacuation order to Police Chief Eric Reinbold. In most municipalities, a police chief was usually the one who issued emergency CodeRed alerts, but in Paradise, this was the town manager's prerogative. It was Reinbold's responsibility to call Gill, who had the final say.

Reinbold thought about how to tell her. A soft-spoken and well-respected Paradise native, he had worked drilling water wells for farmers across Northern California before joining the local police force in 2007. At thirty-five, he was young for the job: The schools superintendent had once been his kindergarten teacher—a fact she mercilessly teased him about. Reinbold had been named chief only fifty-two days earlier, sworn in with a quick ceremony led by Gill on a sunny September afternoon. For the past hour, he had focused on his officers, who were knocking on doors along Pentz Road, instructing residents to evacuate. A sergeant who would have been assigned to represent Paradise at the command post in Jarbo Gap had been diverted to help.

Reinbold hadn't been monitoring Cal Fire's radio traffic. Fire-fighting and law enforcement communicated over separate frequencies. Trying to listen to both simultaneously was akin to playing a country music station over a rock station. But he did know that some of his officers had asked for fire shelters—beach-blanket-sized coverings made of aluminum. Folded into a shoebox-sized waist pack, the shelters deflected heat while preserving a pocket of breathable air. Reinbold had stashed nine of them in his squad car and prepared to speed over to Pentz Road. Now Ladrini's update had left him reeling.

Just after 8:10 A.M., as Feather River hospital began discharging patients, Reinbold called Gill at Town Hall. "Hey, Cal Fire just told me we need to issue a townwide evacuation," he told her.

"No, we aren't going to do that," Gill replied, stunned. "We are just evacuating the eastern side of town." Reinbold repeated what he'd heard from Cal Fire.

"I'm going to call David Hawks," she said and then hung up, her mind racing.

What is happening right now? Gill wondered, pacing the council chambers. Only a few minutes earlier, Mattox—her steady and un-flappable counterpart—had told her about the evacuation along Pentz Road. This alone was shocking, never mind the idea of evacuating the entire town. They had long relied on a system of orderly and well-timed CodeRed alerts as a wildfire progressed. And for good reason. *No place is designed to evacuate all at once,* Gill thought. Not San Francisco with its two bridges during an earthquake. Not Los Angeles with its tangled freeways during a tsunami.

Neither of the retired fire chiefs, who volunteered at Town Hall during emergencies, had arrived. Until they showed up, Gill was on her own. Petrified, she fell back on the town's evacuation plan. Mattox ducked his head into the council chambers.

"I need a second with you," he said, looking at Gill.

For the past hour, his cellphone had chirped with messages from friends in the valley. "Do you need a place to stay?" they all asked, as they saw a black haze solidify above the Ridge. "No, we're fine,"

Mattox had replied. Earlier that morning, he had dropped his daughters off at nearby Paradise Elementary, certain that the town wasn't in danger. He had concentrated on coordinating with his five-member Public Works crew. They were deploying an evacuation trailer stuffed with the construction cones and reflective signs needed to start contra-flow on Pentz Road. But now he had new information to share.

"We need to evacuate the entire town," he confided in Gill around 8:35 A.M., relaying the latest from the police radio frequency.

Gill blinked at him. Pushing thousands more residents out onto the streets in their vehicles could mean sending them to their deaths. She had seen this happen in 2008, when half of the town had gotten jammed on narrow corridors as the Humboldt Fire inched closer. Outside Town Hall, on the Skyway, cars were beginning to back up. As residents drove downhill, they were all trying to take the first exit into Chico, clogging traffic for more than twelve miles. Meanwhile, the four-lane contra-flow hadn't yet been implemented, and vehicles were arriving in Paradise on the two uphill lanes. Gill had already called the Chico city manager several times, begging him to block exits into Chico, which would force traffic onward to places like Oroville. "I'm trying," he replied.

More than a half hour had passed since Cal Fire had called for the full evacuation. Still Gill hesitated. She told Mattox that she wanted to hold off. She would verify the information with Fire Chief Hawks first. He would know what to do. Gill spun on her heel and strode back into the chambers, Mattox following closely behind. Hawks picked up after a few rings. "I'm hearing we are evacuating the whole town," Gill said softly. "And, well, you told me to never evacuate the entire town." "It's only the few zones along Pentz Road, as far as I know," said Hawks, who had recently arrived on the Ridge from his gym in Chico. "You're right, we never want to evacuate the entire town. It's not ideal. But I'm on Pentz Road, and it's bad."

Suddenly, out of the corner of her eye, Gill saw Zuccolillo. He and councilman Steve Crowder, who had just been elected, arrived. They asked her how they could help. "There's nothing to do right now," Gill said, flustered. She ventured that they might check on people gathered at the Paradise Ridge Senior Center and CMA Church. Preoccupied, she paced away. She needed to get the flow of traffic switched on the Skyway.

Soon after, IT manager Josh Marquis showed up. Though four staff members were trained to use CodeRed, Marquis had been designated as the sole dispatcher of evacuation notices. The clerk had called him at home in Chico to tell him to come to the EOC and start issuing CodeRed warnings. By the time his 1989 Jeep Cherokee pulled in to the parking lot, another staffer was waiting for him. "We are not doing warnings. We have to start evacuations right now," she told him.

Marquis cursed. "What are you talking about? I thought we were just doing warnings."

"The fire is coming right for the town."

He parked and followed her indoors, racing past the office where the clerk was recording an updated message for the town's emergency radio station, 1500 AM, about the Pentz Road evacuation. Marquis, thirty-eight, was an hourly contractor who handled everything from PowerPoint issues to keyboard replacements, preferring to keep his distance from the drama that accompanied local politics.

Paradise was responsible for dispatching its own alerts. The town had an independent CodeRed log-in, though it shared the bank of residents' phone numbers and emails with the Butte County Sheriff's Office—a necessary redundancy, in case something went wrong and the town couldn't complete the task. Gill and Marquis didn't realize that the Sheriff's Office had already begun sending its own evacuation alerts. The uncoordinated alerts were scattershot, a problem that had plagued other communities during wildfires.

A year before, the state had been confronted with the grave consequences of not alerting residents in a timely manner. When the Wine Country wildfires blazed through Sonoma County in 2017, only 2 percent of residents had been signed up for its primary warning system, SoCoAlert. A small group of leaders had decided against sending a Wireless Emergency Alert—the Amber Alert–style message that could reach virtually every cellphone in an area—because it was a geographically imprecise tool. They feared it would ping cellphones in unthreatened areas, causing traffic snarls and overwhelming 911 dispatchers. The county's decision meant thousands of residents were left unaware as fire bore down on their neighborhoods.

The state criticized Sonoma County's decision in a scathing thirty-four-page review, calling out emergency managers for a procedure that was "uncoordinated and included gaps, overlaps and redundancies." State senators Mike McGuire and Bill Dodd, whose districts were most impacted, brought a bill to the capitol in Sacramento. They wanted to pass a uniform protocol for how all fifty-eight counties in California issued emergency notifications. The bill passed in September 2018 and stipulated that the new guidelines had to be written by July 1, 2019. As of November 2018, the protocol was still being drafted.

On his computer, Marquis clicked on the zones along the Feather River Canyon and wrote the message: "There is an immediate evacuation order for zones 2, 3, 7, 8, 13 and 14 due to a fire. You should evacuate immediately. You are receiving this message because you are in the affected area." Gill gave her approval, and at 8:44 A.M., Marquis clicked Send. The alert went out to fewer than half of the town's fourteen zones.

Usually Marquis tried to contact only a handful of zones at a time to minimize congestion on the roads. But Mattox was getting impatient. Each minute that passed was another minute that people wouldn't have to make their escape. As Marquis dispatched the first alert, Mattox pulled Gill aside once more. As Public Works director, he carried a master key for the traffic signal boxes downtown. Each contained a computer controller that flipped between green and red. He wanted to switch the stoplights to hold green, he told Gill, to move vehicles downhill faster. Soon he wouldn't be able to cut through gridlock to reach the lights. He felt obligated to leave Town Hall now, before traffic got worse. Gill nodded and briskly turned away.

As Mattox left, he instructed Zuccolillo and Crowder—the lost-looking councilmembers—to drive up the Skyway until they found an intersection without law enforcement. "Stay there and direct traffic," he said. Then, feeling panicked but determined, Mattox drove his Chevy truck to the intersection of the Skyway and Neal Road, just north of where the Skyway cleaved into a four-lane divided highway. He manually switched the stoplights so the oncoming vehicles wouldn't have to stop. The evacuation would hinge on the town's two southernmost cross streets, he knew, where traffic merged. He then headed north to the intersection of the Skyway and Pearson Road, where a probation officer and a sheriff's deputy were directing traffic.

Thousands of vehicles were inching by at 5 mph. Some formed bottlenecks at the bulb-outs. Mattox's wife, a social worker at Adventist Health, was in one of the cars at the intersection, with their two daughters, Sara and Mae, ages six and nine, buckled in the backseat. When she spotted Mattox, she rolled down the window, and Mattox shouted to the girls that he loved them. Their taillights rolled downhill and disappeared in the red glow.

GILL'S ASSISTANT, Colette Curtis, thirty-six, plopped her handbag on her desk and sat down. Originally from the Bay Area, Curtis had moved to Paradise with her husband a decade earlier, hoping to start their family somewhere more affordable. Their two young daughters had been born at Feather River hospital. Regardless of how people felt about the town manager, everyone adored Curtis. With her bright lipstick and sunshiny demeanor, she offered a sharp contrast to her boss's aloofness. She organized Gill's calendar and stood sentry for her. Now, double-checking the six evacuation zones scribbled on her steno pad, she tapped out a Facebook post: "This is a very dynamic situation; roads are closing and opening quickly due to the fire movement. Please be aware of your surroundings and listen to emergency personnel."

Soon after, her office line trilled. Curtis paused to give a live

interview to a television journalist in Redding who wanted an update on the Camp Fire. The smoke was so thick near the northern
city that residents initially believed a wildfire had ignited nearby—
but it was just the ash billowing from Butte County. The chief
meteorologist, who also spoke on the segment, had already commented on the Camp Fire's punishing sprint from Pulga. "Unfortunately, we've got the gusty winds of twenty-five miles per hour
out there and the low humidity of eighteen percent," he had said,
gesturing toward an onscreen map of Paradise. "Those are the two
main factors that we have been tracking during the Red Flag Warning, and unfortunately, they are right on schedule. What makes it
more treacherous is you have all of these canyons and ridges, which
makes it even more difficult for the firefighters. If you look at this
map"—he flipped to a satellite image, which showed a red wave
barreling toward Paradise—"you'll see we've got a really hot fire."

On the program, Curtis explained that the eastern zones along
the Pentz Road corridor were being evacuated. Next door to her,
Gill had just left her office to give Marquis more instructions. At
9:09 A.M., more than twenty-five minutes after he had sent Town
Hall's first alert for Pentz Road, Marquis sent a second CodeRed
evacuation order, which included two additional zones, totaling
eight of the town's fourteen. The message read: "This is an immediate evacuation order for zones 2, 3, 5, 6, 7, 8, 13 and 14 due to a
fire. You should evacuate immediately. You are receiving this message because you are in the affected area." The notice was dispatched
to 2,765 phone numbers. But the telecommunications system was
badly overloaded, and 42 percent of the alerts were never delivered.
(Forty-seven percent of residents hadn't even received the first
alert.) Marquis tried to print the information so Curtis could post
it on Facebook, but Gill stopped him. "We need to evacuate the
next third of the town," she said.

"We can't do it," he shot back. "You can't do it that quickly.
People need time to empty the roads out."

"We don't have a choice," Gill replied.

Marquis tapped the information into the CodeRed program—

but it wouldn't send. The town's Comcast Internet had cut out as the infrastructure melted. The AT&T landlines followed minutes later.

Outside Gill's office, Curtis was finishing the interview. "What should people know right now?" the journalist asked her. "Listen to the local authorities and get your bags ready to go, even if you're not in an evacuation area," she replied. "Be prepared and be aware of your surroundings."

As Curtis was speaking, a car veered into the parking lot of Town Hall. A panicked man emerged, running past Gill's office and into the council chambers. The nearby hillside was on fire, he said, and within minutes Town Hall would be too.

George Morris, a retired fire chief who had recently arrived at Town Hall—interrupting his morning game of golf—assumed the man was nuts. Morris had started his career with Cal Fire in coastal Humboldt County in 1973. He was as solid as a wall, with thinning white hair and weathered pink cheeks from a career spent out-doors. He and Gill lived a few blocks from each other near Butte Creek Canyon, and she trusted him completely. At the EOC, Morris translated Cal Fire's convoluted radio lingo into plain English for her. But it didn't take long for Morris to decide that the newcomer was speaking the truth. Chatter on the Cal Fire radio frequency indicated new spot fires throughout town. Flames had blocked Pearson Road. If dispatchers knew about fires at Bille and Clark roads—in the middle of town, near the Dollar General and Paradise Alliance Church—there were probably hundreds more that they didn't know about. He told Gill that they needed to evacuate the building.

She hurried toward her office, where she saw Curtis, still on the phone with the journalist. Gesturing, Gill asked what Curtis was doing. "Hang up the phone!" Gill implored, pantomiming with one hand, her thumb and pinky extended to look like an old-school receiver, wanting to rip the phone from her hands. She needed Curtis to call the school district and confirm that students were being relocated. "This is a live interview," Curtis mouthed back silently. "I can't just hang up!" Gill threw up her hands and stepped into the hallway. Curtis thanked the journalist for her time, gently placed the phone in its cradle, and stepped out of the office.

"We're evacuating," Gill said. "Grab your stuff and go."

Marquis, who had just walked up behind Gill, paused for a moment before turning to Curtis. "Can you help me with something really quick?" he asked. He led her to the storage room where the expensive police body cameras were stored. "If the building burns down, these can't go down too," he said, jamming them into a box. He looked through the window, struck by the sight. "Why don't you just go," he said to Curtis, his voice strained.

"Are you sure?"

"Just go, I'll be right behind you."

As Curtis pushed through the front door, turning for a last glimpse of the building, she saw that the house behind Town Hall had caught fire. She stared and snapped a photo on her cellphone. Marquis hurried after her, loading the cameras, along with the town's servers and backup tapes, into his Jeep. They climbed into their vehicles and merged onto the Skyway, one after the other. Behind them, the parking lot emptied of staffers' cars.

But Gill and Morris stayed behind. She wanted to sort out whether their local EOC would join the county's in Chico. Where would her staff go when they reached the valley floor? Was the Public Works crew safe? Gill needed to find out. Her phone vibrated with messages from unknown numbers, the area codes from all across the country. She ignored them. Around 10 A.M., her cellphone buzzed with a call from Rick Silva, editor of the *Paradise Post*. Gill knew him well and answered. "I'm hearing there is a townwide evacuation, is that true?" Reeling and numb with shock, Gill gave him the only answer that she knew to be true:

"If residents feel they need to leave, they should."

AS TOWN HALL was evacuating, the Paradise Police Department building was about to burn down.

At 10:20 A.M., another call from Cal Fire's Emergency Command Center rang through. Carol Ladrini answered. "Are you still there?" the other dispatcher said, astonished. "You need to evacuate and get to safety." Ladrini had been joined by two more colleagues, who had helped her answer phone calls, then monitor radio traffic from officers in the field. More than a thousand calls had streamed in from across the county. In Paradise, 132 calls were answered between 6:30 A.M. and 8:20 A.M., when the 911 line had been rerouted to Chico. Alarm companies alerted dispatchers to single houses burning even as entire neighborhood blocks fell.

Family members of the elderly and the vulnerable submitted more than six hundred requests for law enforcement to check on their loved ones. In Magalia, an eleven-year-old girl was home alone from school. Her parents were at work and couldn't pick her up. A quadriplegic on Honey Run Road needed an ambulance. A family on Bille Road was engulfed in ash. Ladrini's tone had become caustic as she faced an unending stream of situations that she couldn't control. She relied on her training, triaging an emergency and providing recommendations based on a script. But over and over, she had to reiterate that help was *not* on the way. Now there was nothing to do but leave. Cal Fire was right—the police station on Black Olive Drive was about to be destroyed, along with the fire station next door. In the police parking lot, a few strangers waited with a shaggy brown dog, hoping for a ride. They had come to the station for help evacuating since they didn't have a car. Ladrini unbuckled her grandchild's car seat and tossed it into the trunk, helping them into her black Mazda CX-9.

Within the hour, Morris and Gill faced a similar reckoning. The retired fire chief feared Town Hall was going to catch fire. They had been spared an hour earlier, when the wind shifted and blew the flames from the burning house in the opposite direction. "If we're going to go, we should go now, because the fence out back is on fire," Morris said, as serenely as if he were asking Gill how she liked her coffee.

"What are our options?" she asked. The only other hope for survival, he explained, would be sheltering in the parking lot by lying flat on their stomachs and letting the fire pass over them.

"Do you have any of those little fire blanket things?" Gill asked. Morris didn't.

"It looks like our neighborhood is gone," he said after a moment, pausing to listen to Cal Fire's radio traffic again.

Gill finally acquiesced. She had wanted to be the last person to leave Paradise, doing whatever she could to help everyone else make it out alive. She couldn't bear the thought of losing an employee—or leaving any staff behind while she saved herself. If it wasn't for Morris, she might have stayed. But she didn't want to be responsible for his death. Gill followed him out of the building and locked the front door. Trembling with adrenaline, she asked Morris to drive them downhill to safety in her red Volvo, leaving his truck—golf clubs in the exposed back—behind. She felt safe with him. "You'll get more work done, too, if I drive," Morris said kindly.

They departed Town Hall without calling for evacuations in the final six zones—1, 4, 9, 10, 11, and 12—along Butte Creek Canyon. Only 4,855 phone numbers had received a CodeRed notification out of a total population of 26,500, which meant that almost 82 percent of Paradise residents never received an emergency alert from *any* of the systems. As they pushed through the flames that leaped along both sides of the Skyway, where traffic was still jammed, Gill tapped out a final series of text messages to the Town Council chat group, littered with misspellings and punctuation errors. She was in shock, her hands shaking. "We are moving eco to chico fore center . . . so is the county." And another: "No one is in our eoc." "George Morris and I were the last ones drove through fire." "Resources are coming." "Traffic is moving."

"Is everyone ok."

For nearly two hours, no one answered.

AT ACHIEVE CHARTER SCHOOL, where Jamie's daughter Tezzrah and Rachelle's daughter Aubrey were in second grade, cars lurched onto the curb in parents' haste to pick up their children. Tezzrah was the fifth-to-last student to leave, waiting for more than a half hour to hear the principal call her name over the intercom, longing for the familiar sight of her father's Subaru. Aubrey's stepmother had picked her up from the same classroom.

Across Paradise, on Edgewood Lane, Travis looked up from his yardwork to see his neighbors Paul and Suzie Ernest straddling their four-wheeler. "Edgewood is blocked by trees," they said. A long line of cars shot by, funneling toward the dead end of the finger ridge, not realizing it was a trap.

And in Magalia, Kevin McKay opened the doors of Bus 963 to two teachers and about two dozen students, their eyes wide with fear. The twin daughters of the immigrant couple who owned Sophia's, the local Thai restaurant. The ten-year-old daughter of a bartender. A seven-year-old whose father was in nearby Tehama, painting the small-town mayor's front door. Their parents commuted to distant communities or worked hourly low-wage jobs that they couldn't walk away from, even in an emergency. They weren't able to collect their sons and daughters in time. The children filed onto the bus, and Kevin closed the doors behind them.

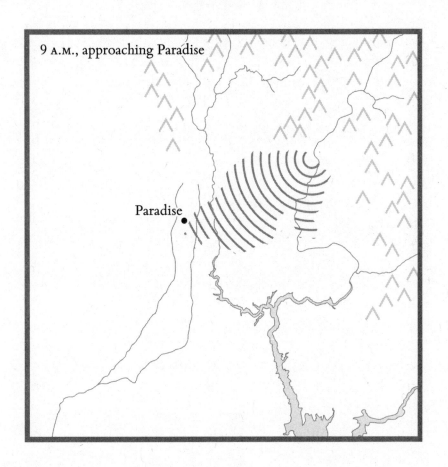

9 A.M., approaching Paradise

Paradise

PART III

CONFLAGRATION

KONKOW LEGEND

All those who had heard the teachings of the good men became conscience-stricken and built the kakanecomes, *the sweat houses, and bowing down therein invoked the Great Spirit, praying for the mercy of Wahnonopem, and that the fruit of the evergreen and ever-bearing tree in the land of the stars, near the Great Spirit, might be showered down to them. But Wahnonopem had veiled his face in his anger and would not hear. He had said that he would send the great fire to destroy his bad children, and his word was the great law upon the earth, in the waters, and in all the skies.*

The good men had told the Konkows that the kakanecomes were sacred, and that no women or children were to go down into them. Only the men who were feeding the holy fire were to bow down before it, with the wickedness in them purified by the fire. But one day when all the people were out on the plain, wringing their hands in their anguish and despair and praying for relief in their suffering, two little boys went down into the kakanecomes and threw some pitch pine sticks upon the fire. The flames flew up to the roof and from there spread everywhere, licking and destroying everything in their way.

CHAPTER 6

ABANDONING THE HOSPITAL

Walking to the parking lot, Chris put his cellphone to his ear and half-listened, groggy. The call was from one of his landscaping employees. Rachelle's phone charger, which he had brought from home, dangled in his hand. Smoke was spilling over the mountains from Concow, the employee said, sounding frightened. She wasn't going to make it in to work that day. Chris fiddled with the phone charger, rubbing the cord between his calloused fingers. He looked up. A thunderhead of smoke was unfurling above the tree line.

"Get out of Paradise," he urged her.

He walked back through the hospital's doors. In the few minutes he'd been outside, a flurry of activity had overtaken the hallways. Fire alarms were shrieking on the walls. A line of patients had suddenly materialized, snaking down the foyer that led to the ambulance bay. Waiting to be evacuated, the patients held tattered maroon binders containing their medical charts and records, the spines marked with stickers denoting allergies and primary doctor information. Anyone who could "walk and talk" was being led out first, including the pregnant woman from Redding. The linoleum floor was streaked with ash from the shoes of people who'd walked in and out of the parking lot.

There were sixty-seven patients to discharge—more than usual for this time of year. Their ailments ranged from pneumonia and chronic respiratory problems to dehydration and sepsis. Employees who weren't involved in patient care, such as secretaries, housekeepers, and students, had already been told to leave. The remaining staff filtered through the hallway, checking on patients. The corridor was a riot of colors in a department that knew only the black scrubs of emergency room nurses: navy blue for the surgical staff, charcoal gray for nursing assistants, bright pink for birthing center staff, maroon for pharmacists. Sweat soaked the cotton fabric, ringing their collars. A couple of them jammed open the ER doors to keep them from swinging shut. Chris couldn't find Rachelle in the line. He stepped back to let a cluster of nurses pass, pushing intensive care patients on gurneys.

"We're running out of wheelchairs," someone shouted.

Six nurses walked by, rolling a hospital bed on which a heavy man with cerebral palsy lay. He was intubated and under an anesthetic called propofol, a milky drug so powerful that it could disable the body's respiratory system. (It was known as the Michael Jackson drug because the pop star died in 2009 from an overdose of the sedative.) If the man were disconnected from his ventilator, he could die. Unfortunately, the expensive respiratory machine wasn't portable. Staff members heaved him onto a gurney and hooked him to an Ambu Bag, a portable breathing device. They did a check: clear airway, untangled lines, the EKG registering smooth waves—all good signs. They continued down the hall toward the exit, throwing their bodies against the wheeled bed to slow it (the gurney had no brakes). Medical helicopters hadn't been able to land in the smoke, so the nurses lifted the intubated patient into one of two ambulances. Into the second ambulance went an elderly woman who had fallen in her bathroom and sustained a brain bleed, along with a paraplegic and a local woman who had given birth to a daughter a few minutes earlier. The obstetrics surgeon had hastened to suture the woman's C-section as the evacuation started. Her husband had scooped up the bundled baby girl—still streaked in amniotic fluid—and driven to safety.

Downhill, the patient financial services office caught fire, threatening the marketing department and human resources center. In nearby homes, propane tanks began to explode, shaking the ground.

As temperatures rose to 2,000 degrees, the liquid gas was boiling inside the metal containers, turning into a high-pressure vapor that ruptured the tanks with an ear-splitting whine and ignited into a fireball, torpedoing the tanks up to 2,500 feet in any direction. Known as a "boiling liquid expanding vapor explosion," or BLEVE, this was a violent by-product of wildfires in the wildland–urban interface, where residents relied on propane to heat their food, water, and homes. A BLEVE was the most powerful nonnuclear explosion created by man. That evening, two firefighters, including dispatcher Bowersox's brother, would sustain serious burns when a 250-gallon propane tank exploded in a burning home in Magalia, flinging molten aluminum and pieces of fence at their necks and faces.

In the hospital, the hallways flickered black for about eight seconds until the backup generator sprang to life. The desktop computers where nurses logged chart notes blinked. Staff began affixing white tape and beige cloth into giant X's on door frames, so that firefighters would know that the rooms were cleared. Others used yellow sticky notes. Roberson, the nurse supervisor, did a final check. The break room, with its six-foot-long folding table and worn green couch, empty. The nurse stations, vacant. The bathrooms echoing, stall doors flung open. Ash from outside rolled down the hallways, searing patients' eyes. With the doors to the emergency bay jammed open, it was getting harder to breathe.

Chris hurried back to Rachelle's suite. She was still in bed, their son propped in her arms. He explained what he had seen. "Go get your mom," Rachelle replied. "Make sure she gets out of the house. We'll be fine. We'll meet you in Chico." Chris planted a kiss on her forehead and left.

A few minutes later, a labor and delivery nurse named Tammy Ferguson appeared with three male hospital staffers. They barged into Rachelle's room, not bothering to knock, lifted Lincoln from Rachelle's arms, and settled her into a wheelchair. Her intravenous line flapped against the armrest. It happened so fast that Rachelle didn't understand at first why she was being moved. Her legs were weak from the anesthesia she had received during her C-section. Her entire body hurt. She reached for her cup of ice water, but one of the men snapped at her not to worry about it. A fast-moving wildfire had reached the hospital, they said. Tammy propped Lincoln on a pillow in Rachelle's lap, preparing to roll her to the emergency bay. The plan was to load Rachelle into an ambulance headed for Enloe Medical Center in Chico.

"Let's go," Tammy said. "There's no time to grab anything else."

Another nurse pushed Rachelle and Lincoln to the front of the evacuation line, where the queue bottlenecked at the open bay doors. Rachelle looked out across the lot just in time to see the only two ambulances pull out and disappear down Pentz Road. There hadn't been enough space for her. Staff members parked their cars beneath the overhang, guiding patients into the backseats. Nurses frantically cleared space in their minivans, pitching car seats onto the hospital sidewalk in a jumble. They lifted Rachelle into a white four-door sedan belonging to David, a fifty-eight-year-old biomedical technician. He wore gray scrubs and a black fleece jacket. A white beard stippled his face. The nurse tucked a blanket and three pillows around the premature baby, tossing Rachelle's IV bag onto the floor mat. Her catheter was still wrapped around her leg. No seatbelt, because of her incision.

"Go with David," the nurse said, slamming the passenger door.

Each juniper bush was a patch of flame. The oak and birch smoldered. David barely paused to introduce himself, speeding through the concrete lot and turning right on Pentz Road. They passed Rachelle's house on the left-hand side. Her driveway was empty. Beige trash bins stood at the curb; the red mailbox flag was down. Everything looked impossibly ordinary—and just out of reach. She hoped Chris had gotten his mother out safely.

Just south of Ponderosa Elementary, David turned left onto Bille Road. He knew that their evacuation was going to be problematic. Four main thoroughfares ran south through town, paralleling each other like the spindly legs of a stool. Pentz Road was situated on the easternmost edge, abutting the rim of the Feather River Canyon. Clark and Neal roads were in the middle, and the Skyway ran along Butte Creek Canyon in the west. Only three roads—Bille, Pearson, and Wagstaff—connected them, cutting across from east to west. About 60 to 80 percent of the town's thoroughfares were access roads that could barely accommodate rush hour traffic on a typical evening. As for the Skyway, studies had shown that its concrete bulb-outs had increased travel times by as much as fifty-four seconds—and that was during normal times. A mass evacuation was sure to make things worse.

At the intersection of Bille and Pentz, a few fire engines were hooked into one of the town's fifteen hundred fire hydrants. They sprayed water at the baking cars, about as effective as a squirt gun. The crews tried to move traffic forward. They waved David through, but about two hundred feet past the intersection, he and Rachelle hit gridlock. David inched his Nissan forward a few feet at a time, punching the Recirculate button on the air conditioner. He and Rachelle didn't talk. The autumn sky disappeared, slowly and then all at once. Puffs of orange tie-dyed the horizon, deeper in hue where homes were burning, lighter where flames hadn't arrived. A toxic mass choked the sunlight, the black smoke suffocating the sky. It was just before 9 A.M., but if Rachelle hadn't known better, she would have guessed it was midnight. She clutched Lincoln. The top of her cotton gown was folded down, her chest exposed. The baby nursed, occasionally mewling, his face contorted. The glowing taillights of cars marked the road in front of them. Rachelle pressed the window gently with a fingertip, then pulled it back in surprise. It was like touching the door of an oven.

David noticed glowing leaves skittering across the pavement and sticking to other drivers' engines, threatening to light them on fire. They must be gathering under his Nissan, too, he realized. He got out to clear the leaves, then moved to help other drivers facing the same hazard. Rachelle sat alone in the sedan. She took off her N95 mask, cupping it over Lincoln's mouth and looping the elastic around his seashell ears. It eclipsed his entire face. She covered her nose with her hospital gown, breathing hot air into the pink cotton. Outside, she saw people abandoning their cars and sprinting by her window.

Her legs were like gelatin. Though the anesthesia and painkillers had started to wear off, Rachelle still felt she had been disconnected from the lower half of her body. After the birth of Vincent, when she'd had her first C-section, she had managed to waddle across the parking lot after being discharged—slowly, but on her own none-theless. After Aubrey, she had opted for the wheelchair. Days later, when she had begun to walk again, it had been at an excruciatingly slow pace, hunched over with a pillow pressed to her abdomen to stifle the throbbing of her incision. Now, after her third, Rachelle couldn't fathom the idea of moving. She knew she should be in pain, but fear overwhelmed any other sensation.

She reached for her cellphone, dialing Chris repeatedly. The call wouldn't go through. The signal dropped. It was a game of chance, like being in a sports stadium or concert arena packed with tens of thousands of fans as they all used their cellphones. Some randomly got connected; others were stymied by the overloaded telecommunications network.

Rachelle's panic rose. She dialed her father. He lived in Fresno and had no inkling that a wildfire had struck Paradise. Rachelle wanted to say goodbye. She thought about her wild spells—as a teenager and then after her divorce, when she had consumed too much white wine too frequently—and figured this might be her last chance to make amends. She held the phone to her ear. The call connected. "Daddy, I just need to hear your voice," she said, explaining that fire was all around. She was alone in a stranger's car and didn't know what to do. Neither did he. "I'll start praying," her father said, mustering his calmest voice. "Take a pillow. Put it up against the window where it is getting hot. Take a deep breath. I love you, baby. I'll pray for your safety."

"Dad, the guy came back," she said.

"Wait—" he said, but the call dropped.

David crawled back inside the sedan, buckling his seatbelt. Traffic had started to move, though barely. For a moment, a gust cleared the haze. Rachelle spotted a familiar white Suburban with an ACHIEVE CHARTER SCHOOL HONOR ROLL bumper sticker a few vehicles ahead. "Oh my god, that's my husband," she said as she leaned forward and pointed out the SUV. "Follow him!" At that moment, the lead car in their lane of traffic caught fire.

Rachelle watched as Chris veered around it and drove into the black puff of smoke. Her children's school bumper sticker vanished.

She wondered about Vincent and Aubrey. She and Mike shared custody, and the children were now with their stepmother. The kids adored their two homes and three stepsiblings—two families, twice the love—but now Rachelle felt panicked. Were they safe?

Rachelle begged David to follow the SUV. But for all David knew, Chris had made it ten feet farther before getting stuck, just like everyone else. Upset, Rachelle pushed back. She knew Chris had escaped. He was invincible. But there were no emergency responders around to tell them whether the wildfire was up ahead or not, whether the road was passable for David and Rachelle—or Chris, for that matter. The closest firefighters were still behind them, at Pentz and Bille. David gripped the steering wheel, staring blindly into the smoke. His chest ached. He felt like the worst person in the world.

"I can't do it," he said finally. "I can't risk you and your baby going into a black cloud that I can't see through. What if there's a choke point?" He explained that he was on medication for a heart arrhythmia. *Please,* he beseeched Rachelle, *calm down.* If something were to happen to him, she wouldn't be able to get out either. They were in this together.

They tried to talk about other things as they waited—his wife, Bonnie, back home on Pinehurst Way in Magalia. He was worried about her. He didn't mention having children, but he told Rachelle about the infusion machines he repaired at the hospital. He was an engineer or something, she couldn't really tell. He helped her slow her breathing, and she helped him slow his. They continued waiting. They weren't sure what they were waiting for. In the meantime, Rachelle texted her father, letting him know that she could see Chris's car.

David pointed out two bulldozers that were parked near the road, in an open field behind a chain-link fence. If the fire overran him and Rachelle, he said, he would gun the Nissan through the fence and park between the two bulldozers. He thought they could be safe there. Rachelle patted his arm—an oddly intimate thing to do to a stranger, she thought, but these were not normal circumstances. And they made an agreement. If David's car caught fire, he promised to take Lincoln and run. "Are you sure that's what you want me to do?" he asked. Rachelle nodded, unable to answer him, because she didn't want to cry, wasn't going to cry.

A CALL RANG THROUGH on Chris's cellphone—another stroke of luck. It was Rachelle's father. He seemed shaken, his voice quaking as he spoke. He told Chris that Rachelle was a few cars behind him on Bille Road. Chris listened as he navigated the pitch-black tunnel of smoke, feeling his way blind. Suddenly, a sheriff's deputy sounded a horn behind him, and he hung up before nearly hitting a downed electrical pole. He didn't know which car Rachelle was in, or he would have tried to backtrack to find her.

After leaving Rachelle and the baby at the hospital that morning, Chris had made the five-minute drive home, through a corridor of houses decorated for the holidays, past scaly-barked ponderosa pines that he had passed thousands of times without truly appreciating them. He had parked his work truck in the driveway, barreled into the house, and woken his mother up. They had stayed up late, chatting until 1 A.M., and she was drained. He told her that she would need to drop him off at the hospital to pick up Rachelle's car and then head down Clark Road to Oroville. Now he fretted. What to pack? What to preserve when your entire world was at risk of disappearing? Not the family photos or birth certificates or expensive electronics. Not the homemade Christmas tree ornaments or the marriage certificate pinned to the fridge or Rachelle's wedding dress, meant to be passed down to Aubrey someday. Not the dollhouse the girl had made from cardboard boxes or the blueprints from Chris's landscaping job. Not anything that made sense. Racked by indecision, Chris stepped outside and pulled his work truck into the garage. The truck's bed, matted with flammable pine needles, would stand a better chance of survival indoors.

Back in the house, Chris grabbed a newly unboxed car seat, the fabric yet unmarred by dirty fingers and food stains. The power flicked off. He grabbed a flashlight, a diaper bag, and a tub of folded laundry sitting near the dryer. He headed back to his mother's sedan. Everything around him quivered in the heat. Pine needles bobbed on the wind until they met the flying ash and ignited. Singed oak leaves and a ginger afterglow consumed the front yard. As they backed out of the driveway, the wind lashed his home with embers, and flames leaped onto the neighbor's place.

His mother pulled in to Feather River hospital. The campus was abandoned. The car seats, tossed out of nurses' minivans to make room for patients, formed an eerie line on the sidewalk. Near the cardiology wing, a few water tenders, each carrying 3,000 gallons of liquid, idled next to some fire engines from another county. Yellow-clad responders were trying to save the facility. Their division leader directed them indoors. The heavy compressed-air tanks on their backs allowed them to breathe inside the burning building. Some doused tiny spot fires near the Birth Day Place. Others crawled atop the roof and sawed two-foot-wide trenches to let the heat and smoke escape before they could breach the wing. Though the walls were fire-rated, they could withstand only so much for so long. *This might be a lost cause,* their captain thought. A few minutes before, the wind had launched a sugar pine cone onto his truck with such force that the windshield had splintered.

Chris said goodbye to his mother. She left in her car; he departed in Rachelle's white Suburban, heading north up Pentz Road and onto Bille Road. It might have made more sense for them to drive together, but they wanted to save their vehicles if they could. Everyone was thinking this way: Paradise was now jam-packed with thousands of cars. This didn't bode well: According to a 2007 report, a third of the town's public roadways were in "poor" condition—considered beyond routine repair. "The conditions of roads within the town affect the ability to provide public access for fire, medical and police protection services," the municipal service review had warned back then. In the coming hours, 179 burning vehicles, many of them stuck on these roadways, would block evacuation routes. The California Highway Patrol would later recover 19,000 more cars, boats, tractors, and recreational vehicles.

Chris had been picturing Rachelle and Lincoln safe, in an ambulance on its way to Chico. When hospital staff had told him that she was "evacuating right now," he had assumed they meant it. But now her father was telling him something else. He tried to call Rachelle, again and again. Each call dropped. He tried to text her: "What kind of car are you in?" "What color is it?" "How is the baby?" None of the texts went through. He called her again. Nothing. He texted her again. Nothing. Red taillights, white headlights, black skyline. The sunshine had been sucked from the sky.

People were abandoning their cars and—thoughtless in their desperation—taking the keys with them, leaving the vehicles as immovable as boulders on the road. They raced around Chris while he dodged and navigated the obstacles in his wife's white SUV. He counted the burning cars—there were fifteen—before pulling off near Sawmill Road. He parked on the shoulder and then sat there, the driver's door cracked. His was the only car that had managed to cut through the gridlock. He fiddled with the radio. The town's emergency station, 1500 AM, was broadcasting outdated information about the Pentz Road evacuation. The only other clear station on air was a conservative talk show.

Chris knew Rachelle would follow him if she could. And yet nobody appeared. He ran back down the street toward Pentz Road Market & Liquor, leaving the SUV behind with his keys dangling in the ignition in case someone needed to move it. Chris watched more cars catch fire. Was Rachelle in one of them? Was the baby?

Smoke greased his throat, his eyes wet with ash and fear. A man appeared out of the smog, his cheeks dark with charcoal. "Who are you looking for?" he asked.

"My wife," Chris said. "My baby. He's six pounds, twelve hours old. I don't know what's going on."

"I saw a blond woman breast-feeding in a car back there," the man said, gesturing at the roadblock. "They turned around and went back." Chris had to believe him. Why would this stranger lie? His family had to be alive. Lincoln—the son that he planned to take trout fishing in the Feather River and hiking in the snow-capped Sierra Nevada—had to be alive. Chris ran back through the jumbled maze of eviscerated cars toward the fire, the ground beneath him gummy with tire rubber and marred with tributaries of aluminum from melted engine parts.

OBSERVATION: A FATAL BREATH

In a tunnel beneath Feather River hospital, Ben Mullin wheeled metal cylinders of compressed medical-grade air out of a storage closet. A cardiopulmonary manager, Mullin had devised a backup survival plan after the Humboldt Fire, when he'd been trapped in Paradise by the traffic. He didn't like the idea of burning to death in his Volkswagen on some winding mountain road.

Mullin, thirty-eight, had been one of the last people to leave the hospital that morning. After the final patients had departed, he and a few others climbed into a colleague's four-wheel-drive truck. They soon realized, however, that escape on Pentz Road wasn't an option. Neither was Conifer Street, the residential lane that looped through the Feather Canyon Retirement Community. (The center had been emptied except for eighty-eight-year-old Julian Binstock and his border collie mix, Jack. Binstock's bungalow had been overlooked in the evacuation.) Mullin and his colleagues tried to find a route to safety even as the concrete rippled with heat. But after encountering flames in every direction, they decided to return to the hospital. Mullin knew a safe place.

Followed by his colleagues, he managed to push open the

first set of sliding glass doors outside the emergency room. But the second set was locked and wouldn't budge, no matter how hard he kicked. He cursed. Then, after noticing a glowing keypad on an exterior door, he ducked outside and tried a combination: 2-4-6-8. It beeped. Access granted. He led the others down the steps to the basement tunnel, which ran beneath the parking lot. At the end of the dark and echoing corridor was the HVAC facility with the boiler and backup generators. Near the middle—where he was—was a housekeeping office and the storage closet with tanks of compressed oxygen and helium. The hallway was about seventy-five yards long and six feet high, meaning that Mullin, at five foot eight, didn't have to crouch. Bundled electrical wires and pipes ran along the concrete walls, ferrying steam, air, and pure oxygen to patients' rooms. Though an industrial-looking door opened from the corridor onto the parking lot, most people, aside from housekeeping, didn't even know the tunnel existed. Mullin was familiar with it only because, as a manager, he monitored gas tank levels for resupply.

There were seven others with Mullin, including the hospital's groundskeeper, an infection control nurse, the biomedical director, and the volunteer director, who had also brought her school-aged son. It was the groundskeeper's last day on the job after forty years; an iced sheet cake that had been ordered to celebrate his retirement was still sitting untouched in a break room refrigerator upstairs. He and the others had been desperate to find a refuge. Mullin had promised that the tunnel was that place. "We can't go out like this," he told them, trying not to think of his sons, five and seven, who were with his wife.

Mullin grabbed two filtration masks—standard hospital

equipment—from a cart, plopping one on the young boy's head. The mask, resembling a hockey helmet, could trap 99.7 percent of the smallest particulate matter in the smoke. The boy's head bobbed under its weight. Everyone else nabbed bottled water, fire extinguishers to douse spot fires, and electric camping lanterns by which to see. Rolling an office chair to the mouth of the tunnel, Mullin instructed someone to sit by the door and keep a lookout for fire engines. If firefighters came, they would presumably be able to evacuate everyone.

In moments of stress, Mullin's mind always went to a checklist: making a plan, then a backup plan. It was ingrained in him after fourteen years as a hospital staffer, many of them as a respiratory therapist on the night shift. Through the cracked door, Mullin saw that a green kitchen dumpster in the parking lot had caught fire. Then a transport van ignited, two cars down from his Volkswagen. A generator box sparked. Even the sign for Feather River hospital on Pentz Road was aflame. Mullin sent a colleague outdoors to douse some of the smaller fires.

Still no sign of help.

Smoke began to billow down the tunnel from the HVAC unit. Mullin walked back to the housekeeping office. He had already wheeled a few cylinders of medical oxygen into the room. Now he unscrewed the tanks. Fresh air filled the space, blowing out the stuffy smog that had begun to gather along the ceiling. The office, he offered, was now a "clean air respite." When the smoke in the tunnel became overpowering, he and his colleagues could gather here and breathe actual air. One task complete, Mullin began hashing out another backup plan. What would they do if the hospital burned and escape wasn't possible? Only his wife, Emily, and a few other colleagues' spouses knew they were hidden in the tunnel. They had called 911, but the dispatcher said resources weren't available to help.

"I would hate for us to burn up and be conscious of it," Mullin said, turning to the forty-six-year-old respiratory supervisor.

He pointed to the tanks of helium. If it came to it, they could build a tent with bedsheets and crack the canister of inert gas. Breathing pure helium would make them drowsy, Mullin explained, because it displaced the oxygen in the air. Within a minute, they would black out. A few minutes after that, their oxygen levels would plummet to fatal levels. Asphyxiation would be quick and painless. His colleague nodded slowly. They wouldn't share the plan unless they had to.

Meanwhile, twenty feet above their heads, unaware of their plight, firefighters were working on the hospital roof.

CHAPTER 7

A BLIZZARD OF EMBERS

Paradise Fire chief David Hawks rolled to a stop in Chloe Court. The cul-de-sac, lined with a crop of cookie-cutter homes, was located on the eastern edge of Paradise. The chief's cellphone pinged with text messages, rolling in all at once, all time-stamped 9:05 A.M. He unfolded his hands from the steering wheel. Finally, some cell service. He paused to reply to the slew of text messages and listen to Cal Fire's radio traffic, which was becoming increasingly frantic.

Over the radio, he heard that embers were landing, or "spotting," in the Butte Creek Canyon, threatening the historic Honey Run Covered Bridge. The Pratt-style truss bridge had been constructed in 1886, for the then hefty price tag of $4,300, as a way of connecting the Ridge to the rest of Butte County. In 1965, when the Butte County Board of Supervisors threatened to tear the deteriorating bridge down, Dr. Merritt Horning—a doctor and co-founder of Feather River hospital—and another local resident had donated money to preserve it, as well as to carve out an adjacent park. It was the only remaining bridge of its style in the United States. Listed on the National Register of Historic Places, it had been closed to traffic for decades, becoming instead a popular spot for rustic-themed weddings featuring twinkle lights and wildflowers in mason jars.

At 9:07 A.M., Hawks listened as a Cal Fire dispatcher told a female caller that no one could come to her aid. The woman was stuck on the basketball court of Ponderosa Elementary with five other drivers; they had driven onto the school grounds when flames sheeted across the only access road. Now they were trapped behind a chain-link fence. The dispatcher instructed her to drive through it if necessary. There were no firefighters available to help; they were across town trying to free other trapped residents. Hawks could hear their reports of using bolt cutters to snip fallen electrical lines. He knew that that was a huge risk: First responders were taught never to touch live wires, since they didn't have the training to know which ones might electrocute them. But representatives from PG&E had assured Cal Fire's Emergency Command Center that the electricity was officially offline. (Records would later show that power had in fact remained on in parts of the downtown grid through 12:30 P.M., when the backup generators rumbled to life.)

Hawks had several missed calls from Town Manager Gill. After speaking earlier that morning, when he was off duty and the wildfire was still threatening Pulga, they had struggled to connect. This time he got through. He repeated what little he knew from listening to the radio. The latest updates worried him. The last time he had seen a conflagration spread this rapidly was in 2017, when the Tubbs Fire had roared twelve miles across Sonoma County in about three hours. Before that, the Cedar Fire had outpaced firefighters in 2003, running thirteen miles in sixteen hours. But this Camp Fire defied the facts on the "rate of spread" card that Cal Fire laminated and handed out to its employees. It was the size of a business card, and Hawks kept his tucked in his wallet. The card defined a "critical" blaze as anything faster than 3 mph. According to Hawks's calculations, the Camp Fire was moving at twice that speed.

Hawks was a kind-faced man with wire-rimmed glasses and a thick shock of white hair. At fifty-two years old, he carried the experience of thirty-one fire seasons and the patience that came with parenting two teenage girls. In addition to serving as the fire chief in Paradise, where he had grown up, Hawks oversaw the northern half of Butte County as a Cal Fire division chief. Summer after summer, he had warned his town's people of the mounting fire risk. He had helped sponsor an annual forum called "Fire on the Ridge" at Paradise Alliance Church, one of the town's designated evacuation centers. In 2017, Cal Fire and the Butte County Fire Safe Council had mailed a postcard invitation to all 26,500 residents of Paradise. Only two hundred people showed up—including the presenters.

In August 2018, Hawks had hoped to shake this complacency with a carefully designed slideshow, which he presented at a Town Council meeting. That Tuesday evening, Hawks had opened by calling for a moment of silence for eight people who had recently died in the Carr Fire. They had perished in an unusual way, caught in a fire tornado that uprooted trees and crumpled electrical towers, flames soaring 400 feet in the air and stripping the earth bare. Fire scientists were still trying to make sense of the freakish meteorological event. After several minutes of silence, everyone's head bowed, Hawks had launched into his presentation with a sense of urgency.

"All right," he said. "I will tell you one thing that I see a lot around town, and it really concerns me. You can't drive a block and not see leaves and needles stacked up in gutters and roof valleys and along the edges of homes. Those are all ember capturers." He pantomimed a small fire exploding, steepling his fingers and ripping them apart. He emphasized the importance of defensible space, then gave an update on another wildfire that was burning to the southwest, in Mendocino County. It had cannibalized a smaller blaze and covered more than 459,000 acres—twice the land area of New York City. The Mendocino Complex had earned the title of largest wildfire in state history, toppling the record held for only eight months by the 281,893-acre Thomas Fire of December 2017.

"Anyone have any burning questions?" Mayor Jones had asked when he was done, commencing public comment with a bad pun. Two nameplates were propped in front of her computer monitor—the first with her name on it, the second with her title. It was another small way the town saved money, by swapping nameplates as people were promoted. Jones had been appointed the year before, having left the Sacramento area in 2014 and moved to Paradise for her husband, who ran a local Farmers Insurance agency. She had entered local politics that same year, winning a seat on the Town Council. Candidates usually ran on their love for the community, not necessarily on their qualifications, but Jones, who had previously served as a district director for the California Department of Transportation for thirteen years, had more experience than most: She had overseen the development of $1 billion worth of new road projects in eleven counties. Her duties as mayor were largely ceremonial: attending ribbon cuttings, giving speeches and interviews, and presiding over meetings like this one.

From the back of the chambers, a man in a gray collared shirt and black pants approached the microphone and introduced himself. "I was just wondering," he said. "In the event of a fire in Paradise, how long does it take to evacuate the town?"

It had been a simple question, but Hawks had not answered it directly, saying that he was "very confident" that Paradise was "very prepared." He had even suggested that the man's question was pointless, saying he couldn't ever imagine that the entire town would ever need to be evacuated at once. He would discourage people who hadn't received an alert from evacuating prematurely, he added.

Now, sitting in the cul-de-sac and listening to the Cal Fire chatter, Hawks thought back to his answer. Oh, how misguided he had been. The reality was that the town couldn't handle the crush of a mass exodus. The town's Public Works department had anticipated the pileup of cars as early as 2002, writing in a memo that only 3,700 vehicles an hour could evacuate on the four main routes. And that was under blue-sky conditions. "We can anticipate that during most fires, at least one of the main roadways will be closed (at least temporarily due to fire, smoke and fire suppression efforts)," wrote the Public Works director. Hawks shook his head. The real-time consequences of the town's poor urban planning were overwhelming.

He looked up from his cellphone to see two ambulances from Feather River hospital veer off Pentz Road. One rolled to a stop in the crisp grass and promptly burst into flames. The second pulled past a white mailbox reading 1830 CHLOE COURT and parked in the driveway. Nurses swung open the back doors, revealing the hospital's most vulnerable patients.

Tammy Ferguson leaped out of the second ambulance with them. Her long blond hair lashed around her head in the wind. She and her colleagues gathered outside the vehicle, lightly touching one another's arms and discussing what to do next. Tammy, forty-two, had bright blue eyes and was dressed in the fuchsia scrubs of a labor and delivery nurse. She emanated a warmth and steadiness cultivated during a long career of coaching terrified women through complicated deliveries. A mother of five herself, she thrived on the adrenaline and the connection with the new mothers and their new babies.

Her first checkup of the day—the beginning of her last twelve-hour shift of the week—had occurred minutes before the hospital ordered its staff to evacuate. Tammy had checked Rachelle's vital signs and made sure her baby was latching as he nursed. She had taken an immediate liking to Rachelle, who was direct and outspoken, unafraid to advocate for herself. Her son was premature, as delicate as a porcelain doll. Tammy thought he was unbearably cute. She didn't feel this way about all babies.

After the evacuation order, Tammy had chosen to stay and help her patients. Jumping into the ambulance next to the woman with the fresh C-section, Tammy had felt a kinship with the patients—all of them longtime residents of Paradise, just like her. When she was thirteen years old and living in Los Angeles, her parents had blindfolded Tammy and her two siblings, spread a map on the kitchen table, and told them to point to a spot. Their fingers found Paradise. It was that easy. Tammy started her freshman year at Paradise High and had lived in town until the fall of 2018, when the four-bedroom house that she rented on Pearl Drive for $1,500 a month had been put on the market. With five children to care for, Tammy couldn't afford to purchase the place and opted for a move to Chico.

Now Tammy watched in disbelief as a paramedic sprayed the

flaming hood of the first ambulance with a fire extinguisher. Bits of foam flecked the grass. A second paramedic assessed patients. Chardonnay Telly, an emergency room nurse in black scrubs, tried to call her father. He was seventy-four and lived in a log cabin in Concow; she hadn't heard from him since the Camp Fire ignited. The sky was a blizzard of embers. Millions of lit matches fluttered from the heavens, and about 90 percent were causing new fires when they struck the ground. The gutters of the house at 1830 Chloe Court smoked from the onslaught. The open concrete cul-de-sac gave the ambulance crews a fair chance of survival, because it wouldn't burn. But nothing was guaranteed. Tammy saw dread flush the face of her colleague Crissy Foster, who had forgotten to kiss her toddler goodbye that morning.

The wildfire moved across the landscape unevenly. Pushed by the wind, it funneled up dry creek beds and whipped down narrow ridges, sparing some homes on the western edge of town, including some in Town Manager Gill's neighborhood. Houses on Pentz Road weren't so lucky. The flaming front was more than five miles wide as it leaped out of the overgrown Feather River Canyon, which hadn't burned once in recorded history. In the cul-de-sac, the air was so hot that the nearby trees were torching. It was quickly becoming one of the town's most intense fire zones.

Ash flecked Tammy's blond hair like fat snowflakes in a winter storm. Each breath was arduous, drying her mouth and stinging her lungs. For a moment, she allowed terror to overtake logic. Should she hide in the surviving ambulance, where there was medical air to breathe? What if the tanks exploded? Should she hide in one of the houses? Or was hiding, in general, a bad idea? She could sprint across town, dodging the flames, until the fire overtook her, but she had never been much of a jogger.

She watched a female EMT crawl through the doggie door of the cream-colored house and click open the garage door. A male paramedic dragged patients inside on their gurneys. The woman with the brain bleed, slung over another paramedic's back, moaned in pain. A man suffering from dementia continued to smile, as if he was enjoying the sky's unusual colors. The first ambulance was now completely engulfed in flames.

The houses across Pentz Road crumpled, the electrical transformers popping. *How quickly the work of a lifetime can disappear,* Tammy thought. She clutched her cellphone, trying to decide what to do next. Like most medical professionals tasked with saving patients under harrowing circumstances, she had an essential pragmatism. She steeled herself against the encroaching fear—something she was well practiced at after years of tough decisions. At seventeen, when she had unexpectedly gotten pregnant, she had peed on a pregnancy test in the bathroom of the Mexican restaurant where she waitressed part time. She hadn't been able to bear the thought of going an hour longer without knowing—acknowledging the crisis brought relief.

She had always been a doer. Against all odds, she'd graduated from Paradise High with a "4.0 and a four-month-old," as she liked to say—and a year of college credits to boot. Most important, the birth of that daughter, Clarissa, had set Tammy on the path to becoming a labor and delivery nurse, just like the woman who coaxed her through labor without judgment or condemnation. She had put herself through nursing school while working full time. Tammy had never given up—and she wasn't going to now. But first she had some phone calls to make.

THE FIRE: LEAPING THE WEST BRANCH

Flames hadn't crossed the West Branch Feather River in more than a century. The canyon was so steep and twisted that it tended to trap firestorms before they reached Paradise. But the wind on this day was powerful. Embers soared atop 90 mph gales like kamikazes bound for a distant land.

Tongues of flame lapped the western slope of Sawmill Peak, zigzagging across the Feather River by dancing on the wooden flumes used long ago by gold miners and more recently by hikers, who trod on them like catwalks. Meanwhile, embers descended on the northeastern tip of Paradise, landing on the tapered roads and snug rows of houses. The structures were ideal kindling, offering landscaping materials, mulch and brittle grass, gutters clogged with debris, unscreened attic vents, open windows and broken windows, even cheap single-pane windows, and bald roofing with missing shingles. Millions of embers tested each point of entry. They needled through the attic vents that allowed a house to breathe and gathered in the rainwater drains where mosquito larvae hatched in the summertime.

This was how the fire spread so quickly: It wasn't a single unbroken front but a hail of embers. Alone, the sparks were too weak to do much damage. A single one could smolder

for hours, and they were miles ahead of the mother conflagration. But if two dozen embers accumulated in a crevice or
piled up on a clump of pine needles, the potential for fire
skyrocketed. A small pile of firebrands could generate forty
times more heat than a languid August afternoon of relentless sun. The superheated air started a chain reaction, broil-
ing so hot that sheer curtains behind a single-pane window
in a nearby home could catch fire and ignite the building
from within. Windowpanes shattered with the temperature
rise, the hot air rushing indoors to embrace the relatively
cool interior. Hundreds of flaming matchsticks swirled over
the furniture, fingering framed family photos like looters,
then incinerating the entire place within minutes.

The spot fires belched poisonous smoke as they reached
for the next home. There were rows upon rows of houses
to devour. Flames raced through the heart of town and
gnawed at structures, a firestorm similar to the conflagration
that annihilated Hamburg, Germany, after the famous Allied
aerial attack of 1943. The flames left the ponderosa pines
standing, too busy torching homes and businesses to scale
into the canopy. But they roasted their exteriors, eating at
their scabby bark. Millions of trees underwent a quiet death
that day too.

The smaller fires merged into a wall of flame and plowed
ahead to gobble untouched areas that other spot fires had
skipped over. More than twenty square miles had already
burned, and the line of flame in Paradise stretched the length
of Manhattan. The coals slammed into houses like red-hot
bricks thrown at 50 mph. Flames reared upward as high as
200 feet. Thick black smoke unfurled in complete darkness.
Fire engines that had just arrived raced along the town's
roads as if there was something they could do.

CHAPTER 8

SAVING TEZZRAH

Jamie had never imagined that flames would blitz into town from the east—maybe from the west, but not out of the impenetrable Feather River Canyon. Heritage Paradise was located across town from where the wildfire had trapped the ambulance crew off Pentz Road. The last Jamie had heard, from the news program playing in the care facility's lobby, the flames were still a comfortable distance away. He had also received a smattering of text messages from various family members warning that the wildfire was closer than expected. Jamie didn't consider himself an expert, but he figured that if the wildfire advanced to Heritage, which was on the Skyway, flames would have had to take out the heart of downtown—and that was unthinkable.

But then a stranger in a blue polo drove his Harley-Davidson into the parking lot and pounded on the care facility's front door, telling Jamie's boss that downtown was exactly where the wildfire was burning. He warned that the flames were heading their way and offered to help out. Jamie's boss instructed the man to help triage residents, organizing the wheelchair-bound patients under the rain awning, away from the falling embers. An evacuation from a place like Heritage would have been difficult under ordinary circumstances. Now, pressed for time, they needed every bit of assistance they could get.

Jamie stood outside the entrance of Heritage, trying to calm his boss as chaos unfolded around them. They were good friends and had a friendly rapport, jokingly calling each other their "work spouse." Her cellphone hadn't pinged with a CodeRed evacuation alert, but she was sure one would come soon. Heritage would be ahead of it. Another administrator had already ordered four medical transport vans to carry their elderly charges downhill.

In the parking lot, staff tried to keep all fifty-six residents calm as they waited for rescue. The patients were mostly oblivious to the closing distance between themselves and the wildfire. A change of clothing and jumble of medications dangled in plastic bags from the handles on each wheelchair and walker. One woman had begged a nurse to run back inside for her purse, where she had hidden a box cutter in case escape wasn't possible. As a child, she had suffered severe burns and knew she'd rather end her life than face flames again. The nurse thought the woman might have forgotten her prescription inhaler and obliged.

Now, as Jamie's boss talked with him, her phone jingled. One of the van drivers was calling to say that California Highway Patrol officers stationed downhill had turned the medical vehicles away. The Skyway was a snarl of traffic, each lane gridlocked in the same direction, and the officers didn't want to risk disaster by allowing cars uphill. This made sense: The road was built to handle sixteen hundred vehicles an hour—not nearly forty thousand—and they needed as many downhill lanes as they could get. Jamie's boss exhaled sharply, frustrated, then implored the man with the Harley-Davidson to weave around the lanes of stalled traffic to the bottom of the hill. Maybe he could persuade the officers to let the vans through. If they didn't arrive, the patients would be burned alive.

Fire was a particular threat to convalescent homes, whose residents were often too frail or medically vulnerable to evacuate quickly. Compared to the general population, elderly adults were twice as likely to die or be injured in a blaze. The number of senior citizens in Paradise made the threat exponentially greater. "Newlywed and nearly dead," residents often joked about the town's skewed demographics. Retirees had sought the sanctuary and affordability that Paradise offered, along with vital medical services and small-town safety. About 25 percent of residents were older than sixty-five, compared to 14 percent statewide, and the disability rate was nearly twice the state average. But few government programs had been created to help the elderly and infirm survive a natural disaster—even though wildfires were a known threat. Instead, it was up to ordinary civilians to step in and make sure their neighbors got out.

This meant that some folks got missed, as Julian Binstock and his dog had been near Feather River hospital. At Ridgewood Mobile Home Park, a retirement community off Pentz and Merrill roads, three women were left behind. The husband-and-wife property managers had knocked on every door, but some residents hadn't answered. Every second mattered, and the managers couldn't wait. Flames tore through the pastel-colored trailers, which were bordered by small gardens mulched with redwood chips. The trailers were decades old, manufactured long before the U.S. Department of Housing and Urban Development enacted building regulations for mobile homes in 1976. They fell like dominoes, their particleboard and thin aluminum incinerating within seconds. There was little space between one flaming trailer and another, and soon all were ash.

Butte County had devised two programs that might have helped the elderly population. The first was the In Home Support Services program, or IHSS, which helped older and disabled residents live independently by sending social workers to check on them in their homes. The Ridge had an enrollment of 960—a sliver of the thousands of residents in Paradise and Magalia who had reported to the U.S. Census that they had a disability of some kind. Moreover, state and federal laws protecting private health-related information prevented the county's emergency services officer from accessing the list of IHSS clients until a natural disaster warranted it. By then it was often too late to help them. As the Camp Fire galloped from Pulga, the program activated a phone tree. Staff individually dialed each of the Ridge's 960 IHSS clients to ask if they had a relative who could assist them or if they needed law enforcement to pick them up. They got through to only 215 clients. They left 247 voice-mails and were unable to contact 498 clients whose numbers were unlisted or disconnected.

The county's second program was the Special Needs Awareness Program, or SNAP, which had been founded in 2008 and had an enrollment of thirty-seven hundred people. Registrants voluntarily submitted their contact information to let the county know they would need assistance during a disaster. The Sheriff's Office had a map of their addresses so that officers would know who might need extra help, if they had time to locate their homes. But SNAP was more of a voluntary educational program that emphasized personal preparedness than a tool widely used by law enforcement. As part of the program, Butte County posted a sample emergency preparedness plan on its website. The twenty-page document, drafted by San Diego County in the wake of the 2003 Cedar Fire and 2007 Witch Fire, included tips on what to pack in a go-bag and how to establish a support team. It didn't include any phone numbers to call for help. SNAP also issued reflective window plac-ards to residents, so that first responders could spot their homes during a disaster. But at the time of the Camp Fire, it was not clear how many people—if any—had actually placed the placards in their windows. Many first responders didn't remember seeing them.

At Heritage, the residents were lucky. They had Jamie and his colleagues. The facility's recent renovation had given staff members new energy and inspired focus on patients' quality of life. The owner had hired a professional chef and encouraged staff to spend more time with residents. He wanted to make sure they felt seen and valued in a way that—because of their age or health conditions—they often weren't by society. He scheduled bingo tournaments with prizes like beaded necklaces and scented lotions. In good times, these thoughtful attentions made residents happy. In bad times, they made them stick together.

Jamie's boss thought out loud, talking with Jamie as she deliber-ated how she might save their elderly charges if the vans didn't ar-rive. If it came down to it, she said, staff could shelter residents in the tiled showers, or perhaps the walk-in freezer. Another option could be to walk patients off the mountain, navigating their wheelchairs and walkers down the Skyway, until someone offered up a spare seat. Jamie told his boss that he wouldn't leave her side until they knew everyone was safe.

While they entertained her hopeless plans, the motorcyclist was attempting to persuade the patrolmen at the blockade to release the vans uphill—in what was looking like a losing argument. Staff began lining up their own cars and loading residents in twos and threes, just as nurses had done at Feather River hospital, in a makeshift evacuation. Jamie offered to lift the heavier patients. As a maintenance man, he wasn't trained in moving the medically frail, but he tried to mimic the nurses, gently supporting the elderly under their arms. His wife would have known what to do. Erin's arms were so toned from moving patients that Jamie had nicknamed her the Pull-up Queen.

A man driving a tall Dodge truck pulled in to the parking lot, offering two extra seats. The wildfire was closing in, he said. Staff members slid two of the immobile patients onto the truck's bench seat in the spacious back, stacked some wheelchairs in the truck bed, and asked the driver to drop the patients off at a Chico facility. Slowly, over the course of the next half hour, all of the residents were packed up and sent off, except for the seven least mobile. Several of them were obese and couldn't move because of their size. Some were recovering from hip and back surgeries and couldn't bear extra weight on their limbs; the staff depended on mechanical lifts to transport them. It would have been difficult and dangerous to shove them into tiny vehicles.

On the roof, a few members of the maintenance crew were hosing down the shingles in hopes of saving the building. "Where are we in this?" Jamie's boss called up to them.

"I think we have defensible space," the lead maintenance man shouted back. "We can wait here."

"I don't think we have a choice," she said, throwing up her hands. She hoped the vans would arrive soon. In the distance, the fire was a mushroom-shaped monster, its dark smoke puffing from the base and mottling the sky. The rupturing of propane tanks ricocheted through the air.

AT 9:45 A.M., Jamie's boss was about to usher residents back indoors to shelter in place in the walk-in freezer or tiled showers when she saw a white van pull in to the parking lot. A second van followed, then a third and a fourth. They were, she thought, like Moses parting the Red Sea, if the Red Sea had been an ocean of parked cars. Somehow the motorcyclist had gotten the patrolmen to let them through. She heaved a sigh of relief, the knot in her throat releasing. It felt like a miracle—the residents would be okay. With the help of the van drivers, staff finished loading the last wheelchairs and walkers.

His job done, Jamie headed into the living room adjoining the facility's lobby, where his daughter was waiting. He had picked up Tezzrah from Achieve Charter School just before 8 A.M., managing to drive from Heritage to her school and back before the streets filled with vehicles. He had considered leaving town with the seven-year-old but brushed the thought away. Heritage needed him.

Tezzrah had known there was a wildfire nearby. She was sharp, peppering her father with questions about the blaze's proximity. "It looks big," she had commented. Jamie tried to stay quiet. It was impossible to ignore the claustrophobic blanket of smoke, but he didn't have any answers.

Now, about to face his daughter again, he realized he still didn't. For the past hour and a half, Tezzrah had waited in the living room with an administrator's young daughters, twelve months and five years old. A nurse had brought her dog to the facility, and the girls petted him as they watched cartoons and ate orange Tic Tacs. Through the closed doors, the children could hear some of the residents with dementia shrieking in their confusion. Their wailing startled even the nurses. "I'm not leaving without my recliner!" one woman had shouted. "I'm a little concerned we aren't going to make it out, so we aren't going to look into the recliner right now," an administrator had answered, nimbly buckling the patient into the medical van. As they retrieved supplies from indoors, staff members tried to hide their fear so they wouldn't further terrify the patients—or the little girls.

When Jamie finally reappeared, Tezzrah burst into tears. She could imagine the flames engulfing her hair and her clothes. She felt things deeply and had always had a dramatic flair. Though Mariah and Arrianah sometimes irritated her by stealing her favorite art supplies and patterned washi tape, she missed her sisters now, as well as her mom. Erin had always been the parent to pick her up after school—not Jamie, who did drop-offs—and the break in routine had to mean that something was very wrong.

On Facebook, Jamie had seen photos of the approaching wildfire posted by Ridgeview High, a continuation school for at-risk youth. In the pictures, snapped from an old lookout platform in Magalia, an orange tsunami scaled the Feather River Canyon. Jamie wouldn't have believed the images if someone had described them to him. He called Erin as he finished packing up the last patients, instructing his wife to evacuate with Mariah and Arrianah north through the mountains—not through Paradise. Known as the "Upper Skyway," the highway had been paved in 2013 at a cost of $21 million on a recommendation from the Butte County Civil Grand Jury. The former gravel road was the only evacuation route that avoided Paradise. After following its length, Erin could hook south to Chico. It was an hourslong drive, and she was apprehensive, never having driven the route before. In the background, Jamie could hear his younger daughters crying. "I guarantee you that all of Magalia will be going that way," he said. "Just follow the line of cars.

"As soon as you can, let me know where you're at and if you're okay," Jamie continued. "Finish packing up. Paradise is on fire right now, and so are parts of Old Magalia. There is no way for you to come down here. I will do my best to keep Tezzrah safe. I love you."

It had been their last conversation before Jamie's cell service cut off. Around 10 A.M., as the final medical vans departed, he walked

Tezzrah out of Heritage and settled her in the back of his Subaru Outback. One of his colleagues was going to hitch a ride with them. She waited with Tezzrah in the car while Jamie ducked back inside the facility to grab snacks and supplies, the weight of his daughter's eyes on his back. He jogged through the empty hallways. Emergency lights blinked on and off and radio stations blared news of the destruction outside. Was it even real? Jamie had witnessed scenes like this only in movies. His stomach roiled, as it often did when he was anxious. He felt a wave of nausea rise in him; he had a weak stomach at the best of times.

He headed to the basement, where he found bottled water and snacks in a maintenance locker. Grabbing an armful, he thought of his seven-year-old. She would get hungry or thirsty—those, at least, were problems Jamie could solve. Tezzrah was his first child, the one who had taught him the meaning of promise and parenthood. From the second Jamie had witnessed her dark eyes open for the first time, he had vowed to protect her. He remembered how small Tezzrah had been, how she fit perfectly in the crook of his arm: as compact as a football, her thicket of black hair soft against his skin. Tezzrah believed the world was a safe place because Jamie always made it safe for her. How much longer could he continue to protect her?

Jamie hurried. He knew Tezzrah would panic if he was gone for too long. "This is really happening," he said to himself as he cut back through the dimly lit building, skirting the abandoned medical carts that scattered the hallway.

OBSERVATION: PARADISE IRRIGATION DISTRICT

High above Paradise, on a narrow plateau in Magalia, the town's water treatment plant was nestled in a cove of ponderosa pine. The Paradise Irrigation District had constructed the plant in the neighboring community because of its proximity to Magalia Reservoir and Paradise Lake. These pristine pools collected the rainwater that flowed downhill. The irrigation district was consistently water-rich, receiving more rainfall per year than Seattle, sometimes up to ten inches a day in the winter. The rugged topography wrung storm clouds out like rags. Other water districts in Butte County had to neutralize agricultural runoff and industrial waste; by contrast, Paradise's gravity-fed system was blessedly simple. Workers filtered out dirt and debris, added a bleach sterilizer, tested the water, and then sent it downhill to Paradise. In a state dominated by battles over water—who owned it, how to transport it, whether to divert it—the Ridge had gotten lucky. The water in Paradise even tasted different. Everyone said so. It was crisp and clear, as if it had been bottled at a mineral spring.

Water district workers Ken Capra, sixty-one, and Jeremy Gentry, thirty-nine, had spent years laboring at the plant. They worked in twenty-four-hour blocks: one day on, two days off. At night, they slept on a pull-down Murphy bed at the plant, receiving partial pay while they slumbered. It didn't make either man a fortune, but it did provide abundant free time to do the kinds of things—bass fishing, woodworking—they couldn't have done with a traditional day job. For Gentry, it meant extra time with his sixteen-month-old daughter.

On Thursday morning, he had just finished a twenty-four-hour shift and was preparing for another—a rare doubleheader because staffing was short. Below the plant in Paradise, Gentry could see a heaving column of smoke floating west. Something was clearly wrong. But when he turned to his computer to check the pressure and water levels for Paradise, he discovered that the plant's connection to the five water tanks downhill had been severed.

By the time Capra showed up for his shift, Gentry had tripled the flow in case firefighters needed extra water to battle the flames, as he had practiced the week before in a tabletop wildfire drill. About 12,000 gallons of water rushed down the slope every minute—more than 17 million gallons in all. Normally, they'd run 6 million gallons in a day. But in town, the 10,500 plastic service lines that connected homes to the thick underground main had already been damaged by the wildfire. The pipes leaked. It was like dumping water into a bucket with no bottom. Paradise's water distribution system was separated into six zones, and all were on a slope except for one, which was fed by a 3-million-gallon tank near Clark Road and the Skyway. This tank was the first to fail. Water gushed out of the plastic distribution pipes. As the water pressure dropped, ash and toxic sludge from burnt-out homes were sucked back into the main lines, contaminating them.

Alarms flashed on Gentry's computer screen, signaling a loss of contact with the rest of the system. He and Capra had no landline, no television, no Internet, no cell service. On the emergency frequency of their handheld radio, they could hear trapped firefighters screaming for their colleagues to rescue them. Gentry and Capra could do nothing but listen to the panic of grown men who believed that they were going to die. "We have live power lines coming down on some people here on Wagstaff," a division chief shouted. "We need to open this intersection. I got people trapped all through here!" Then came the sound of a fire engine, its driver reporting over the radio that flames had overtaken the cardiology wing. *The hospital,* thought Gentry.

Out the window, he and Capra saw half a dozen Cal Fire contract pilots dip Bambi Buckets into Magalia Reservoir, the huge helicopter pails dangling at the end of long cables. The aircraft droned laboriously as they departed, carrying the water downhill. Capra and Gentry's bosses had told the men to leave if they ever felt unsafe, but the treatment plant seemed the best place to be. The metal roof certainly wasn't going to burn. Besides, they had plenty of water.

CHAPTER 9

THE LOST BUS

The doors of Bus 963 closed. Kevin wrenched the steering wheel, turning onto Pentz Road and peering through the dark smoke. Cinders tumbled from the clouds, igniting thousands of small fires along the roadside. Kevin planned to cut across town to reach Clark Road—the second-largest thoroughfare in Paradise, capable of accommodating nine hundred cars per hour—then head to Oroville. Traffic was piling up on Pentz Road, and he didn't want to get stuck.

Behind him, the twenty-two schoolchildren in the bus were silent, an eerie contrast to the regular din of his route. These were Paradise kids, not the hardened Magalia students he usually drove. When the wildfire had been reported, he was the bus driver closest to Ponderosa Elementary and had offered to help out. The children, too small to see over the tops of the seats, were nearly invisible in his rearview mirror. Kevin spotted a golden yellow beanie and a blue tie-dyed baseball cap. He didn't know their names; they didn't know his.

"Who are you?" Mary Ludwig, fifty-one, a second grade teacher, had asked when Kevin had first pulled up at Ponderosa Elementary.

She had never seen him before, which she found odd. Mary had taught in the district since 1994 and thought she knew every bus driver. She was friendly with a lot of people in Paradise; she and her nine siblings had grown up there. She was the second-youngest, and the first to be born in Paradise. Her father had been a teacher at Biggs Elementary in the valley. Everyone knew his raucous brood. It was hard for the Ludwig kids to go anywhere—the Fosters Freeze, Stratton's Market—without someone striking up a conversation.

Mary had shiny chestnut hair and a warm personality. She was a small-town girl who had married a small-town boy, moved to the big city of Chico for five years, then returned to escape the traffic and noise. She didn't hesitate to pull complete strangers into tight hugs or share a funny story from her classroom. She liked crafting creative lesson plans and had recently read *James and the Giant Peach* to her students in an English accent, which she admitted wasn't very good. Their latest unit had centered on understanding the momentum of a peach. The children had tossed balls down a knoll behind the classroom to study how slope affected speed.

Kevin looked at Mary. He was new, he said, but had grown up on the Ridge too. Mary's eyes watered; she explained that her eyes felt too scratchy for contacts and she had been forced to wear her prescription sunglasses in the dim gymnasium. The children had been kept indoors, eating applesauce cups and watching an episode of *The Magic School Bus* until Kevin arrived. An aide had handed Mary a stack of emergency paperwork as she walked outside, guiding the remaining two students from her classroom toward the bus. Chunks of burnt bark rained down on the playground. Firebrands landed in her hair, singeing it. She tried to leave the students with Kevin, as the other teachers had. But Kevin shook his head. "I need you to come with me," he implored. Someone needed to look after the children as he drove.

Mary hesitated. She preferred the comfort and protection of her silver Toyota Highlander, parked by the music room. She wanted to drive home and check on her teenage son, then—if she had time— pick up her daughter from work at Kmart. She had done her duty as a teacher, and there was no need to get on Bus 963. But Mary also knew that if her own kid were boarding a bus with a new driver during a natural disaster, she would want their teacher to be with them. Not wanting to be on the bus alone, she went to wrangle Abbie Davis, a first-year kindergarten teacher, into joining her. They had talked only a few times, connecting over a shared love of Dansko clogs. Abbie had a cherry-red pair; Mary wore canary yellow.

Abbie, twenty-nine, was petite and dark-haired, with thick, expressive eyebrows that betrayed every emotion. She and her twin brother had been raised by a single mother, and she intimately understood the poverty and hardship that many of her own students faced. She took pride in her classroom, spending too much of her own money to make it feel like a second home for the children. She had dedicated most of the summer to decorating, adding fake grass to a reading nook to make it feel like the outdoors and making sure each student had a wall hook for their backpack. Recently, she and her boyfriend, Matt Gerspacher, twenty-nine, had decided to get married. They had grown up together. On the first day of school, he had sent her a bouquet of flowers. Abbie knew she had gotten lucky. Now, as she boarded the bus, she hoped desperately that they would both survive to see their wedding day. Mary clambered up behind her. "You'd better be a good driver," she told Kevin.

In his rearview mirror, Kevin watched the playground at Ponderosa Elementary disappear in the distance. He turned onto Wagstaff Road, where flames were roaring along the edges. Stumbling upon the blaze shocked Mary. The air was stifling, greased with carcinogens from burning household products. Embers lunged sideways on the downdraft. Kevin called her and Abbie to the front, pointing out the fire extinguisher and first aid kit. He gestured to the two emergency exits and emphasized that they were not going to leave the bus unless they absolutely had to. It was the safest place to be. Mary squeezed his arm, thanking him. He told the teachers to take attendance and pair older children with younger ones. "And handwrite three copies as you take roll, so each one of us has a manifest of the kids in our care."

"Why?" Mary asked, bewildered.

"If something happens, authorities need to know who was on this bus." Kevin fixed his eyes out the windshield, focused on the blinking brake lights up ahead.

Mary and Abbie nodded, then moved toward the back of the bus, following orders. Rowan Stovall, who had just turned ten, comforted her seatmate. "You'll see your mom and dad again," she told the kindergartner. "The bus isn't going to catch on fire. We are going to be okay, I promise." Her mom, who affectionately called the fourth grader Rowboat, had gotten stuck in Concow with her boyfriend. She hadn't been able to pick her daughter up from Ponderosa Elementary, but the girl felt safe with Mary, her beloved former teacher. The two had developed a special relationship after Rowan was held back a year in Mary's class. Rowan clutched her seatmate's hand.

"Is it ten P.M.?" a boy in a flannel shirt asked, tugging on Mary's shirtsleeve as she passed. He was confused; it was so dark outside. Another boy was in a panic, ripping at his hair as he babbled about how his "ninety-four-year-old" cat was going to burn up. Even more worrisome were the ones who didn't speak at all. *How do I distract the children and comfort them at the same time?* Mary thought. She knelt beside a tiny girl in a zipped fleece jacket, asking her name for the manifest. The girl was so terrified that she couldn't remember her last name. Mary rubbed her back. Across the row, she saw a backpack resting on an empty seat. A kindergartner had curled up beneath the bench, cocooning herself from the unfolding nightmare.

KEVIN WENT OVER different scenarios in his head, trying to figure out the best way to steer forward down Clark Road. An RV rammed in front of him, cutting the bus off. *How dare you,* Kevin thought, seething. *Can't you see there are children on board?* He was not going to panic. He knew children were sensitive to the energy of those around them. He could see the kids' hysteria escalate whenever Mary or Abbie took a break to stare out the windows, or record videos on their cellphones, or call their loved ones to say goodbye. The women's voices warbled with fear. Mary's son hadn't evacuated soon enough and was now trapped on deadly Pearson Road, which dropped into a gully known as Dead Man's Hole for its lack of cell service. Children, more innocently, called it the Pearson Dip. Abbie worried that her fiancé, Matt, who was refusing to leave their house on Filbert Street until he saw the bus pass by, might die because of his stubbornness.

Kevin flicked on the ceiling light so other drivers could see the children in the back of the bus. He summoned Abbie and told her that she would be his scout, pacing down the bus aisle and calling out new spot fires along Clark Road so he would know when to change lanes and keep some distance from the wildfire. He was a quick study and learned to read the arc of Abbie's eyebrow and the tilt of her head, the subtle ways she signaled the presence of flames, not wanting to speak aloud and scare the children. Meanwhile, Mary continued scribbling down their names.

A countywide plan passed in 2010 had included a scheme for reducing fire hazards on evacuation routes like Clark Road. The Butte County General Plan for 2030, approved by the county's Board of Supervisors, called for clearing vegetation from roads and creating alternative evacuation routes. An earlier study had recommended widening the northern portion of the Skyway—where Jamie's wife, Erin, was trying to leave Paradise—to four lanes. But funding for these projects wasn't allocated; that part of the Skyway was never widened.

In the late nineties, more than a hundred "fire safe councils" had popped up around California to supplement the limited efforts of local governments in places like Butte County. Though some regions, such as fire-scarred San Diego County, benefited from the help of dozens of councils—which cleared brush for disabled residents, ran woodchipper programs to pulverize dead trees, and plowed fuel breaks to halt a wildfire's assault—others, like Butte County, did not receive nearly the same attention. The Butte County Fire Safe Council, founded in 1998, comprised one full-time employee and three part-time employees. Its annual budget was less than $500,000, and it relied mostly on state grants. The council received no dedicated funding, though the Board of Supervisors did contribute $40,000 in 2013 to prevent it from going under. "We're filling the gap from local government on people's preparedness, and they do little to help us," the council director, Calli-Jane Deanda, would later explain. In August 2018, the nonprofit had received half a dozen grants totaling more than $3 million.

This funding had prevented layoffs, but there had not been enough time to enact many other improvements around the county itself. The Town of Paradise had its own council, which ran a program to educate schoolchildren about fire risks. Residents were urged to keep cloth go-bags with important documents and keepsakes by their front doors. The tactic was proposed so often that one of the council's fifteen volunteers had been nicknamed the Bag Lady. Led by chairman Phil John and fire chief David Hawks, Paradise's council met at Atria Senior Living on the second Wednesday of each month, often presenting to a half-empty room. John operated a $4,000 budget cobbled together from grants; Town Hall contributed no money. His was a lifetime appointment, because no one else wanted the role. He enjoyed dressing up as the council's mascot, Wildfire Ready Raccoon, who shared John's birthday: April 26. The beloved critter had his own trading cards and picture book, and he was present at every Gold Nugget and Johnny Appleseed Day, dancing in the parade and giving out hugs. John had also recorded an educational parody rap called "Wildfire Ready in Paradise," set to the tune of "Gangsta's Paradise."

"The name's Wildfire Ready, I don't live in the city," the song went. "I live up in the mountains where the trees are pretty. I got friends that are people, I got friends that are dogs. I got friends that live in houses, I got friends in logs. If you're livin' near the forest, there's some things you got to know, how to keep yourself ready, in case a fire starts to grow. I'm ready. Are you ready?"

They had not been ready.

. . .

TWO OF THE school district's assistant superintendents materialized out of the smoke and knocked on the glass door. Kevin startled, then opened it for them. Their truck had caught fire in the
parking lot of Ponderosa Elementary and they had decided to proceed on foot. It was faster than driving anyway. Boarding the bus
for a few minutes, they warned Kevin to avoid Paradise Elementary—
an evacuation center and for years the town's only elementary
school—because it was already on fire. Then the two got off to
continue their walk. They planned to help direct traffic.

Mary had attended high school with one of the administrators.
She was comforted by the coincidence of running into him. But
the blaze was everywhere, scorching the mountains and hillsides
with an unprecedented fury. The red and blue spin of police lights
ricocheted past as officers drove into ditches and around fallen trees,
rushing in response to reports about a cluster of people, including
Ben Mullin, the cardiopulmonary supervisor, trapped in the basement of Feather River hospital. They were also trying to track
down a woman who had gone into labor in the Fastrip gas station
parking lot. Mary pointed out the first responders to the children
in a bid to distract them. "Look at those brave men coming to help
us!" she said. She recognized the face of a sheriff's deputy whose
son had been in her class the previous year. He couldn't be harmed,
she thought, because his wife and two children needed him. It was
the kind of illogic that made sense in moments like this—people
with families couldn't die. The children screamed through the
locked windows: "Thank you, Officer Brody! We love you!" Their
noses left smudges on the glass.

Beloved landmarks passed by: Paradise Alliance Church, Mountain Mike's Pizza, McDonald's, Dollar General. The Black Bear
Diner, with its carved wooden bear propped out front, holding a
sign reading WELCOME TO BEARADISE. The familiar sights brought
with them a sense of hope. "Who likes pancakes?" Mary yelled,
smiling broadly and raising her hand. A smattering of small palms
followed. Kevin commented that he also kept a wooden bear statue

in front of his house. The children laughed gleefully, because they knew that couldn't be true—there was only one "Welcome to Bearadise" sign! The sky broke, the velvet black fading to light gray. Then the darkness closed in again. The kids watched as flames catapulted onto the roofs of the Black Bear Diner and the McDonald's, then spread to the KFC restaurant next door. Mary fell silent. So did the children.

As they turned onto Pearson Road and passed the intersection of Black Olive Drive, an officer directed the bus south, away from the Skyway. They had been one block away. For a short distance, they moved easily, without stopping. Looking out the window, Mary recognized the teacher's aide from Ponderosa Elementary running along the roadside, heading toward Kmart. Mary had felt like the bus was flying, but the aide was outpacing them. Paradise's isolated lanes and loops had become choked with cars. To the north, Rachelle and Chris were trapped on Bille Road, while Jamie was trying to weave through traffic on the Skyway. Kevin had taken a different tack, trying to escape Paradise along routes that only a native would know—but he was turned away repeatedly by law enforcement officers with out-of-town uniforms who claimed to know better. Kevin gritted his teeth and thanked them, only to curse under his breath in the next traffic snarl. The bus was pushed off Pearson Road to smaller streets: south on Foster, east on Buschmann, south on Scottwood. Luckily, a text had made it through from Kevin's girlfriend, Melanie, letting him know that his family was safe in Chico. She had gotten a hotel room for his son and mother. A small mercy.

Roe Road appeared before them. It was dangerously narrow, its sides flanked by dead brush and ponderosa pines. Efforts to regulate vegetation here had been met with opposition; residents had feared this would lead to the logging of healthy trees. The pines had become so prized that they couldn't be chopped down without a special permit—which cost nearly $60, with an extra $23 per tree—and even then the offending landowner had to agree to plant a new sapling, at least 15 gallons in size, within the year. Cutting down more than five large trees required an appearance before the Town Council.

All morning, Kevin had referred to the timber in Paradise as fuel—a phrase that Mary and Abbie had never heard. To them, trees were a source of beauty. Mary described the town's ponderosa pine groves as the "rainforest" of Paradise. It's what made the town feel like home. But looking ahead, Kevin's word choice made sense. Roe Road was claustrophobic. It was harrowing on an ordinary day because the line of sight was so limited. Now it looked as if the brush could ignite at any moment.

A Paradise Police officer flagged Kevin down. "Do you have kids on this bus?" the officer said, peering up at Kevin as he cranked open the driver's window. "I'm about to shut this road down, but you go first. Get out of here."

"Hey, man, Roe Road is highly overgrown," Kevin argued. "I'm worried about getting through there."

"Just go," the officer replied. "There's no other way out."

Kevin halted in the middle of the intersection of Scottwood and Roe, trying to leave a football field's length between the bus and the car ahead. The pause also gave him the opportunity to attempt a getaway: He tried, very slowly, to pivot away from Roe Road and take a different route, against the officer's recommendation. The drivers behind him laid on their horns, livid that he wasn't moving forward. They wedged their vehicles into the clearance, and the patrolman directed a few more cars forward into the intersection, trapping Kevin in place. In the confusion, an elderly driver scraped the back of the bus, jostling the children from their seats. Kevin was stuck. The decision had been made for him: The only way out was forward.

Mary, who had been helping calm a student, recognized the road. She walked to the front of the bus. "What the heck, Kevin!" she said, her voice cracking. "Why are you taking us down Roe Road?" She begged him to go a different way. "You know it's a death trap," Mary said. "Please do not take us down this road." Kevin gripped the steering wheel. They didn't have a choice, he said.

Abbie interrupted, saying she thought some of the children were in shock. She didn't know what to do. Mary switched places with her, sitting with a young girl who usually had a lively personality but had turned strangely morose. Kevin continued to stall in the intersection, Roe Road before him. The children grew drowsy, some on the verge of passing out, nauseated by the carbon monoxide and exhaust fumes. Hours had passed since they'd last had food or water. The bus was unbearably warm. Kevin kept his eyes locked on the tunnel ahead. The canopy ruffled, ready to catch flame. Mary squeezed the girl's hand once more, then walked back to the front of the bus, sliding into a seat with Abbie. She was weary and depleted.

For a moment, the two women found solace in each other. "Look out the window, Mary," Abbie whispered. "I don't think we're going to make it." They clutched hands, imprinting tiny half moons on each other's skin with their nails, Abbie's fist as small as a songbird. She revealed that she had already lost one fiancé in a riverboating accident—and now the man who had offered her a second chance at love was refusing to leave town for her sake. What if he died while waiting for her? Together, she and Mary prayed aloud. Mary wondered whether the school district might later fire her for this public show of faith. Perhaps, she thought, they would understand that this was a special circumstance.

"Please," they pleaded, "let the smoke kill us first."

CHAPTER 10

THE BEST SPOT TO DIE

A fierce gust of wind shook Sean Norman's SUV and sent waves of fire sheeting across Wagstaff Road. The Cal Fire captain peered through his windshield. Just ahead, he made out the silhouette of a Chico Fire Department engine. The crew was spraying water on a cluster of flaming homes. Norman crept forward a few more feet, then rolled down his window. He knew how hard it was to go against the instinct to save burning buildings— but it was no longer their priority. "Hey!" he shouted, until he caught the firefighters' attention. "Don't try to put that house out. See if people are in there and get them out." They nodded, leaning over the hose to roll it up. Norman closed his window, muting the howl of the wind. The Cal Fire seal plastered on his door, which normally made traffic a nonissue, now brought him little advantage. No one was moving, period.

"I'm cut off by fire on Wagstaff and the traffic is completely gridlocked coming down south on the Skyway," Norman said over the radio to his supervisor. The supervisor called back, asking him to cover the northern half of Paradise and the outskirts of Magalia— an area that incident commander John Messina was calling Division Hotel. The made-up designation helped everyone understand which area of the foothills firefighters were reporting from on the radio. It was a tough assignment, because in the midst of the unfolding hell, some residents still hadn't received an emergency alert or accepted that they needed to evacuate. Norman was one of thousands of firefighting personnel that had responded through the state's mutual aid system and were being sent into the burn zone to persuade people to leave their homes behind—without losing their own lives in the process. The calculus of such a task sometimes felt overwhelming.

"Don't die out there," Messina had warned him.

The gravity of Norman's job had hit him hard earlier that summer. A thirty-seven-year-old city fire prevention inspector had died while evacuating residents from the Carr Fire in Redding, leaving behind a wife and two small children. The fire tornado had repeatedly flipped his 5,000-pound Ford F-150 truck down the road, then dumped it into the woods. For months his colleagues had searched for his lost helmet, which had blown away, to give to his family. They never found it.

Death was a risk that came with the job—but it was a job that Norman, forty-seven, had wanted since he was five years old. His grandmother had worked as a secretary for the San Francisco Fire Department's arson division, and as a child he had loved visiting her at the station. He had studied the firefighters as they scrubbed the engines to perfection, the red doors gleaming in the sun. After his family moved to Sonoma County, Norman had taken to biking to local fires whenever he heard alarms sound at the volunteer station downtown. He would stay and watch the firefighters work until the flames dwindled, then wave at them as they departed in their engine. The idea that he could someday work among them was thrilling. His childhood fascination had never faded—terrifying his father, a psychologist, and mother, a preschool teacher. Once, after training as a volunteer firefighter in high school, he had rushed into a burning house near his mother's preschool. She had been outside with her students and witnessed the act. "How do you just *do* that?" she asked him later that evening. "That's what we do," he answered earnestly.

His job had gotten more difficult in recent years. Norman couldn't imagine anything worse than what he had seen a year ago. The colossal Thomas Fire, which had ripped through the chaparral hillsides north of Los Angeles in 2017, had burned well past Christmas and brought firefighters to the brink of insanity; the mud flows that followed buried a fire-scarred community on the coast and killed twenty-three people. Some of the bodies were submerged so deep beneath the muck that they were never recovered. It had made Norman worry about the fate of his own town of Grass Valley, in the foothills southeast of Paradise. He and his wife had moved from the Bay Area to the hamlet of thirteen thousand in 1997. Having grown up on two and a half acres, Norman couldn't imagine ever living in a suburban neighborhood of tract homes. Priced out of Sonoma County, like so many other families, they had looked to the wildland-urban interface for a piece of land. It was only after a trip through Grass Valley on a work assignment that Norman decided that this town was the place. Several miles away from their new home was Rough and Ready, where the PG&E-caused Trauner Fire had burned down a historic schoolhouse in 1994.

Grass Valley was sixty miles northeast of Sacramento, neighboring the vast wilderness of the Tahoe National Forest. Property was cheap and plentiful. The town was similar to Paradise in topography and forestry, nuzzled against the overgrown Sierra Nevada and often buffeted by menacing seasonal winds. Fire officials cautioned that Grass Valley met the conditions for a historic burn. The movie theater in the small downtown even played fire preparedness trailers before the main film screening. Still, Norman savored the isolation. His five-acre property on Oak Drive was secluded and shaded by pine; it meant his son and daughter could have the kind of childhood he and his wife had imagined for them. His six-year-old, Patrick, was obsessed with toy fire engines and the outdoors, while his four-year-old, Katie, was fixated on horses. She wanted to own as many as their barn could hold. "What job are you going to do to pay for all of these horses?" he would ask her. "Mom will pay for me!" she would taunt, knowing her mother was more likely to cave to her demands.

Now, as he drove farther into Paradise, the wind pounded and pulled at Norman's SUV as if trying to lure it off the pavement. He forced his way toward Bille Road, then uphill toward Magalia. A few houses had caught fire, distracting him. People were pouring out of them. They ran, leaving their cars behind; without electricity, they couldn't remember how to get the garage doors to open manually.

In the doorway of one home, Norman saw an older woman silhouetted against a patch of light, crawling on her hands and knees. He swerved off Wagstaff Road, parked in her yard, and ran to her, pulling the woman to her feet to help her into the back of his SUV. She was light in his arms, her face wizened, her wrists notched and delicate. She didn't say a word. Norman settled her inside and closed the door, then tried to get his bearings in the choking smoke. Anytime the smog cleared, giving him a chance to see, flames would appear, the neighborhood around him warping like a demented funhouse mirror. He pictured his children, happy at home, clinging to this thought like a talisman. They needed a father who came home at the end of the day.

"Don't make a mistake," he repeated to himself.

BEFORE FIRE OFFICIALS and legislators in California began calling record-breaking wildfires "the new normal," widespread fires in the United States were not that normal at all. There had been one historic siege in 1910, when indiscriminate logging led to the "Big Blowup," a fire that sparked among the dead timber and burned up more than 3 million acres in the Northern Rockies, killing eighty-five people—many of them firefighters—and forcing evacuations across Idaho, Montana, and Washington. But after that, wildfires largely faded from the landscape, helped in part by the legacy of President Theodore Roosevelt, whose devotion to conservation and public lands had led to the creation of the U.S. Forest Service. The agency had taken an aggressive stance toward wildfires, pushing its "10 A.M. policy," which called for every wildfire to be doused by the morning after it ignited.

Across the country, federal and state agencies responded in kind, shoring up resources to help protect the land. The California Department of Forestry was founded in 1905. Though its mission was to safeguard and manage the state's timberland, before long it was also providing emergency services, contracting with local communities as a way of subsidizing its existence. The agency's "ranger" title was swapped for "chief." Green and khaki uniforms and wide-brimmed hats were replaced with the navy blue garb of urban fire departments. On the agency's patch, a red flame was added next to the green conifer. Even legislators embraced the shift, supporting the agency's emergency responder role by funding an armada of engines and all-terrain vehicles, as well as bulldozers, air tankers, helicopters—and even inmates, who were paid $1 per hour by the state to help fight fires.

The rarity of out-of-control wildfires meant that millions of homes began to freckle the wildland-urban interface. Power lines traced their advance, crisscrossing the land. But as development increased, so did the potential for fire. The power lines were vulnerable. Dead and diseased timber piled up. Residents required protection. And so California legislators promised to forestall and conquer fire rather than accommodate it. In 2006, the agency rebranded itself with a new name: Cal Fire. The change could not have come soon enough: Over the previous four decades, fire season had increased by seventy-eight days. The dangers would grow even worse in the decade that followed. "Climate change has impacted California significantly," Cal Fire director Ken Pimlott cautioned in a 2016 memo. Temperatures had increased, weather patterns had changed, and plant and animal species had migrated. Not to mention urban development, which was still spreading unchecked. "We are averaging 25 percent more fires than the five-year average," Pimlott concluded.

Across the state, 189 communities, including Paradise, were listed in Cal Fire's "very high fire hazard severity zones"—the livelihoods of 3 million people at risk of complete erasure. In 2017, conflagrations set new state records for size and destruction. The damage was worse the following year, with flames threatening Yosemite National Park, torching mansions in Malibu, and rippling along the outskirts of Redding. By 2018, Cal Fire's expenditures had skyrocketed to $947 million, more than ten times its costs in 2010. That summer, the agency exhausted its nearly $450 million budget within the first two months of its fiscal year. The wildfires began to seem less like outliers and more like an ominous new reality.

Conflagrations had increasingly defied firefighters' efforts to control them. They were more powerful and erratic, and they were striking communities in ways that had once seemed unfathomable. From 2013 to 2018, blazes scorched 12.7 million acres in California, razing 50,615 structures and killing 186 people. At a news conference in August 2018, outside the Emergency Operations Center in Sacramento, Governor Jerry Brown told reporters that the state was in for "a really rough" ride. Firefighters were working in triple-digit temperatures to stamp out flames across California.

"The predictions that I see, the more serious predictions of warming and fires to occur later in the century, 2040 or 2050, they're now occurring in real time," Brown said. "It's going to get expensive. It's going to get dangerous, and we have to apply all our creativity to make the best of what is going to be an increasingly bad situation."

NORMAN STEERED AROUND the corner, only to run into more fire. He backed up and got lodged on a planter box, his SUV butting up against the front door of a house. He couldn't see much through the windshield. Suddenly a wooden PG&E distribution pole fell, denting his hood with a massive bang. The elderly woman in his backseat gasped, her arms quivering. Norman didn't know what to tell her. He was beyond offering comfort. He decided to gun over the power line, which was supposed to be de-energized. But then the line arced, still live. Norman cursed. He drove through a few more front yards, tearing up the dirt and grass, his vehicle getting pounded with rocks, picket fencing, debris. At least the car wasn't engulfed in flames. His mapping system was updating slowly—too slowly, worthless. Norman wasn't familiar with this part of town, wasn't sure where to turn.

Then somehow he ended up on Oak Way, a half-mile access road that ran north, paralleling the Skyway and connecting Bille and Wagstaff roads. He hadn't even known the street existed. He made his way toward the Skyway and deposited the woman in the parking lot of a Walgreens, where firefighters had jimmied open

the store's door, grabbing fire extinguishers off the shelves and creating a temporary refuge area for residents. Like their terrified
charges, other people across the Ridge were sheltering in their cars
at a grocery store, a reservoir, a vacant storefront, a thrift store, an
intersection—any place where they might be spared from roasting
in their vehicles. "The fire is right behind me," Norman told the
firefighters. "Put people inside the Walgreens. Otherwise they are
going to die in their cars."

Other residents were running down the Skyway, terrified, devastated, furious. Some seemed to be unfazed. A man wearing a
backpack walked his Labrador on a leash, as if he were hiking along
a trail on a warm fall afternoon. An old woman dragged a suitcase,
the fabric fraying on the pavement. A woman tugged at her young
son's hand, urging him forward. Too many people for Norman to
help at one time. The decision whether to evacuate was deeply
personal, with people often misjudging the danger they were in,
especially when they lacked good information. Orders from government agencies got everyone moving faster. Without them, people tended to act according to their experience in the last wildfire,
relying on an erroneous sense of knowledge.

Psychologists have identified three distinct phases in a human's
response to disasters. The first is denial. Previous wildfires had
brought a series of near misses to the Ridge, and the memory of
these experiences was powerful, bringing with it a false sense of
security. Denial was partly why people called friends and family
members or scrolled indefinitely through their Facebook news feed
instead of taking action. Others focused on their belongings, wast-
ing valuable time cramming their vehicles instead of getting out.
Facing the unknown, they found security in their possessions. Psychologists have found that at least 75 percent of people in a catastrophe remain frozen in this state of inaction.

After denial came the second and third phases—deliberation
and action. Though the Town of Paradise had practiced evacuating
before, people often made irrational choices when it mattered. Like
passengers in airplane crashes—who often ignore closer exits to follow the crowd—drivers mindlessly headed in a conga line toward

the Skyway, not thinking of alternatives. They ignored other open roads and paused at red lights and stop signs, even though police certainly wouldn't have ticketed them for blasting through the intersections. Others chose not to evacuate at all, overestimating their abilities and chances of survival. This sense of exceptionalism has come to be known as the Lake Wobegon effect, a term coined by a physician in the 1980s after he noticed that hundreds of elementary schools across the nation claimed their students were above average—which, statistically, could not be true. The elderly were particularly guilty of this thinking, because input from others did little to sway a lifetime of survived experiences. Disliking change and the hassle of an evacuation, they often underestimated a wildfire's risk.

In many cases, it took a specific moment—witnessing a neighbor's house burning down or receiving a call from a family member with orders to evacuate—for a resident to think and react rationally to the crisis at hand. How quickly a person arrived at this phase depended on genetics, experience, and training. Town Manager Gill got stuck in denial for nearly half an hour, debating the merits of the evacuation orders. Norman, on the other hand, had trained—and repeatedly practiced in real life—how to move through the three phases quickly. First responders like him understood the psychology of disaster response, overriding their own fear instincts to help others out of theirs.

Now the captain steered back to Oak Way to see if anyone else needed help. The Walgreens, he realized, was situated on an unburned island in the center of town; the flaming front had wrapped around the area in a U shape. But as the embers got bigger and gained energy, Norman knew, the space between the two arms of the fire would close. He had already told the firefighters to scope out other buildings where people could take shelter. As he drove down Oak Way, flames arced toward his vehicle, sucking up oxygen. He reversed and inched forward in a desperate game of Frogger, trying to prevent his vehicle from melting in the radiating heat. The engine in his SUV would fail without oxygen. He didn't have much time. Norman tried to make out the landscape over his dented hood. *Don't get trapped,* he told himself. *Don't be stupid.*

He drove over a few more front yards, smashing through a white picket fence. It was the wild kind of stuff he had dreamed about in his fire-obsessed youth—only this was not a dream and his survival wasn't guaranteed. He looked up and saw a couple in the street. The man was kneeling on the ground near a bush. Near him, the woman was pacing. A parked car filled with their possessions sat in the driveway, and a second car waited on the street. Three houses away, the fire was devouring a shed, raging in their direction. *What the hell are they doing?* Norman crawled out of his SUV and ran toward them. "You have to go," he shouted, his voice disappearing in the cacophony of flying tree branches. "Get out of here right now."

"We can't leave," the man said. "We lost the keys to our car."

"You don't have time," Norman yelled again. "No more waiting, you gotta go." He'd take them himself if he had to.

"Do you have an electric saw?" the man asked.

"No, I do not carry a Sawzall in my vehicle. If you're not going to leave, I'm going to," Norman replied, glancing back to make sure his escape route was still viable. He was fighting not flames but psychology. The couple didn't understand the peril they were in. "You have two minutes to leave," Norman said, knowing there was nothing he could do, really, to force these people to abandon their vehicle.

The man ran into a nearby garage and reappeared with a saw. He worked deftly, cutting off the components around his steering wheel and hot-wiring the car. Norman watched him, impressed. Even he didn't know how to do that. "Okay, we are going," the woman said. They took off, one vehicle tailgating the other.

Norman carried on down Oak Way, kicking in the front doors of houses. He didn't want anyone to be left behind. But the residents on this block appeared to have gotten out. The flames had begun to surge again, the wind carrying dead songbirds, shingles, tree bark, and other debris on its breath. Norman returned to his SUV. He reversed, driving back the way he had come. He hadn't gone a mile when he ran into the same couple.

They rolled down the window: "We ran out of gas! Can you help?"

OBSERVATION: THE SHERIFF'S DAUGHTER

Butte County sheriff Kory Honea, forty-seven, had reversed traffic on the Skyway and was helping direct vehicles downhill when he looked up and saw his daughter. She was standing in the intersection, moving motorists forward. Twenty-four years old, Kassidy was following in the law enforcement footsteps of her parents. Her mother was a dispatcher; her father was the county's top law enforcement officer. She had joined the Paradise Police Department as an officer six months earlier and was the only woman on the force. She could hold her own. Sheriff Honea was so proud of her. They had the same clear blue eyes and wide smile, even the same initials. Sometimes one would borrow the other's gold name badge.

A Winnebago trailer passed, then a gray SUV, a red pickup. Kassidy windmilled her arms, beckoning them forward. She was in navy blue; her father was in tan and green. They both wore N95 masks over their mouths. There was comfort in recognition and fear in understanding, particularly in moments like this.

The sheriff took his mask off for a moment. A call had crackled over his radio. Some of his deputies were trapped, and he needed to check on them. "I'm gonna go, my guys need me," he told his daughter. He wanted to say so much more, unsure whether he would ever see her again. "I love you, kiddo."

"I love you, Daddo," she said back.

THE FIRE: SIEGE IN SIMI VALLEY

More than 460 miles to the south, a new wildfire sparked. Near Simi Valley, on the boundary between Los Angeles and Ventura counties, the Woolsey Fire began so quietly that the hazy ash might have been, at first, mistaken for air pollution. California was now beset by flames along the north and along the south, each region overflowing with its own miseries and diverting resources from the other.

The massive pressure gradient that spread across the state's backbone tried to equalize, sagging over Southern California and building above the mountains. It fed the winds, which fanned the Woolsey Fire through the chaparral-covered Santa Monica range at the same time that they fed the Camp Fire through the pine-stubbled Sierra Nevada. The Woolsey Fire torched small ranches, movie sets, and the Malibu mansions of celebrities, models, and millionaires. It steamrolled through trailer parks, uprooting power poles and trees on the way. A phalanx of private firefighters arrived to protect the wealthiest homes as it tore a path to the Pacific Ocean, the blue-green waters reflecting the destruction. Fire engines were staged on beaches normally scattered with bright beach towels and foam surfboards.

Confused seagulls careened through the sky, dive-bombing into the sand. Along the Pacific Coast Highway, hot winds gusted over the cars, the people as vulnerable as celluloid dolls.

CHAPTER 11

"THE SAFETY OF OUR COMMUNITIES"

The line of vehicles barreled down Edgewood Lane and past Travis's house, clouding the air with dust. Fire roared through the nearby gully, and a pack of drivers had turned off Pearson Road, near the mobile home park on the corner, to avoid it. Paul and Suzie, still across from Travis on their four-wheeler, watched with worry as the vehicles passed. Travis checked for alerts on his phone. The traffic was an unusual intrusion: No one traveled on Edgewood unless they lived there.

The remoteness was part of the lane's appeal. It made the area feel far from the hustle and bustle of civilization, though the convenience of grocery stores and shops was actually only a few miles away. Neighbors on Edgewood recognized one another. In the early evenings, one woman strolled with her pets—not just her three dogs but also the goat, the horses, the vicious goose, and even the family bird, which perched on her shoulder. Another neighbor was known for his baked tofu. They watched out for one another, alerting the rest of the road if they saw a black bear pawing at an unlocked garbage bin.

Time had passed so quickly, Travis realized. Only a few minutes earlier, the neighbor across the street had pulled her clean underwear from the clothesline and evacuated safely. Her jeans and blouses shook on the line, forgotten. By the time Paul and Suzie had done a reconnaissance mission down the street on their quad, the mouth of Edgewood was choked by flames. The couple had already parked their cars—a blue Toyota 4Runner and a gold Toyota Corolla—in Travis's driveway, as they always did during a wildfire. His home was safer, after all; the man was religious about keeping invasive weeds at bay and maintaining the ring of defensible space around his and his wife, Carole's, bungalow. In the back of Suzie's sedan, her parakeet, Lucky Penny, flitted about its cage.

With Edgewood blocked, Travis knew the best option would be to escape on their four-wheelers. He left his packed Subaru in the driveway and headed for the garage, pulling on the emergency cord so he could heave the door open without the help of electricity. Then he backed out his beloved vintage truck, in which his Polaris quad was stowed under a golf cart cover in the bed. Some models of the four-wheeler sold for upwards of $15,000. Carole often joked that it was Travis's "show pony" because he spent so much time washing and polishing the sides and buffing the leather. He kept a towel draped over the seat so that it wouldn't get dirty. On this morning, he held the towel over his mouth as a buffer against the smoke.

Farther down Edgewood, Beverly Powers, sixty-four, and her partner, Robert Duvall, seventy-six, idled in their vehicles. They had loaded their two trucks and travel trailer with camping equipment, just in case. But the couple had taken too long to pack and found themselves trapped at the intersection of Edgewood Lane and Marston Way, about 200 feet south of Travis's home. At a house near theirs, sixty-three-year-old Ernest Foss sought comfort in his dog, Bernice, as his caretaker—his thirty-six-year-old stepson, Andrew Burt—pushed the bedridden musician in a wheelchair out the front door, dragging the older man's oxygen tank behind them. On the street, Joy Porter, seventy-two, braked, part of the caravan of cars. She and her son Dennis Clark, forty-nine, were horrified to realize that they had chosen a dead-end route.

Travis pulled down the ramps on the truck bed and unloaded the four-wheeler. It was nicer than Paul's; to jump-start the engine on his, Paul had to throw the quad into neutral and complete the circuit by jamming its innards with a screwdriver. Travis had watched him jury-rig the engine hundreds of times. For seven years, the men had ridden the craggy trails around Paradise together. Their favorite course was to the south, on the 1,500-acre property held by a prominent local family. Travis and Paul had received special permission to explore the property's farthest corners; in return, they often picked up discarded beer bottles or cleared downed branches. The afternoon rides were relaxing, the men lost in the forest with the sun on their backs. Travis loved the crisp autumn air and the musky scent of pine after the season's first rain. He knew every ditch and gully in the land surrounding his eight acres.

Now he and Paul brainstormed in the driveway about where to head to safety. There was a grassy field not too far away that might offer refuge. Paul also knew of a slot canyon with a waterfall that gushed through its center. Out of the corner of his eye, Travis saw his next-door neighbors, Jeanette and Mike Ranney, emerge from the smoke. Mike wore Nomex pants and leather boots, his long brown hair in a ponytail down his back. He was a contract wildland firefighter, a regimented and detail-oriented sixty-two-year-old who knew how to take care of himself. He was carting his collection of silver coins and uncut opals, as well as one of their outdoor cats in a large duct-taped Tupperware bin. Jeanette, fifty-three, carried their other kitten in a cage meant to trap skunks.

Travis stalled his quad. "What are you guys doing?" he asked.

"I don't know," Mike said. "I'm going to make a run through the fields, try to reach Libby Road to the west. Maybe we can outrun the fire." If not, he explained, he and Jeanette would shelter in the unnamed stream behind Travis's house. It wasn't much of a waterway—steep and narrow, bordered by brush—but Mike thought it might offer the best shot at survival. Travis and Carole had once wanted to dam a small pond off the stream, but the mating calls of red-legged frogs were so deafening at night that they had changed their minds about encouraging them.

"Do your best," Travis said. There were no police cars, no sirens—just the roar of the approaching wildfire. The sky flashed orange as Edgewood Mobile Home Estates exploded about 800 feet away. Flames soared 120 feet high, a wall of fire about three-quarters the height of the Arc de Triomphe. Travis knew they had to leave. The cats yowled in their carriers. He shouted goodbye, not sure he would see Mike and Jeanette again, then navigated his four-wheeler over the narrow bridge that crossed the creek below his house. Paul followed with Suzie, her arms cinched around his middle, chin tucked against his shoulder. Embers the size of dinner plates somersaulted through the air, their edges glimmering orange. Gray pines, one of the most flammable tree species, ignited in staccato bursts.

A few yards down the trail, Travis heard the wildfire's approach from the west. It rumbled, chewing through the pine needles and slamming dry leaves against the tree trunks, as loud as a jet plane. *How did it get there?* he thought. Suzie clung harder to her husband. They gunned the four-wheelers, weaving through the forest at 10 to 12 mph. Manzanita scraped at their ankles. Travis, realizing that the flames were blitzing in the direction of both the grassy field and the cavern, tried desperately to think of another place to shelter. Suddenly it came to him. About two miles south, there was a sloping, lava-cap-covered bluff overlooking the Sacramento Valley. Seeded with grass, it had little vegetation or timber that might burn.

"Follow me," Travis mouthed over the din, motioning Paul and Suzie forward.

His voice was hoarse from the smoke. Ash stung his eyes as he whipped past trees that seemed to be bending with the wind, a testament to the fire's strength. The trio navigated another half mile, then parked their quads side by side on the bluff. Travis pointed downhill, at a reddish rock outcrop that might shield them. Paul had packed a coat and a blanket, and he tucked the extra clothing around Suzie, trying to protect as much of her skin as possible. He also reached for the small Igloo cooler strapped to the back of his quad. It was stocked with the last tomatoes from his and Suzie's garden, a checkered napkin, a canteen of water, and some strawberries. The couple had figured they would need to eat lunch eventually. Paul grabbed it by the handle and the three of them hurried downhill, trying not to catch the toes of their shoes on the craggy terrain. Feet crunching on the dehydrated grass, they skirted around rocks the size of bowling balls. Travis stumbled over a dip and slammed to the ground, hitting his head and bruising his left leg. Without stopping to catch his breath, he scrambled to his feet again, pressing forward in the darkness.

His cellphone kept ringing. He had no idea how he still had cell service. He answered each call, explaining first to his father, then to his son, and then to his daughter-in-law that he might not make it. He also spoke to Carole, who was stuck on the Skyway with all the other evacuating drivers. She was tracking her husband's movements from a GPS app on her own phone and had called 911 on his behalf. Travis, Paul, and Suzie hadn't been on the bluff longer than ten minutes when they heard the whine of trees as they boiled: The flaming front was bearing down on them.

"We were trying to outrun the flames to the north, but they've surrounded us now. I don't know what to do," Travis told Carole. Sitting in her car, she realized the tires on her sparkly blue Subaru Impreza were flattening in the heat. "Hopefully I get out of this," Travis said. "I love you."

Finally the three reached the boulder, the fire close behind. From a distance, the outcropping had appeared bigger than it actually was. About the size of an armchair, the boulder was the perfect size for two people—not three. They crowded together behind the rock, hunching to protect their heads. The wall of flames blasted forward. The wait was terrifying.

"This is gonna be a bad one," Paul said.

MORE THAN 150 MILES to the southwest, conversation was buzzing at PG&E's newly minted Wildfire Safety Operations Center. It had opened earlier that year at corporate headquarters, a thirty-four-story skyscraper located on Beale Street in San Francisco's Financial District. The center, housed in a room on a top floor, was staffed 24/7 by meteorologists, engineers, and other safety experts, who studied fire risk across the company's service area. They needed to decide whether or not to black out portions of eight counties as a matter of public safety.

The center had opened in May 2018 to much fanfare. The media were rarely allowed access to PG&E's impenetrable fortress, which the company had occupied for more than a century, but for this occasion, journalists had been granted a widely publicized tour. Patrick Hogan, the utility's senior vice president of electric operations, had been quick to pin the impetus for the new center on climate change. "Extreme weather is increasing the number of wildfires and length of [the] season in California," he said, failing to mention PG&E's vulnerable grid. "We must continue to adapt to meet the challenges created by this 'new normal.'"

Just before lunchtime, the analysts noted that the day's devastating winds were losing some of their intensity. Perhaps the shutoffs might be avoided—though PG&E was hardly in the clear yet. Meredith Allen, forty-eight, was the senior director of regulator relations, responsible for making sure the company met the guidelines set by the state and federal government. A Benedictine College graduate, she had a law degree from the University of California, Berkeley. Allen worked hard and had already been up for hours. She had emailed four department heads from the California Public Utilities Commission with an update at 4:16 A.M. "We are continuing to closely monitor weather conditions," she'd explained, adding that the potential shutoff was "expected to occur between 0600 and 1000 hours, or possibly later."

Across the room, analysts sitting at computer keyboards studied the sixteen screens on the wall, which relayed everything from weather forecasts and pressure gradients to satellite images and footage from wildfire cameras. A few screens displayed maps that showed, in red and purple swirls, the elevated levels of fire risk across the state—high and extreme danger. Directly behind the employees, through a row of windows overlooking the bay, sunshine glinted on the choppy water. But no one was paying attention to the view. On the front wall of the room, in big blue letters, a sign read NOTHING IS MORE IMPORTANT THAN THE SAFETY OF OUR COMMUNITIES!

PG&E adhered to the Incident Command System, like Cal Fire and the Town of Paradise. This meant that in a building next door, the company's Emergency Operations Center—activated when something had gone terribly wrong—would soon open as well. Employees were already being diverted from their normal jobs to staff the EOC; they would work in twelve-hour shifts until the Camp Fire was contained. This room was white and looked to be freshly painted, with a hundred computer terminals organized into rows. Off to the side was a meeting space labeled the "Huddle Room." In the EOC, everyone wore a vest of some kind to represent their role: red for meteorology, hazard-orange for external visitors, black for the CEO, and green, of course, for finance. Though the existence of the center projected order and control, in reality there was very little the utility could change or contribute at this point.

By the end of the day, they would need to have some idea of what had caused the Camp Fire—and what role they had played in it. PG&E was required to submit an "electric incident report" to the public utilities commission within two hours of an issue occurring in its system. The rule applied to any event that might have been caused by its infrastructure if it had resulted in a fatality, inflicted damage greater than $50,000, or drawn significant public attention. From January 1, 2017, through December 20, 2018, the utility submitted 131 reports, 86 of which were related to fire.

Streams of data flooded across the computer monitors. The electrical grid indicated that more than 34,200 customers had lost power because of the Camp Fire. At 10:36 A.M., PG&E had turned off natural gas service to Paradise, affecting 12,000 customers. Even more ominous—possibly implicating the company—an outage had registered on the Caribou-Palermo transmission line at 6:15 A.M. The timing seemed to correspond with the ignition of the Camp Fire. Company leaders hoped to get a helicopter up to examine the line soon.

It was nearly noon. PG&E had already missed the two-hour window for issuing its report.

.   .   .

THE FIRE REACHED THEM, a flash of brilliant light in the black sky. Travis watched in horror as Paul and Suzie vanished into the flames, screaming. *I'm next,* Travis thought, sprinting downhill. He crouched behind another boulder—just in time. The main fire front passed overhead, the rock taking the brunt of the heat. Bits of flaming grass swirled around him. Voracious tendrils of fire reached for his body, singeing his clothes.

Travis leaped through the remaining flames to the scorched area that had just burned. The edge of the lava cap was a dark line in the distance, marking where the canyon dropped off. The Paradise Ridge had been formed after Mount Lassen exploded 65 million years ago, sending layers of lava more than a hundred miles south. The magma had calcified into a series of gently sloping ridges. Then, during the Ice Age, geological upheaval had tilted the land. Across the Sierra Nevada, powerful rivers carved V-shaped valleys in the foothills. The Feather River and Butte Creek whittled away at the table that would become known as Paradise, etching its steep-walled canyons for millennia with the patience known only to water. Travis was now standing near the edge of another ridge. On a good day, he could see past the county seat to Table Mountain near Lake Oroville—but not on this morning.

The blackness was disorienting, a bank of smoke rolling around him. Orange suffused the air in patches, delineating land and sky. Tall pines stood out against the supernatural glow of the wildfire, swaying from side to side, embers flickering in their branches. Travis noticed that Paul's quad—about a hundred feet uphill—was starting to smolder. He couldn't imagine that his friends hadn't survived. They would need the vehicle to ride to safety. Besides, Paul loved that thing. Travis hobbled up the incline, intent on putting the fire out. The plastic milk crate where Paul stashed his gear was already melting. Travis snuffed the flames, knocking the screwdriver that Paul used to jump-start the engine onto the dark, rocky ground. He paused to record a video of the scene on his cellphone, because he didn't know what else to do. On the screen, the world became smaller and more manageable. It was 11:29 A.M.

Travis was terrified to return to where he had left Paul and Suzie, terrified of what he would discover. He walked back downhill toward the boulder. They were still there, a heap of limbs and blackened clothing. Paul had tried to shield his wife's body with his own, but Suzie's pants had melted off and her shoes had liquefied. Several inches of her golden hair, streaked with silver, were charred. Paul's new leather boots, which he had bought two weeks earlier, were plastered to his shins, his feet baked into their soles. The yellow and orange shoelaces were shriveled on the ground, nestled among buttons from Suzie's shirt. They were talking to each other quietly, Travis realized.

Paul's arm twitched. He was trying to put out the fire near them with the strawberries from their cooler. "Are you okay?" Travis asked, approaching. They were alive—for now. He stared at Suzie's feet. He could have sworn she'd been wearing shoes when they left Edgewood Lane. Travis knelt and gathered ice packs from the Igloo cooler, placing them on his friends' hot skin. Seeing Paul contort with pain, Travis ran uphill and ripped the seat off his neighbor's quad. He brought it back, settling Paul in the recliner so he wouldn't have to lie on the hot lava cap, which was sharp as razors. Strawberries scattered the ground, and he fed Paul some of the sweet fruit, trying to pull him back from shock. His gibberish wasn't making sense. Tears streaked his face; Travis had never seen him cry before.

The couple were too injured to drive. Suzie couldn't stand on her burned feet, and when Travis reached for Paul's hand, the skin slipped off in his palm. He would have to leave them in order to find help. He knew he was their only hope: If he stayed, first responders wouldn't know how to locate them. Travis looked at his friends through the smoke, deliberating. Suzie caught his eye and whispered: "Don't forget about us. Promise you'll come back."

11 A.M., engulfing Paradise

Paradise

PART IV

CONTAINMENT

KONKOW LEGEND

Over fields and valleys, across the dry streams and the mountains, the flames scorched the dry, parched earth, burning the trees and melting the rocks, with the people fleeing in terror before them. But the flames were faster, and everything that was alive—the game and the wild beasts and even the birds in the forestland and all the Konkows but two—was destroyed.

Peuchano, so named from his great sufferings, was a kindly, pious man, and he and Umwanata, his mate, had always thanked the Great Spirit for his kindness to them, and he remembered them even in the great fire. The flames came roaring toward them like wild beasts, but they rolled away on every side as if pressed back by an invisible hand—the hand of Wahnonopem, the Great Spirit.

CHAPTER 12

THE LONGEST DRIVE

The sky crashed around Chris as he searched for Rachelle, the smoke swallowing his windshield and whistling through the vents of his Suburban. He'd decided to get back in the car and head to Chico, hoping he'd find his wife there. Now he puttered forward on access roads, trying to ignore the storm surging outside. The plastic taillight covers wrinkled from the heat. He called everyone he thought might have information about evacuees, including his sister-in-law, who was a nurse at Enloe Medical Center in Chico, and a close friend whose wife processed medical records at the same hospital.

Chris's phone pinged with an incoming call. It was Stacey Zuccolillo, the wife of Rachelle's ex-husband. She was sitting on the Skyway, three children crammed in the backseat of her Toyota 4Runner: her daughter, Aubree, along with Rachelle's Aubrey and Vincent. She connected the call to her Bluetooth speaker so the little kids could hear his voice. They were excited to meet their baby brother later that evening. Chris told them their mother and Lincoln were okay. "On their way to Enloe," he said firmly. He couldn't tell them the truth: that he didn't actually know where they were or whether they were safe. Before Stacey could switch the call off the speaker, though, Chris paused and continued in a different voice: "Hey, Stacey? I actually can't reach Rachelle. Can you help me find her?"

Aubrey's and Vincent's eyes grew as big as moons. They looked at each other, shaken. They knew that they weren't supposed to have heard this news. They had just left behind their councilman father, who was directing traffic on the Skyway, and now their mother was missing. What if they never saw either of their parents again? Stacey promised to help Chris as soon as she could, but there was little she could contribute while trapped in Paradise. People ran alongside her SUV, their backpacks bouncing with each stride. In the distance, there was orange light in all the places orange wasn't supposed to be. Even during their short call, the fire had lurched closer.

Chris tried to dispatch a text message to Rachelle: "I wish you were with me."

DAVID FLIPPED HIS NISSAN around and headed east toward Pentz Road, pushing against the direction of motionless traffic. He and Rachelle didn't make it far, easing into the parking lot of a liquor store near the intersection of Bille Road, where they had a little more space. The market was white-walled and spare, its name printed in block letters above a Bud Light poster. It offered little protection—but at least it hadn't caught fire. Firefighters spritzed water over the cars parked on the road, but the water was quickly running out—the irrigation district's water tanks had bled dry. In the span of a few hours, 9.5 million gallons of water had disappeared. The engine crews continued trying to siphon water from the dried-out hydrants.

Firefighters handed out bottled water to those snared in gridlock, reaching through the partially opened windows with a smile. Their faces were smeared with sweat and ash—and something less tangible. Rachelle had never seen grown men look so shell-shocked. She accepted one of the bottles gratefully, pouring the liquid into her mouth. Her tongue was scratchy and dry. Lincoln gazed up at her, his blue eyes unfocused. With his narrow nose and blond fuzz, he reminded Rachelle of her daughter. David hopped outside again, kicking away the pine needles and leaves gathering beneath the tires, and then drove the car in a circle around the parking lot. By nosing into southbound traffic on Pentz Road, he hoped to be among the first to leave the convenience store when traffic started moving. Maybe they could find another route.

They waited.

David noticed Rachelle's IV bag askew on the floor mat. He picked it up and tried to get the saline solution and medication flowing again: He traced the cord, made sure the clamps were open, and hung the IV from the rearview mirror—but there was no pump to force the pain medication into Rachelle's veins. David's job didn't involve patient care, and he wasn't sure how to proceed. He left the car again. "Is there a nurse here by chance?" he asked, speed-walking over to a group of people huddled on the shoulder of Pentz Road. An older woman raised a gauze-wrapped hand and said she was a retired nurse. She had only one eye, David registered with surprise. She walked with him to his Nissan and leaned over Rachelle, examining the line. She managed to get the solution pumping again, but before she could plug the port back into Rachelle's arm, police officers interrupted them.

"Everybody get back in your cars and follow us—now!" they roared, leaving no room for argument. The fire was approaching again. The one-eyed woman waved goodbye, turning back to her own car. David and Rachelle followed the syncopated blue-and-red lights north on Pentz Road. They were among the first in line. One of the officers turned west onto Wagstaff Road and pointed them toward the parking lot outside Kmart. It was the biggest store in the shopping center, which also housed a Dollar General, a Save Mart supermarket, and an Allstate insurance office.

Firefighters had turned the parking lot into a temporary refuge area, or TRA, a place to shelter when escape was cut off. First responders had learned from previous fires that last-minute evacuations were often flawed—and more likely to kill residents than if they simply waited in a safe place for the flames to pass. The tactic had been tested by Forest Service ranger Edward Pulaski more than a century earlier, during the "Big Blowup" of 1910. Forced to shelter in an abandoned mineshaft as fires raged across Idaho, he and forty-five others had lain facedown on the ground to avoid inhaling the superheated gas, which could roast their lungs and sear their throats shut. According to legend, Pulaski had hunched near the entrance with a loaded gun to prevent the panicked firefighters from fleeing to their deaths.

In the past decade, sheltering in place had become a matter of policy in another country familiar with wildfires: Australia. Citizens were offered a choice: shelter in place or leave early, otherwise known as "stay or go." (An alert during the 2020 bush fire siege, for example, read: "You must take shelter before the fire arrives. The extreme heat is likely to kill you well before the flames reach you.") The government ran educational campaigns that trained Australians to fireproof their homes and safely hunker down within them. Residents were encouraged to purchase sturdy leather boots and non-flammable clothing and select a room with at least two windows or exits.

But it wasn't a perfect system. After the 2009 Black Saturday bush fires killed 173 people in the state of Victoria, Australia had reexamined its shelter-in-place policies. Many of the victims were found within their ruined homes or on their lawns, having made a last-ditch effort to escape. At least fifty-seven people had hidden in bathrooms—among the worst spots to seek shelter, because there are generally fewer doors and windows to serve as vantage points or exits—and thirty-seven of them had died. The government was sharply criticized for not having adequately articulated the dangers of staying at one's home. A report published by the Victorian Bushfires Royal Commission emphasized that leaving early was always the best option.

In the United States, firefighters urged residents not to defy evacuation orders. But in the event that early evacuation wasn't possible, some communities did have designated public TRAs—like an athletic field or place of worship—that were less likely to burn. In the 2003 Cedar Fire in San Diego County, thousands of people gathered at Barona Casino after escape became untenable. During the 2008 Tea Fire in Santa Barbara County, residents sheltered in a college gymnasium. As the Woolsey Fire was surging across Malibu, people were congregating at Pepperdine University.

Paradise had two designated gathering points: Paradise Alliance Church and a parking lot outside the 765-seat Performing Arts Center. In addition to those spots, thousands of people were also assembled at ad hoc shelters, including the parking lot outside Kmart, where David and Rachelle were headed. People packed the lot outside the big-box store, tipping their heads back to study the dark seam of smoke in the sky. Shopping carts were flung on their sides on the pavement, tossed by the wind. Dogs jerked at their leashes. The Kmart smoldered, threatening to ignite under showers of embers, but the lot was strangely orderly, even peaceful. People were polite, waiting obediently for instructions. As firefighters knew, people panicked most when they were alone, feeling trapped and helpless, separated from others. The crowd of locals might have been stuck at Kmart, but they were stuck there together—and with friends and neighbors they recognized.

David pulled onto the road shoulder before the shopping center's entrance. He didn't want to get trapped in the lot with his vulnerable passengers. While he explained to a firefighter that he needed to evacuate to Enloe Medical Center in Chico, Rachelle stared out the window and tried to bank her diminishing hope. She adjusted Lincoln in her arms, remembering the evening she and Chris had told his daughters about the baby. Rachelle had nearly spilled the news while they were in Missouri for her stepdaughter's college graduation. She had managed to hold back, not wanting to diminish the young woman's accomplishment. In private, Chris had already started joking about the timing. He would roll his eyes in mock despair: As soon as he had finished raising his two children, he discovered another one was on its way. They had waited to share the news until his daughters were back in California, making the announcement over a white-tablecloth dinner at a fancy Italian restaurant. The girls had squealed, overjoyed at the prospect of a baby brother. It seemed like so long ago now.

David yanked the sedan out of park, pitching forward. "We're going to follow him back to the hospital," he said, with a weary smile. Returning to the Feather River campus felt like defeat, but at least they could find medical care there. As they headed south on Pentz Road, Rachelle scanned the wreckage outside for her home. Many of the trash bins had liquefied, now just blue and green stains on the curb. At her address, only a brick chimney towered in the air. There was nothing else left. The boxes of baby supplies stacked in the foyer, gone. The dresser and crib set, the throw rugs shaped like baseballs, gone. The game room, dubbed Motel 6, which Rachelle had hoped to convert to a master bedroom, gone. Four generations of her family history, lost. The sight took her breath away. She leaned forward to record the scene on her cellphone, the landscape passing in a blur of brown and gray. The grass was scorched in Rorschach splotches. At the end of the driveway, miraculously, the bins were intact.

"It's all gone," she said. "Holy shit."

David exhaled sharply, but she didn't register the sound. For a moment, Rachelle was gone too. She was remembering the weekend trips with her children, drives with Chris to the aquarium in Monterey or the science museum in San Jose. Their annual summer vacation to Pismo Beach, where the ocean lipped the horizon and Aubrey stuck her toes in the chilly surf. Chris's move to Paradise for her, certain he would hate it. The ease with which the community had won him over with its green-layered mountains and small-town geniality. Vincent's PeeWee team, undefeated for two seasons, which surprised no one, because Paradise was a football town and indoctrination started early. The wedding dress that she had carefully preserved, packed in a box and stored in the attic above the garage, Aubrey dreaming of wearing it someday. All the things Rachelle had sacrificed—her own intuitions, her own feelings—so she could put her family first.

It hadn't been easy. Something had been off in her first marriage, to Mike, but she'd ignored it for a long time. She loved her husband and adored their children. How could anything be wrong? She posted photos on Facebook from the Sacramento Zoo and at the Aquatic Park Swimming Pool in Paradise. With her "two favorite boys" on Father's Day: Mike, wearing a baseball cap embroi-

dered with WORLD'S GREATEST DAD, cradling their son in his lap.
When Rachelle campaigned for a seat on the local school board in
2012, Mike had helped make the signs, stenciling her name in
white against a blue background. He attended every debate, smiling at Rachelle from the audience. Both Rachelle and Mike were
ambitious and loved debating politics, sparring from separate patio
chairs on their deck overlooking Butte Creek Canyon. Always
scrappy and good on a budget, Rachelle had asked a friend to style
her hair into an elaborate up-do for the Election Day victory party.
She and Mike had been so happy.

Rachelle had taken the divorce hard, struggling to regain control of her finances and find a new place to live. After moving into
her grandparents' home on Pentz Road, she had started buying
cheap white wine by the box to take the edge off. She stashed it in
a desk drawer at work. Soon she was showing up drunk to school
board meetings. After one particularly embarrassing session, Rachelle resigned her seat and checked herself in to a rehabilitation program. She was determined to get better. Slowly, in fits and spurts,
she did. During her stint in the Santa Cruz program, Rachelle
purchased a green mug from a local coffee shop. In white letters, it
read IT'S ALL BETTER IN HERE. Years later, Rachelle still sipped her
coffee from the mug every morning, a daily reminder of how far
she had come; it was a permanent fixture in the cupholder of her
Suburban. Rachelle's father had always told her that it didn't matter
how many times she fell on her face—what mattered was that she
got up again. Rachelle had taken his advice to heart: She maintained her sobriety and was open about having hit rock bottom.
She had looked toward the future, set goals, and built a new life
with Chris.

Now she was looking at that life in ruins. The dogwood tree in
what used to be their backyard was a blackened stump. Every
spring, Rachelle had mailed its cream-colored buds to her family
back in Fresno. She remembered how the tree had sheltered her on
that night the previous spring when she had clutched the pregnancy test and learned that a baby was coming. Her son was finally
here, but he had come into a world that Rachelle didn't recognize.

Everything she had worked for had burned to the ground or been lofted into the air, spilling into the atmosphere in poisonous puffs.

NEARLY SIX HOURS AFTER leaving Feather River hospital, David pulled back in to its parking lot. While they had been stuck in traffic, the fire had swept west across Paradise, just missing them. Under the emergency room overhang, doctors were triaging other people who had tried—and failed—to evacuate, like Rachelle, and preparing to move them to the helipad. The lower wing of the building was on fire, and so were the outbuildings: marketing, human resources, the central utility plant. The cardiology department had been lost. One of the emergency room nurses had broken her foot while trying to outrun the blaze, and she hobbled in an orthopedic boot. The Toyota Tundra driven by the ICU manager had toasted like a marshmallow, its sides crisped brown. Staff arranged straight-backed chairs and couches from the waiting room on the helipad, an open space circled by beige landscaping rocks where they might be safe. The rows of seats faced each other, as precise and orderly as they had been indoors, but they'd traded the fluorescent hospital lighting for an ashen sky.

The backup generator had caught fire earlier in the day, and without power, the toilets inside the hospital had clogged and flooded. A scattering of plastic buckets, where people could squat to relieve themselves, sat cordoned off by sheets. Deputies pushed elderly women from a nearby nursing home across the parking lot in wheelchairs. Thin white sheets were draped over their heads and shoulders like keffiyehs. Beneath one sheet, a woman wore a red crocheted hat and cradled a goldendoodle in her lap. Ben Mullin, the cardiopulmonary supervisor who had been sheltering in the tunnel, preparing to breathe noxious gas, had emerged when he noticed new people arriving at the hospital seeking medical assistance. He recognized David's white Nissan and went over to give him a hug. David broke down in his arms, gasping and sobbing.

A pediatrician rushed over to Rachelle and listened to Lincoln's heart and lungs. The hours-old baby blinked and tried to wriggle away from the cold stethoscope. The pediatrician instructed Rachelle to stay in David's sedan—he didn't want the baby to inhale any more smoke. Rachelle was too exhausted to protest. They had spent hours trying to escape, only to end up right back where they had started.

Meanwhile, Chris was circling the parking lot at Enloe Medical Center in Chico. He had finally made it downhill, but he couldn't find a place to park. Every evacuee with the slightest ailment had shown up at the hospital seeking help. Chris's twenty-four-year-old daughter lived three blocks away, so he gave up and drove to her house. She pulled him into a hug and guided him into her kitchen, where Chris drank a glass of cold water and tried to catch his breath. His phone lit up with a text message from a friend, who had received word from Rachelle. "Don't worry, they're at Feather River hospital," it read. "Everything is going to be OK."

Chris blanched. How could Rachelle and the baby still be in Paradise?

OBSERVATION: SARA'S CALL

Cal Fire dispatcher Beth Bowersox recognized the voice on the line as soon as she picked up the call. Her neighbor Sara Magnuson had called 911 hundreds of times during her slow slide into dementia. At one point, she was ringing the emergency line ten to twelve times a day, complaining of chest pain or claiming a burglar was breaking into her house. The seventy-five-year-old loved Christmas and kept twinkle lights strung along her eaves year-round; one strand, draped across her front door, tapped the frame when the wind blew, making it sound as if someone was working the lock.

From across Drendel Circle, Bowersox would watch Sara retreat into her home, appear in the driveway ten minutes later, and go back inside again, apparently at a loss as to why she had come outdoors in the first place. Her husband, Marshall, had run a jewelry business from their home until his death in 2013. Without him, Sara hadn't known how to take care of herself. She saw and heard things that no one else did. To the irritation of neighbors, she regularly set off her car alarm in the middle of the night to fend off imaginary "Nazi" dogs. At Sara's request, PG&E workers had stopped by several times to check her gas meter; she had reported that her gas was leaking, though it never was.

Sara had few family members willing to help out. She had last talked to her brother, who lived in Phoenix, eight months earlier. He found Sara "difficult to get along with." After she'd been caught driving without a license, her car had been confiscated, and she couldn't afford to recover it from the impound lot. In recent months, Bowersox had beseeched Lauren Gill to put a conservatorship over Sara. She was a danger to herself. But as much as Sara's antics annoyed Bowersox—and everyone else on their cul-de-sac—she also felt sympathy for the elderly woman, who spent so much time alone. Looking out the front window of her bungalow onto Sara's home, Bowersox couldn't help but think of how scared she must feel navigating the world every day.

Earlier that morning, their neighbors had twice tried to evacuate her. "I'm fine," Sara had insisted, fierce as ever. The septuagenarian's signature look was tight leggings and low-cut tank tops that enhanced her cleavage. Bowersox had always admired Sara's confidence. But it meant that Sara didn't understand that an out-of-control wildfire was at her fence line. She refused to leave. By the time she could process what was happening, it was too late. Now she called the 911 line for help. "Nobody's left," she whispered to Bowersox. The power had been turned off, and Sara's house was dark. The woman was curled up in her bathtub, afraid.

"Sara, I need you to get out," Bowersox told her urgently. "I need you to run. Start walking out of the house. Maybe you will see a car or a fire engine. But you cannot stay in your house. You need to leave." Sara wouldn't—or perhaps couldn't—follow Bowersox's directions. She might not have understood. Soon after, the line failed.

Bowersox labored to breathe, her shoulders heaving. Tears filled her eyes. She stepped away from her pod. She

couldn't bear to answer another call. People called 911 expecting first responders to save them, but more and more on
this grim morning, there was nothing they could do. After a
few minutes, Bowersox wiped the tears from her cheeks. She
had no choice. Gulping back sobs, she put on her headset
and answered another call from another person she couldn't
help.

CHAPTER 13

NO ATHEISTS IN FOXHOLES

The ambulance crew huddled in the darkened garage of the house on Chloe Court and tried to come up with a plan. The family who owned this house had just moved in; the space was strewn with stacks of cardboard boxes and a tangled nest of bicycles. Amid the clutter, three patients rested on foam backboards on the concrete floor, sheets tucked around their bare feet. Red triage bands around their wrists marked them as the patients in the most serious condition. Tammy planned to help them, but she needed to say her goodbyes first. She climbed back inside the remaining ambulance to make her calls, settling next to the man with cerebral palsy, who was too large to move. Gazing out the ambulance window, Tammy saw that a house on the cul-de-sac had caught fire.

Tammy called her twenty-four-year-old daughter, Clarissa, the baby she had birthed the winter of her senior year. She had nursed her while studying for final exams and still managed to walk down Om Wraith Field with her class at Paradise High's graduation. Tammy had looked for her daughter's chubby face in the audience, hoping that the baby's future would be different from her own. She had married the girl's father at a park in her native Los Angeles the summer before, borrowing a white dress for the occasion. Her pregnant belly had pushed against the fabric. Tammy's family hosted a potluck afterward. It was her mother, after all, who had insisted they marry. A recently born-again Jehovah's Witness, she would support her daughter in making this transgression right, though she refused to do anything more than help with the wedding.

Once they were married, Tammy had moved into a one-bedroom apartment with her new husband, who was two years older, and Clarissa. The bedroom barely fit their double bed and the baby's crib. Tammy worked after school and on weekends at a restaurant called La Comida; her husband worked at Round Table pizza. They struggled to pay the bills. Tammy had once called her mother, crying: "I don't have money to go get cough syrup for Clarissa." Her mother maintained that she wouldn't help. Later that evening, though, Tammy found the medicine and some bagged groceries on her front porch.

Tammy had tried to foster a closer relationship with her own daughter. She was more generous and open with Clarissa, treating the teenager and her friends to pedicures, and when Clarissa entered high school, taking her to the midwife's office so she could make educated decisions about her body. Now, in between sobs, Tammy told Clarissa over the phone how much she had loved her and how hard she had tried. "I'm not going to make it out of this, and I'm so sorry," she said, begging her daughter to leave Paradise without packing. "You need to leave now. The time you thought you had—you don't have it. Please just leave."

Next, Tammy called Savannah, who was about three years younger than Clarissa. After her birth, Tammy had felt the loss of her youth—she had skipped straight from high school to a household in which she was a wife and mother of two. At twenty-four, Tammy was working three jobs: as a housekeeper, an obstetrics technician at the hospital, and a cocktail waitress at the Crazy Horse Saloon, a nightclub in Chico known for its strong cocktails and mechanical bull. (Wednesdays were Buck Nights.) Her husband had gotten a new gig at a car parts store. Tammy would bathe their girls and read them books, then leave for the nightclub after they had fallen asleep. She would return home at 2:30 A.M., only to wake up a few hours later for her 7 A.M. shift at Feather River Hospital. She and her husband, rarely home at the same time, drifted apart, their relationship reduced to that of roommates. It wasn't long before they separated.

She was sorry about the divorce, Tammy told Savannah now, and for the chaos that came afterward—starting with the irascible man she had met at the saloon after splitting from their father. Clarissa and Savannah were ten and seven when she found out she was pregnant with his son, Brayden (now fourteen), then his daughter, Allyson (now thirteen). Tammy had stayed in the relationship for more than seven years; she didn't know how to survive without him. And yet he was emotionally abusive, which Savannah had taken especially hard. By the time Tammy was thirty-six, when she found out that she had gotten into her dream nursing program at Butte Community College, her relationships with both her boyfriend and, more important, her daughter Savannah had seriously decayed. Another father figure followed, another pregnancy. Tammy graduated from nursing school with four-month-old Brooklyn in the audience. That month, she sent out announcements celebrating the birth of her fifth child, Clarissa's high school graduation, and her nursing school graduation.

Tammy had wanted to be a better mother, she told Savannah. She apologized for all the ways she knew she had failed and asked Savannah to be there for her younger siblings. She had to go, Tammy said, but she was proud of her kids. "I love you," she repeated, until she was certain that her second-oldest daughter felt it.

And finally, with a nervous sigh, Tammy called her own mother, who was on vacation in Oregon. They worked in proximity—her mother was a nursing supervisor at Feather River hospital—but in recent years, their relationship had grown even colder. After Tammy's stepfather died, her mother had remarried and become more involved with her church. These days, she and Tammy barely talked. Still, Tammy remembered the pride in her eyes when Tammy had been given the Daisy Award the previous year. The prize recognized one nurse every trimester for dedication and professionalism. Tammy had been nominated after coaching a first-time mother through a difficult labor. "I have never felt such genuine love and compassion from a complete stranger before," the mother wrote in her nomination letter. "Tammy's love for her job was so evident in her every action towards me. Her motivating words got me through something I was beginning to feel was impossible. My favorite phrase she used was 'Get mad and push her out!'" Tammy had won the award after just three years; there were nurses who had been on the floor for three decades and never received it. At the informal ceremony, hospital officials clapped and cheered while Tammy scanned the sea of faces for her mother. "I hope that I've made you proud," she told her mother on the phone. She said she was sorry for how much she had put her through.

In each of her conversations, Tammy apologized, again and again. For the family outings she had canceled, the whim of a vindictive boyfriend derailing an entire afternoon. For the relationships she'd had with men who emotionally abused her and her children. For the broken family that her children had had to endure—for the hectic schedules, the shuttling of five kids to and from the houses of three different fathers. For the nursing study flashcards that Tammy always kept in her Jansport backpack and pored over during her son's baseball games. For the recordings of medical terms she'd subjected her children to as she tried to memorize them during rides in her Suburban. For the years that Tammy had been there, but never really there, as she worked her way through nursing school.

It had taken Tammy a long time to realize that she didn't deserve to be treated as if she was nothing, that even with the stretch marks on her stomach from five pregnancies, she was beautiful and kind, worthy of love. She wished she had met her current boyfriend sooner. He was a local police officer who took Tammy out for steak on their first date and treated her with respect. Funny, with the best smile, he was also a good role model for her children. She was happy that she had finally achieved her dream of becoming a nurse, she told her family, but she was sorry that whole decades had passed before she had done something meaningful with her life.

. . .

THE SKYWAY WAS a ribbon of hope—at least in the minds of residents, who stubbornly clung to its promise. If they could *just* edge their vehicles onto the town's widest road, they believed they might stand a chance of survival. Jamie should have been safe here—but why was traffic not moving? He crawled forward a few more excruciating feet, nosing downhill. One lane over, he recognized a few of the colleagues who had departed Heritage with patients in their backseats. They had gotten a head start, leaving an hour before Jamie, but they hadn't made it any farther. Their faces glowed, illuminated by their cellphone screens, as they scoured social media for information. They glanced up only when electrical transformers burst with a bang atop distribution poles. Everyone was on edge.

The nurse seated next to Jamie fidgeted, trying to make another call. Earlier that morning, her elderly uncle had picked up her sons from school and brought them back home to the house they shared on Buschmann Road. He hadn't answered calls since then. *Probably asleep,* she thought. Her uncle had been one of the few to defy evacuation orders during the Humboldt Fire in 2008, preferring to spritz the house with a garden hose than suffer through the onslaught of traffic. His stubbornness had saved their family's longtime home—but now the stakes were higher. She didn't think he was taking this fire seriously.

She would later discover that her instincts were correct. Her uncle was fast asleep in his loft above the garage. Using a flashlight to see, her three young sons were stamping out spot fires on the lawn. They couldn't rouse him from his bed.

The nurse had been born into one of Paradise's oldest families. Her father, who owned a roofing company, had shingled half the town's homes—or so he claimed. She had met Jamie and Erin through mutual friends, and over the past decade, they had grown close. Her husband was one of Jamie's supervisors at Heritage. After Tezzrah was born, she had been one of the baby's first visitors. She called the girl, who was about the same age as her youngest son, her "second daughter." Now, though, the nurse was considering leaving Jamie and Tezzrah. She needed to make sure her sons were safe. Police had already forced her husband downhill with their two-year-old daughter. In his rush to pick the toddler up from daycare, he had accidentally taken his wife's key fob, stranding her with a car she couldn't unlock.

Now she was stuck with Jamie. He tried to calm her, but his assurances weren't doing much good. She couldn't stop shaking; he could tell she was making Tezzrah nervous. "I'm so sorry, Jamie, but I gotta go," she finally said, swinging open the door. "I can't just sit here." She turned to say goodbye to Tezzrah, then lurched out of the car, the door latching shut. They watched as she teetered toward the road shoulder, tripping and stumbling in the darkness.

Other people cracked their car doors, too, stepping outside to see for themselves why traffic wasn't moving. Jamie smiled at his daughter to distract her. Tezzrah stared back, her brown eyes serious. Minutes later, a sheriff's deputy tore past them on foot, hollering: "Get out of your cars or hunker down! The fire is coming fast!" His voice was muffled by a respirator. A gold star glinted on his chest.

Jamie clambered out of his Subaru. Two of his colleagues from Heritage called to him over the gridlock, asking what he thought they should do. "He's law enforcement, and if the fire is coming fast, then let's just go," Jamie yelled back. Heart pounding, he pulled the Subaru off the road and yanked his keys from the ignition. He grabbed Tezzrah's backpack, filled with the bottled water and snacks. Tears streamed down the seven-year-old's face as Jamie helped her from the backseat. "Daddy, are we going to die?" she asked for the second time that morning.

Jamie took a deep breath, then knelt on the concrete to look his oldest daughter straight in the eyes. "Tezzrah, you are not going to die today. I promise you that. I am your father, and I will lay down my life for you if I have to. Even if I have to carry you off this mountain, I will get you out of here." He pulled off his puffy black hoodie, blotted at his daughter's cheeks, then wrapped the jacket around her, telling her to cover her face with the sleeve so she wouldn't inhale too much smoke. Then he stood up, enfolding her small hand in his. "Do not let go," he said with a squeeze.

As they ran, Tezzrah lagged behind, unable to keep pace with a grown man. The pavement was slick with pine needles, and she skated over them in her sneakers. A nurse from Heritage ran alongside them, clinging to the facility's chef, a woman named Jill Fassler. The nurse held on to Jill like a drowning woman, until Jill lost her patience. "You've gotta stop dragging me down," Jill said, snapping at the nurse. "Let's just keep moving." Jamie, one lane over, tightened his grip on Tezzrah's hand as he tugged her along. They darted past a woman leaning on her walker and two people weighted down by oversized birdcages.

A few minutes later, they hit the intersection of Wagstaff Road. Flames roared down the hillside, threatening two gas stations: American on their right, Chevron on their left. Firefighters had snapped open the lock on the front door of Needful Things, an antiques store with stucco walls and a metal roof, to create a temporary refuge. Jamie had made up his mind to reach Walgreens, farther away from the explosive gas stations, but they didn't get far before an engine crew sent them back to the antiques store.

"I'm not going in there," Jill said, balking at the front door. "It's crammed with hundreds of people and animals. We won't have a chance if we get stuck in a building like that." Jamie couldn't see a better option. He left her outside and guided the nurse and his daughter indoors, away from the din of the wildfire. The store was cold and dark, with concrete floors and black cabinets cluttered with tea sets and porcelain dolls. The store was not meant to hold so many people. Items kept getting elbowed off the shelves. *If only the owner knew what was happening in his shop right now,* Jamie thought.

They pushed their way toward the back of the store, through a doorway leading to the break room. It was packed with old DVDs and boxy floor fans. An antique Chico road sign—EAST 9TH—rested against the wall. Dust had gathered in the corners, and the air reeked of Pine-Sol. Tezzrah eyed a gold necklace, studded with jewels and locked inside a glass counter. While she wondered idly if it was the most expensive item for sale, Jamie grabbed water jugs from a cooler and examined some tightly rolled Persian carpets. If they needed to, they could douse themselves with water and crawl under the rugs for protection.

Hearing a familiar voice outside, Jamie lifted the metal delivery door on the loading dock to reveal Jill standing outside. She had nowhere else to go.

"I thought that was you!" Jamie said, smiling. They chatted for a half hour, surrounded by the gold jewelry and an onrushing wild-fire. Tezzrah ate a Lunchables snack from her backpack, and Jamie used the bathroom. He had almost begun to believe that this was a safe place after all—until firefighters shouted for everyone to move again. More than a hundred people stampeded out of the building. Jamie helped Tezzrah to the front of the store. In the parking lot, he saw that evacuees were crowding onto two passenger vans, ordered from Chico by first responders. Each was built to accommodate no more than sixteen people. As seats filled, people were forced to stand. With surprise, he saw his colleague the nurse scramble on board—the last one to enter before the doors swished shut.

Jamie and Jill looked at each other, trying to decide whether to follow on the vans' return trip. "I don't want to get on that bus," Jamie said. "I'd have more control if I was behind the wheel." "Then let's go back and get our cars," she replied.

The wave of fire hadn't yet overtaken the lanes of abandoned vehicles on the Skyway, though it was close. Embers whirled between the rows of parked cars as tiny patches of flame whooshed bigger. Jamie sprinted uphill, wheezing in the smoke, with Tezzrah and Jill about twenty-five yards behind. It dawned on him that he would have to carry Tezzrah. On a gravel clearance near Reliance Propane, Jill slowed down and called out to Jamie, suggesting he forge ahead without them. He'd get uphill faster that way. Then he could pick up his Subaru and swing back around to pick them up. "If you trust me enough, I'll stay here with Tezzrah. We're holding you up." Drunk on adrenaline, he agreed. "Whatever you do, do *not* leave her," he told Jill. "Even if they try to make you get on the bus, don't do it. I'll find you." Then, kneeling in front of his daughter once again, he handed Tezzrah her backpack and gave her a tight hug. As he turned to leave, he saw Jill wrap her arms around the seven-year-old and squish the girl's face into her soft abdomen to protect her eyes from the flying ash and debris. Jill, in turn, watched Jamie's figure grow fuzzy in the smoke until he vanished into the unknown.

Every few minutes, a firefighter would walk over to ask if they were okay. Jill would explain that they were waiting for someone. "You're okay here for now," one of the firefighters told her. "But I wouldn't stay much longer." He pointed to the propane factory across the street, giving her a meaningful look.

Folded into Jill's arms, Tezzrah couldn't see much. She perked up at the sound of each passing fire engine, asking if it was her father. Ten minutes passed, then twenty, then half an hour. Hot air swirled around them, and explosions ricocheted in the distance. Paradise was a hunting town, and ammunition supplies were discharging as they caught fire. Occasionally, a driver managed to extricate their vehicle from the abandoned gridlock and make their way downhill. "I hope my daddy is okay," Tezzrah said. "He is going to be just fine," Jill replied, hoping she was right.

A firefighter approached them for the final time. "Hey, the last round of vans is leaving in five minutes," he said. They couldn't wait much longer.

BACK ON CHLOE COURT, three propane tanks ruptured in the neighbor's yard, hissing like rattlesnakes and shooting flames ten feet high. On the opposite side of the yard, tires detonated beneath the burning ambulance. Her calls complete, Tammy felt flushed with purpose. She needed to take care of her patients. Looking up, she spotted Paradise Fire chief David Hawks's SUV parked in the cul-de-sac. She tumbled out of the ambulance and sprinted toward him, her fellow nurses in tow.

She wanted to yank open the passenger door of his vehicle and climb inside its soft interior, where it seemed safe. Hawks made it clear that that wasn't an option. "You have to defend the house," he said firmly. He was trying to speak while listening to the radio, as he had been doing for the last half hour. His tone was distant; he was about to leave Chloe Court to help his colleagues, who were trapped with more than 150 residents at the intersection of Pentz and Bille roads. His mind was elsewhere. Still, Tammy looked to Hawks for answers, as if he had any.

"That house is your only chance of survival," he continued. "I want you to rake up pine needles. I want you to spray down the roof with the garden hose. I want you to fill up buckets of water and fill the bathtub."

"What if the house catches on fire?" Tammy asked.

"Listen, I don't know what's going to happen," he said. "I can't tell you if we'll make it. This fire is crazy. But I do know that your only chance of surviving this fire is protecting that house, which will protect you."

"We are going to die here."

"You are not going to die here," he replied. "If the house catches fire, or you catch fire, find a rocky place and lay low. Roll around to put the flames out."

"I don't like the options you're giving me," Tammy said, smiling helplessly. The absurdity of the situation was almost funny.

"There aren't any good options," Hawks replied. "But you cannot just sit here."

Something inside Tammy clicked: There was no decision to make, she realized. Suddenly businesslike, Tammy turned away from Hawks's SUV and walked toward the house with her fellow nurses. She looked back. Hawks rolled through the stop sign and turned north on Pentz Road.

IN THE BACKYARD, Tammy shuddered as she grabbed another handful of pine needles, the bristles sharp against her palms. Her fingers smelled like air freshener. *At least they'll have a clean yard when they get home,* she thought. Nearby, her fellow nurse Crissy beat at the wooden patio with a broomstick, smothering the fiery wreckage as it landed. There weren't many tools in the garage: a three-pronged rake, a leaf blower, a few buckets. With the fire chief's departure, they had to figure things out on their own, opting to split the property into sections, with paramedic Mike Castro, twenty-eight, standing on the roof as lookout.

Born in Los Angeles, Castro had been raised by his grandparents on a 20-acre almond orchard in the town of Orland. He had decided to train as a paramedic after an off-duty medic had resuscitated his friend following a jet ski accident. (The friend later died from his injuries.) Now Castro tried to stay calm. The stucco house had green hoses hooked up to three different spigots, each yielding a trickle of water. He barked out orders when he noticed new flames flaring after a gust of wind. His colleagues sprayed the embers with water. Castro learned to sense the wildfire's movement. The stilling of the air was the most foreboding. It portended a wind shift, the high pressure inverting the low pressure, then the arrival of sparks from a completely new direction.

On the ground, his crew flipped lawn furniture, dog beds, and anything else even remotely flammable onto a concrete patio in the neighbor's backyard. Firebrands had struck the other paramedic in the face, freckling his cheeks with second-degree burns. The paramedic tried to pray, when he could remember to, and recalled with grim amusement his father's favorite maxim: "There are no atheists in foxholes."

A spark pirouetted into the flowerbed, and pediatrician David Russell stamped it out. He nearly collided with Tammy. "Hey! It's good to see you," he said. "Sorry it's under these conditions. But you're doing a great job." Russell was soaking wet, having sprayed himself with the hose so his khakis and soft-shell jacket wouldn't catch fire. Even his shoes were sopping. He gave Tammy a hug, dampening the front of her pink scrubs. He towered over her, lanky and impossibly thin, bordering on skeletal. "What are you doing here?" Tammy asked, incredulous. He had left Feather River hospital in his truck—not with the ambulances. He had tried to escape, he explained, but his vehicle had run out of gas. He pointed to its mangled frame, just behind the crippled ambulance. The thirty-four-year-old father of two had a third child on the way. His wife was thirty-nine weeks and four days pregnant. Where his family was, he wasn't sure. "Well, keep it up," he told Tammy, distracted.

She returned to the garage to comfort the patients, whispering to them that they were going to be all right. She held the hand of the brain bleed patient. The elderly woman seemed to be fading: Her face was limp and gray, her frizzy hair forming a halo around her head. The C-section mother tried to lighten the mood by cracking jokes, even as fear overwhelmed them. "We'll be out of here soon," Tammy promised. "My boyfriend is a police officer. Maybe he'll be able to get to us." She had left a voicemail on his phone earlier; he was resting after working the night shift and hadn't picked up.

While Tammy had been with Chief Hawks, the paramedics had scoured the house for supplies. The scene was eerie, as if the homeowners were going to return any second to finish putting away their groceries. A crossed-out shopping list lay askew on the counter. Stuffed animals were scattered across the carpet, and finger paintings were stuck to the fridge. One EMT had decided to fill the bathtub in the guest bedroom with cold water, in case the garden hoses lost pressure. Castro had different priorities. If they were going to spend the night here, he would need a drink. The paramedic selected a red wine blend from the well-stocked rack off the kitchen and set it aside. "Thank you so much for letting us in your house," he wrote on a scrap of paper. "We sought refuge here. If you find this, thank you. We love you. P.S. We stole a bottle of wine."

As Tammy soothed her patients in the garage, two men materialized out of nowhere, instantly recognizable as newcomers among the charcoal-faced ambulance crew. Unbeknownst to those on Chloe Court, Hawks had radioed for help before leaving—and members of the Butte County Search and Rescue, or SAR, team had heard the call. On their way over, the pair had seen a man—shirtless, shoeless—rocketing through traffic on a rusty bike. He seemed to be strung out on some kind of drug. He'd probably be the only one of them to make it out alive, the men had joked.

"Who are *you*?" the older of the two paramedics asked, puzzled by the duo's sudden appearance. "I'm with SAR," one of the newcomers responded. "Heard over the radio that you guys were trapped. Man, it sounds like war up here. I'm not in the military, but those propane tanks, and the ambulance tires, and the oxygen tanks in the back . . . they sound like a fighter jet on the Fourth of July."

The two men were stockbrokers in Chico. They volunteered for law enforcement in their free time, exchanging their tailored suits for green uniforms when needed. The program trained them to dangle beneath helicopters, swim across surging rivers, and wield a gun. When a disaster happened "off pavement," as they referred to unincorporated Butte County, they were sent in. It was only their second wildfire, but the two men busied themselves helping out. The water main near the front door of the house had burst, flooding the grass and depressurizing the hoses. The ambulance crew handed up buckets of water to the roof, which Castro used to drench the fuming shingles. By now, nearly a dozen people were seeking shelter in the cul-de-sac. A system had emerged out of the chaos, and everyone was feeling more confident.

Within the hour, Hawks returned in his SUV. "How are things looking?" Tammy asked, hopeful. "Not good," he replied. His face was knotted with worry. He had been trying to direct a helicopter to Chloe Court to rescue them. The aircraft only worked if it could beat the air into submission, though, and with the dangerous winds, the air would likely beat the helicopter down instead. At this point, Hawks said, the safest option was to leave. As he talked with the ambulance crew, a fire engine from Chico pulled up to save the house that had become a shelter, summoned by Hawks's earlier requests on the radio. The mood instantly lightened, the possibility of evacuation now seeming a surety.

Hawks instructed Tammy to secure the two frailest patients—the quadriplegic and the man on a ventilator—into the one remaining ambulance. She and the other two nurses would ride in SAR's Ford truck, along with the C-section mother and the brain bleed patient. But before they split up, the group paused to snap a selfie on Tammy's pink iPhone. In the photo, Chardonnay, the emergency room nurse, peers in from the left, her dark hair slicked back in a ponytail. Below her, a respirator shields all but Tammy's blue eyes. A man from the Search and Rescue team grins behind her, alongside two of the ambulance crew members and pediatrician Russell. The walls of the house are barely visible above their heads, the metal ladder they'd used to scale the roof crossing a corner of the frame. Their faces are relaxed, their smiles huge— a portrait of relief.

Tammy climbed inside the truck with her colleagues Chardonnay and Crissy, eager to drive downhill. The driver headed toward what Cal Fire's Emergency Command Center had announced was now the safest spot to shelter in Paradise. The flames had already steamrolled past, after all, and diligent landscaping work had made the area less likely to burn again. Crissy looked at Tammy, confused, as they pulled in to a familiar parking lot. "We are supposed to be at Enloe," Crissy said, looking around the gray campus of Feather River hospital. "I don't want to be here. I don't want to be here!"

OBSERVATION: TEN THOUSAND FEET
ABOVE THE FIRE

Nearly ninety miles south of Paradise, in the foothill town
of Auburn, a California Highway Patrol plane rose into the
air. Brent Sallis, thirty-eight, piloted the Australian-made
GippsAero GA8 Airvan, while Joe Airoso, thirty-six, peered
out the window. It was a clear morning, a touch windy, but
nothing exceptional. Even their radio was quiet, unlike most
days. Sacramento Police had requested their help tracking a
murder suspect as he zigzagged across the state capital. They
planned to hover above the grid of suburban homes, relaying
the man's whereabouts to officers on the ground. But as they
lifted off from the municipal airport and headed south, an
impenetrable fog seeped over the horizon behind them.
From forty miles away, the smoke plume punched through
the fall sky. Soon Sallis and Airoso were redirected to Paradise to assist with a massive new wildfire.

They whirred north, taking a half hour to reach the
Ridge. Below them, Highway 70 carved through the valley
and flames hiked up the buttes. As they neared Paradise, the
smoke condensed, thick as milk, as it scudded along the
mountaintops. They snapped on oxygen masks to breathe,
then climbed to 13,000 feet—twice as high as they normally
flew—to avoid the turbulent updrafts and minimize the ash

filtering into the cabin. To Airoso, it looked like a bomb had gone off. "There was this huge column of smoke that was not moving, like a solid, physical object in the sky," he would later recall. "It had dimension, black at the base and gray and white at the top. It smelled like everything was burning. . . . It looked like an apocalypse."

Their eight-passenger aircraft was used for aerial intelligence. Faster than a helicopter and heavier than an air tanker, the airvan contained heat-sensing technology that was invaluable to firefighters on the ground. A thermal camera hooked to the aircraft's belly could cut through the dark strata of smoke, offering a God's-eye view of the terrain below. The camera worked by detecting the heat signature of the landscape. Radiation registered as a gray-scale gradient on Airoso's computer monitor. The hotter the object, the paler it was.

Sallis steered the aircraft north. Seated behind him, Airoso toggled the camera's joystick and studied the images flickering on his screen. Blowing embers looked like pinprick stars in the night sky, twinkling almost merrily. The computer program superimposed a map on top of the camera footage, so Airoso could see the name of every street in Paradise. His radio was set to Scan, switching between the frequencies of Cal Fire and local law enforcement agencies like Chico Police and the Butte County Sheriff's Office.

But where is the fire? Airoso wondered.

On the screen, he saw traffic obstructing the four escape routes out of Paradise. The fire was everywhere—in canyons, along roadsides. The town was a maze, and in their panic no one could find the exit. On Clark Road, the wildfire had blackened the hillsides down to Highway 70 near Oroville, leading officers to hold back traffic needlessly. They surmised that the blaze must be below them, when in reality they needed to send the vehicles in exactly that direction, over the safe burnt-out land.

Meanwhile, thousands of people blocked the Skyway as a wave of white heat pushed toward them. Some of the cars on the northern stretch had already caught fire, kernels of white-hot silver on Airoso's screen. Nobody was getting around the molten barrier. Thousands of feet below him, he could hear his 94 Highway Patrol colleagues communicating over the radio, their voices tight with desperation. The biggest problem, Airoso saw, was Highway 99. Three special response teams had been assigned to shut down every exit into Chico, but they hadn't yet released vehicles onto the northbound side of the highway. They were worried that forcing traffic south to Oroville in both directions would cause a fatal collision—they didn't realize no one was driving into Paradise at this point.

Airoso talked with the Highway Patrol's dispatch center, trying to guide field commanders. They were distracted by the radio traffic. Airoso tried to raise his voice above the babble, but he was barely audible over the other conversations. At last, Highway Patrol division chief Brent Newman, fifty-two, who oversaw ten counties and four major highways across Northern California, overheard Airoso's directions. He knew his high-ranking status would make everyone pause what they were doing and actually listen. "Just call it!" he bellowed to Airoso over the radio.

The line fell silent.

Airoso spoke into the sudden hush. "Let the lanes go on Highway 99. And send more people down Clark or Pentz."

Below, as the flaming front advanced, a police sergeant at the back of the line on the Skyway was uttering his last prayer. Suddenly he saw the brake lights in front of him fade as traffic lurched forward. Gridlock broke and thousands of vehicles began flowing out of Paradise.

"Chief Newman," Airoso said over the radio, watching the movement below with relief, "mission complete, sir."

THE FIRE: PARTICLES OF POISON

By early afternoon, the wildfire had taken out the southern half of Magalia. It blasted the town sign standing on the roadside—MAGALIA, POPULATION 11,500—bubbling its green surface and leaving the text illegible. The blaze circled the historic one-room Community Church and scorched the three crosses staked out front. Along an elbow stretch of the upper Skyway, the waxy cones from McNab cypress trees—adapted to high fire risk climates—burst from the heat, their seeds freed after decades deprived of fire. The local Subway, the Dolly-"O" Donuts, the hardware store, all leveled. Flames skipped over the town's only grocery store, Sav-Mor, and Jaki's Hilltop Café. The wind blew ambient heat down residential streets, drying the homes so drastically that they ignited under the assault and coughed toxic smoke. Already, 20,000 acres across the Ridge had burned.

More than 3.4 million metric tons of carbon dioxide and noxious gas puffed into the atmosphere, the equivalent of all the pollution produced by the state's factories and traffic in a week. An invisible siege of water vapor, carbon monoxide, and carbon dioxide dispersed in the pall, mixing with the toxins from cleaning solutions, insulation, plywood, carpet, furniture, electronics, rubber toys. The fine particles measured less than 2.5 microns—about one-fifth the size of a particle of dust—and lodged themselves inside every living creature, causing irritation with each rasping breath.

The soot rolling off the wildfire poisoned people from the inside out, burning eyes, inflaming throats, and stinging lungs. The carcinogenic chemicals crossed into the bloodstream and stressed the heart. Scientists had only recently begun to study the long-term health impacts of wildfires, running tests on housecats that had been trapped in fire zones. They had found that cardiac arrest for humans was as much as 70 percent more likely to occur on smoky days. And the amount of wildfire smoke in California was only expected to double in the next century, along with the number of deaths due to chronic wildfire smoke exposure. In the United States, smoke claimed twenty thousand victims annually.

But the fire was undeterred. Roiling and fuming and seething, the blaze clawed north.

CHAPTER 14

PARADISE ABLAZE

The brake lights ahead of Kevin dimmed, and traffic moved forward. Roe Road, with its drooping oak and pine boughs and tangled brush, lay ahead. Glancing in the rearview mirror, he saw the two teachers huddled together in a single seat. He didn't like seeing them so upset. "All right, girls, we've got a job to do!" he hollered.

Mary, her prayer finished, darted forward and crouched by his seat. Her eyes were bloodshot from the six hours she had spent in the smoke. Even inside the bus, the air was hot and ashen. Earlier, Abbie had told Mary that she was worried they would lose a kindergartner to smoke inhalation or carbon monoxide poisoning. Maybe Kevin would have an answer. "Hey, Kevin," Mary said, "the kids are starting to pass out. They can't breathe. What should we do?"

Kevin slithered out of his polo and yanked his undershirt over his head, tossing it at Mary. She held the undershirt gingerly. "Tear it into twenty-five squares, so we each have one," Kevin explained, pulling the black polo back over his belly. He put his foot on the brake and showed her how to rip the cotton undershirt into rags. "We'll douse it with some water, and the kids can use them as masks."

Mary shredded the thin shirt and handed the cloths to Abbie, who pulled a half-full plastic bottle from her purse—the only water on the bus. She dampened the masks before dispensing them to the twenty-two children and instructing them to hold the fabric over their mouths. "I want you guys to suck on the rag a little bit," said Mary. "It'll make your throats feel better. But I have to warn you, I don't know when bus driver Kevin last washed his undershirt!" The students giggled.

Abbie walked down the aisle with the last dregs of water, giving each child a sip. She would have loved some herself: Her chest ached, her head swam, and each intake of breath scratched her throat like gravel. But she had to save as much as she could. Someone might need it more.

Mary took over to give Abbie a second to rest. As she walked down the bus aisle with the bottle, she tripped, spilling the precious water. "Kevin, I need to get off this bus to get more water," she said as she returned to the driver's seat.

"I'm not letting you off. It's way too dangerous out there," he said, eyeing her in the rearview mirror.

"I don't care, Kevin," Mary replied. "We need water."

He knew she was right. He opened the door with a huff, and she descended into the darkness.

Feeling her way along Roe Road, Mary bumped into a tall, rangy figure. He was a young man in his twenties who had abandoned his car to see why traffic wasn't moving. Inked with tattoos, he appeared to be part of the populace that many locals had taken to calling the New Ridge: harder-looking adults who lived in trailer parks on the outskirts of town and were known to struggle with addiction.

"Do you have any water?" Mary asked, uncertain how he would react. "I have twenty-two kids on a bus, and we need it badly."

His smile was bright in the gloom. "Let me check my car," he said, cutting back through traffic.

A few minutes later, he reappeared with two plastic bottles. They were crunched and nearly empty, as if they had been rolling around his floorboards for a year, but they contained some water. His generosity felt staggering. "Thank you," Mary said, squeezing his arm.

She returned to the bus, climbing through the door to Kevin's whoops and cheers. It hadn't gone far: Every ten minutes or so, the bus would shudder forward an inch, if that. They had been trapped on Roe Road for more than an hour, though they were only half a mile from merging onto Neal Road, which promised a direct route to Oroville—and safety.

Mary and Abbie were resoaking the rags when Kevin noticed an older man hosing the shingles on his mobile home. He asked Mary to hop off the bus to fill up their three water bottles one more time. Mary complied, feet crunching across the desiccated grass. Reaching the man, she held out the three bottles and asked if he would fill them. "Of course," he replied. She returned to the bus, doling out the water, then rushed back for one last refill. "How many kids do you have?" he asked. She told him and he ducked inside his home, returning with a half case of water bottles. He handed the flat to Mary without a word.

"If the bus catches fire, can we come huddle with you?" Mary asked.

"Sure," the man said, splashing more water onto his roof.

"Mary," Kevin yelled to her. "Get back on board!" The bus was creeping forward, and though they weren't going that far, he didn't want her out of sight. Mary sprinted back. They were careful not to spill any of the precious liquid again. Mary took the rows behind Kevin; Abbie took the other side. They drizzled more water into the children's tiny mouths, which opened and closed like goldfish's. Their lips were chapped from the smoke, and their faces were pink with exertion. The students did whatever was asked of them without complaint, though they were exhausted. The fire outside had heated the metal bus up like a pizza oven. Mary estimated that it had to be at least 100 degrees. The children were sweating through their clothing. One little boy had undone the pearl buttons on his flannel shirt, exposing his pale, bare chest.

A few rows back, ten-year-old Rowan Stovall couldn't tear her eyes away from the bus window. The days of fishing for bluegill at the Aquatic Park, baiting them with dandelions, seemed like a thing of the past. She feared there would be no more collecting crystals or skipping rocks at Paradise Lake with her mother, no more karate or horseback riding lessons. Rowan was tough. She never cried when she skinned her knee or bit her lip—she had been raised with male cousins—but she couldn't hide her emotions if someone hurt her feelings. She loved animals with a tenderness that her mother found humbling: She doted on their three cats and tracked the speckled fawns that munched on their lawn in the evening. Now, as Rowan stared out at the burning forest, she saw a deer trapped by a burning log. Its spotted body stumbled forward, then slumped to the ground, overtaken by flames.

ACROSS TOWN, snared on narrow Oak Way, Cal Fire captain Sean Norman turned onto the town's paved bike path, which paralleled the Skyway. Other drivers were making the same decision. The Paradise Memorial Trail was carved into a five-mile strip of donated land. The route had once been used as a rail line by the Diamond Match Company, transporting timber to the valley floor until the Southern Pacific Railroad halted operations in 1974. One of the few remaining depots was located in Paradise Community Park.

In September, the Gold Nugget Museum board had gathered at the historic site for a ribbon cutting ceremony, celebrating the conclusion of a $1.3 million trail renovation. A length of the trail had been renamed in honor of Luther Sage "Yellowstone" Kelly, an early Paradise settler and the subject of a 1959 film starring Clint Walker. The museum had installed thirty-one plaques on the trail to commemorate other noteworthy pioneers like him. At the ceremony, the audience had been filled with the smiling faces of families with young children. Now those same families were waiting in roasting cars and the trail had become an impromptu evacuation route. Norman followed the wide path until he hit a parking lot, merging back onto the Skyway near the Walgreens.

The area Norman was responsible for was mobbed. He needed to get people moving downhill. He drove northeast for a half mile, checking on the intersection of the Skyway and Wagstaff Road, where officers had told people like Jamie and Jill to abandon their vehicles. This decision, Norman realized, had trapped everyone to the north behind a blockade of cars. The gas stations on each corner were overrun with people. The wildfire tossed hot embers down their shirts, and the skin on their arms was red and swollen— the telltale sign of a burn injury. One woman patted her curled blond hair, which smoked. Dark clouds gusted overhead. Norman retreated to the Walgreens, past a car that had caught fire with its driver still inside. The blaze had torched the houses behind the drugstore. The dumpster, stuffed with cardboard boxes and wooden shipping pallets, was an inferno.

At Walgreens, the engine stationed in the parking lot was running dry. Norman sent some firefighters inside to rip extinguishers from the store walls. "What do we do?" asked a local cop. "Just keep sending people in," Norman replied. "Do not let this building catch on fire." About a hundred people were now sheltered inside. He recognized a pair of buggies surrounded by men dressed in orange instead of the standard-issue firefighter yellow: two teams of prisoner hand crews that had managed to beat their way uphill against traffic. This crew was from the Washington Ridge Conservation Camp in Nevada City, near Norman's hometown.

Minimum-security inmates accounted for as much as 40 percent of California's firefighting force, saving taxpayers about $100 million annually. The inmates were paid $2.00 daily, plus $2.00 hourly when they were dispatched to a wildfire—a fortune compared with the wages for jobs inside state prisons, which paid 8 to 37 cents per hour. And for each day in the program, they received a two-day reduction in their sentences. Though critics said that the program essentially amounted to slave labor, the prisoners were an invaluable source of manpower, with 218 crews in 29 counties across the state. One Cal Fire division chief likened them to the Marines of the fire service.

The bands of prisoner firefighters were led by two captains whom Norman knew well. He often went kayaking and whitewater rafting with them. They directed the crews to start clearing brush and digging firebreaks, ringing the Walgreens with a bare patch of dirt. As soon as the roads opened up, one of the captains told Norman, they planned to transport civilians downhill in the crew buggies.

Norman nodded. He could help by opening the road for them. Checking first for any straggling civilians, he peeled out onto the Skyway in his SUV and began to ram the deserted cars near Walgreens, trying to push them off the pavement with brute force. Metal struck metal with a grating keen. If his airbag deployed, it could give him a bloody nose, knock out his front teeth, or worse. Rethinking the decision, Norman called out to an engine from another county. The command structure was now in disarray, and a branch or division supervisor like Norman could step in with a new assignment for a crew at any moment. "Start ramming cars off the road," Norman yelled at the engineer. The man looked at him, perplexed, as he gripped the steering wheel. This wasn't in his job description. "Do it!" Norman shouted.

As he gave orders, the intersection swelled with a fresh wave of people. A school bus ferrying a group of evacuees from Paradise Alliance Church had crashed into a ditch, he learned. Some of the passengers had medical issues, and they begged Norman for help. He sorted them into strangers' cars. Speaking into his radio, he pleaded with the Emergency Command Center in Oroville for a bulldozer or some other piece of heavy machinery to smash through the traffic. The out-of-county engine had managed to open one lane on the Skyway, but it wasn't enough. "I need an additional bus to the Walgreens," Norman said into his radio. "We are sheltering people as best we can."

He directed a thin stream of vehicles forward through the open lane, parceling people off into backseats, one by one. A few unruly drivers wouldn't stop honking. "I have a fire-resistant suit," one man said, pulling up next to Norman. "So I'm gonna go ahead."

"Okay, great. I don't care. I need you to put this lady in your truck."

"I don't even know her!" the driver protested. His dog snarled in the back of the pickup.

Norman leaned over and met the man's eyes. "I don't give a shit.

She's a human being. I need you to put her in your truck." The man acquiesced, grumbling.

As the final passengers were loaded into the crew buggies, Norman followed the out-of-county engine uphill to make sure the rest of his division was safe. He knew there were people to the north who needed his help. By this point, only first responders were heading uphill; everyone else was being sent to safety on the valley floor. The engine crew planned to douse the dozens of vehicles that had caught fire on the Skyway. They pushed past Wagstaff Road, where Jamie and Tezzrah had been sheltering at Needful Things, and punched through the gridlock as Norman had ordered. At the tip of the Skyway, the thoroughfare connected to Clark Road before climbing to Magalia. The Y-shaped intersection was flanked by a strip mall. Norman noticed that about 150 people were sheltering in the mall's bowl-shaped parking lot with his colleague, Cal Fire engineer Calin Moldovan.

A Romanian refugee, Moldovan, thirty-four, had fled his then Communist home country as an eleven-year-old. His parents sought political asylum in the United States, settling with the boy in the Sacramento suburbs. For years, Moldovan had worked in construction, importing stone from Brazil and Italy and installing countertops and fireplaces in custom-built homes, before fulfilling his lifelong dream of becoming a firefighter. He had taken a major pay cut, but it had been worth it: Moldovan wanted to help people, as others had helped him. As a child, he had once seen a straw-thatched home in his village catch fire. Neighbors helped push furniture out the window and dump buckets of water on the blaze. "It was a very third-world-country way of tackling a fire," Moldovan recalled. "I remember the chaos, and how everyone helped. Something inside of me was drawn to it." Now Moldovan worked out of Station 33 in Magalia—which, according to reports, had just caught fire. He had tried to reach it, but with flames to the north and south, his route had been blocked.

Moldovan was forced to stay in place and direct reluctant drivers into the parking lot for their own safety. Some of them now hunched against a cluster of buildings. In the strip mall, there were two metal-roofed structures under construction, a coffee shop and Optimo Lounge, a nightspot known for its midpriced Chinese food and live tribute bands. On holidays, the restaurant served patrons free chicken chow mein until midnight to offset the liquor. To the south was Fins, Fur & Feather Sports, a fishing and hunting store that sold ammunition. Across the street, on the corner of Clark Road, was a Fastrip gas station. To the north was a gigantic propane yard. Looking around, Moldovan realized that the intersection was ringed by highly flammable businesses. If any of them caught fire, they could kill everyone in the parking lot.

Kneeling on the concrete, people prayed. One woman was lying on the ground in the fetal position, sobbing inconsolably. She watched as her home across the Skyway burned down, the walls falling inward as the roof collapsed. "Why aren't you stopping it!" she moaned, staring daggers in Moldovan's direction. "Why aren't you doing anything?" Another woman was going into diabetic shock; a Paradise Police officer found some shortbread cookies for her. A county employee flashed her badge, demanding that Moldovan open up the Skyway.

"Why aren't you getting us out?" a man added.

"I can open the road, but you won't make it very far," Moldovan said, thinking darkly of the blockade of cars downhill near the Walgreens.

He'd just heard on the radio that one lane of traffic had been cleared by a fire engine, though, and hoped to send ten cars toward it at a time. But Captain Norman's arrival put an end to Moldovan's plan. The sides of Norman's vehicle were warped from the heat—proof of the peril that lay between the parking lot and the open lane farther downhill.

Regardless, Moldovan was thrilled to see the captain. Norman had a higher rank and more experience, which meant Moldovan was no longer in charge. Norman instructed Moldovan to force his way into one of the locked buildings so they could shelter the 150 people inside. "We'll break every door if we have to," Norman said. The concrete parking lot would act as a moat, cleared of vegetation, debris, and any other flammable materials. That, along with the new shopping center's adherence to updated building codes, might be their salvation.

The legislature had indicated as much in the early 2000s, noting that certain structural features, like wood shake roofs and unscreened attic vents, made houses more vulnerable to wildfires. In 2008, California had become the first state in the country to pass building codes with the wildland-urban interface in mind, mandating fire-resistant construction in new houses. But it wasn't an instant fix— the rules didn't apply to existing homes. In places like Paradise, where nine out of ten houses had been built before 1990, the change had little effect. Residents couldn't afford to retrofit their homes—simply replacing single-pane windows in a house could cost upwards of $10,000—and the state offered no incentive to do so. On average, the homes in Paradise were three decades old. Only about 350 houses in the previous ten years had been built to the new standards. (Of them, 51 percent would survive the Camp Fire, compared to 18 percent of the 12,100 homes constructed before then.)

Defensible space might have protected some of these neighborhoods—but not everyone cleared the mandatory 100 feet of land around their homes, as Travis did so zealously at his bunga-low on Edgewood Lane. Some retired or disabled residents weren't physically capable of the work, nor could they spare the money for a landscaper. Other folks simply couldn't be bothered. Cal Fire mailed warnings to residences in violation, but the agency rarely issued fines; there was little muscle to enforce them. (From 2009 to 2018, the Butte County Cal Fire unit inspected 29,776 properties. It issued one citation.) In Paradise, a legal mandate didn't even exist. When the Town Council adopted a new nuisance abatement ordi-nance in 2011, the five members had inadvertently eliminated the requirement for defensible space. They hadn't read the document closely enough to realize that someone had neglected to copy and paste the language into the new legislation. (Nine years later, as he updated citation forms in 2020, Messina would be the first to no-tice that the ordinance had disappeared from the municipal code and would rectify the omission.)

Now, as Norman and Moldovan shattered the glass doors of the Optimo Lounge with firefighting tools meant for digging fire-breaks, Norman wished there was even more defensible space around the Y-shaped intersection. He glanced up. More than a dozen people were recording him on their cellphones. *What a weird impulse,* he thought, wondering if this was how his wife and children would witness his final moments. He herded people inside the empty building, as if they were boarding an airplane or a lifeboat: the elderly and infirm first, followed by families with children.

Norman sighed. He was exhausted, his body as weary as if he had been awake for twenty-four hours. On the Skyway, he saw officers unloading a semi truck of Pepsi, preparing to use the truck as a makeshift shield in case the flames slunk closer. Above their heads, a helicopter shuddered in the wind, its Bambi Bucket of water nearly scraping the tops of the ponderosa pines as it headed for a nearby gas station. With a whoosh, the pilot completed his drop—but in the thick smoke, he missed his target. Not realizing the error, everyone cheered.

KEVIN CRANKED OPEN the bus door. "Do you need a ride?" he hollered at a young woman, who was looking lost on the side of Roe Road. The twenty-year-old preschool teacher gratefully boarded Bus 963. Her car had run out of gas a few blocks back, she said, and she no longer had a way out of town. She slid into a seat in the back, passing rows of children who had gone silent, uninterested in the presence of a stranger. They were too worn out to care.

The intersection with Neal Road neared. As the car in front of him turned, Kevin finally got a glimpse of the crossroads. Vehicles were crammed into every lane. Panicked drivers wouldn't let the bus merge. Everything ahead of them was ablaze: houses, trees, shrubs. An inferno of vegetation. If Kevin didn't kick the bus into gear, they were going to get caught too. In the back of the bus, Mary clutched her inhaler. Abbie closed her eyes and thought of her fiancé. Kevin gritted his teeth and pictured the twenty-two children running for their lives, scattering into the forest in every direction.

Just then, a truck cut around the bus and blocked a lane of traffic on Neal Road. Kevin accelerated into the space and made a wide turn onto the evacuation route. The truck belonged to the Ponderosa Elementary principal, who had been tailing the bus for miles to make sure they got to safety. Kevin swung Bus 963 onto the road and hit the gas. "We're moving!" he exclaimed, incredulous. It didn't matter what Kevin's friends said: Being a lowly bus driver meant something to him. On this morning, Kevin felt it. As he rolled downhill, the black sky lightened slightly.

Abbie turned to look out the window. They were passing a familiar property on Fawnridge Court—the home of her future in-laws, where she had enjoyed many holiday meals and Sunday dinners. She spotted her fiancé's truck parked in the driveway, and his father standing alongside it in a reflective yellow vest. Neither of the men was budging until they knew Abbie was safe. She had argued with Matt about it on the phone earlier, begging her fiancé to leave. "Nope, not doing it," he had replied. To see him now felt like the greatest gift. Abbie waved at him, awash in emotion.

Kevin laid on his horn, cracking the driver's window and yelling at Abbie's fiancé and future father-in-law: "Let's go! Let's go! Let's go!"

OBSERVATION: THE MAN IN THE TRASH TRUCK

Dane Ray Cummings covered Waste Management's route through Magalia and Stirling City. Every Thursday for the past eight years, the fifty-year-old had stopped at 328 homes, completing the route twice. First for the trash—driving it down to the Neal Road Recycling and Waste Facility, which everyone in town called Mount Neal—and then for the yard waste and recycling. On this morning, Dane had only 110 houses left on his route when his boss called. "Get the hell out," his boss said, his voice crackling on the phone. The wildfire had hit Magalia. But two days earlier, Dane had received a brand-new truck that drove like a Cadillac and, in its newness, seemed fairly fireproof. He wanted to finish the route to check on his customers.

Dane was proud to be a garbageman. The gig paid good money, around $70,000 a year—enough to support his two children, ages five and fourteen, and pay alimony. It was such a great job, in fact, that Dane celebrated his hire date— August 8, 2010—along with his parents' and kids' birthdays each year. The Waste Management customers adored Dane. At Christmas, they left batches of foil-wrapped coconut cream balls and peanut brittle atop their garbage bins. Sometimes the residents waited for his truck to show up so they could shake Dane's hand and chat as he loaded their garbage. They knew all about his divorce, which had ended with Dane getting full custody of his children—though as a single parent, he didn't always have a good handle on them. He cared enough to know all about his customers' lives too. Now he saw it as his duty to make sure they had all safely evacuated.

Dane knew that power had gone out on the Ridge around 10 A.M. If residents couldn't manage to open their garage doors, some of them wouldn't be able to leave. Margaret Newsum was facing that exact problem, as he discovered after checking about fifty houses. She was waiting in the doorway of her yellow home, wearing a pink T-shirt and leaning on her walker. Margaret was ninety-three years old but "never missed a lick," as Dane put it. She and her husband, Buck, had bought their home on a cul-de-sac in Magalia in 1957. They had been married for forty-five years when he died of lung cancer in 1990. Margaret was lonely afterward but stayed involved with her church and visited the salon twice a week to get her hair set. She had never eaten much sodium and had stopped driving at ninety. She attributed her longevity to these factors. "You make your life what you want it to be," she liked to say.

"What the heck are you doing here?" Dane asked her.

"My caretaker went down to Chico this morning, and she can't get back up here." She had tried to call others for help, to no avail.

"Do you have any family?" he asked.

When she said no, Dane helped collect her three prescriptions and her walker. Then he threw Margaret over his shoulder like a sack of potatoes and, huffing and puffing, carried her back toward the trash truck. Two young men spotted them and ran onto Margaret's front lawn to help, lifting her into the front seat and buckling her in place.

"It's against your company rules to have a passenger!" Margaret protested, not wanting to get Dane in trouble.

"I don't care," he replied. "I'm not going to leave you here to die."

Before climbing into the driver's seat, he hung her walker off the back of the trash truck.

CHAPTER 15

PROMISE

Stumbling up the lava cap, his friends motionless behind him, Travis paused to unlock his cellphone and call 911. There was just one problem—they were in the middle of nowhere. Travis knew he wouldn't be able to describe their location to a dispatcher. And with no streets nearby, Cal Fire wouldn't be able to send a crew. Travis needed to backtrack to find help and guide them to Paul and Suzie, but he also didn't want to leave his friends behind. Taking one last look at them, blurred smudges in the darkness, he felt the weight of his promise to Suzie. He started the engine of his four-wheeler.

Travis drove through coils of smoke toward his house. The quad juddered over cracked boulders and fossilized branches. The wooden bridge over the unnamed creek had turned to ash, so he splashed through the water, mud sucking at his tires. As he headed uphill, Travis realized with a jolt that his house was somehow still standing. He could see its outline through the smoke, growing sharper as he wheeled closer. His neighbors Mike and Jeanette were dousing flames in the gutters and ripping off chunks of siding. Embers had burrowed inside one wall and ignited the insulation. "They're going to be so mad you're ruining their house!" Jeanette shouted at her husband. But they needed to destroy the side of the house to save it—and themselves.

For the past hour, they had hunched behind the southern edge of Travis's home, using the structure as a windbreak. They had planned to submerge themselves in a creek, but the fire had moved too fast. It was just as well—the stream water was close to boiling. From behind Travis's house, they saw the wind blast through like a napalm explosion, cutting through the brush and shattering the windows. "We would've been wiped off the face of the earth if we had gone to the creek," Mike now told Travis. Their house had been destroyed, he said, so they were trying to save his. Travis explained that Paul and Suzie had been badly burned. He called 911 with Mike by his side.

"Have the units go down South Libby Road as far as they can," Mike told Travis. The dirt road ran parallel to Edgewood Lane, and it would help the crew get as close to the couple as possible. As a contract firefighter, Mike spoke the language of firefighters, and the dispatcher listened. "I'll be waiting on my four-wheeler," Travis told her, then stepped back onto his quad. He was anxious to return to his friends; he couldn't imagine how terrified they must be, crippled and abandoned as fire burned around them. Mike turned up the volume on his radio, straining to listen to Cal Fire's frequency. A team was being dispatched to South Libby Road.

As Travis left to intercept the crew, he crossed paths with a man driving from the other side of Edgewood, near where the street ended. The fenders of his green Jeep Cherokee were melting, and the bumper had liquefied. His clothes were wet, pinpricked with holes where burning embers had struck him. He rolled down his window. "Whatever you do—do not go down—to the end of the road," he stammered to Travis. He and his friends had tried to escape, he explained, but they had been "trapped like rats." His speech came in disjointed bursts. He had survived only by abandoning his Jeep and following a fox down the banks of Dry Creek, submerging himself in the shallow water. When he had returned to his car forty-five minutes later, the worst of the firestorm was over. His Jeep was still running, and his chihuahuas, Romey and Jules, were alive in the backseat. His friends hadn't been so lucky.

Travis drove about 200 feet down Edgewood Lane to see for

himself. The man must have been exaggerating, he thought. Perhaps people were being rescued now and he could flag down some first responders for Paul and Suzie. Up ahead was a jumble of cars, burned to their frames and submerged in a sea of molten aluminum. Their tires had evaporated, leaving the rims flush against the gravel. Even the door handles had burned off. Travis didn't see any firefighters—or passengers. He slowed down, peeking in the blasted-out windows.

Upright in the seats, staring blankly ahead, were five skeletons. At 1,500 degrees Fahrenheit, they had been cremated in their vehicles—the Camp Fire's first fatalities. Another skeleton was crumpled facedown on the ground near an eviscerated Winnebago.

Travis recoiled in horror, bile rising in the back of his throat. He whipped his four-wheeler around and sped toward South Libby Road to wait for the firefighters, thinking only of Paul and Suzie.

AT FEATHER RIVER HOSPITAL, the early afternoon had brought a sense of peace. The wildfire clawed west across town, leaving Pentz Road blackened—but open. Hospital officials urged people to leave. They didn't want patients trapped in Paradise overnight, and they worried that the other buildings might still catch fire. The emergency department was smoldering. More than six hours after they had first loaded patients into cars, first responders again organized a line of evacuees into sheriff's vans and staff cars, preparing to send everyone downhill at last.

A sheriff's deputy slid a blanket-covered gurney into the back of one van. The body underneath was cool and lifeless, tagged for a mortuary in Chico. The eighty-six-year-old woman with the brain bleed, who had sheltered in the garage on Chloe Court, had died on a backboard on the helipad. Elinor "Jeanne" Williams had lived with the love of her life, Robert, in a rental house on Pentz Road. The walls of their home were papered with so many family photos that in recent years her eleven great-grandchildren had sent only wallet-sized pictures, since they knew she didn't have space for anything bigger. Jeanne had raised three daughters and worked as a housekeeper at Feather River hospital for nearly three decades. She had lived a long and full life before falling and hitting her head while using the bathroom, sustaining the brain injury that had sent her to the hospital. Jeanne's family wouldn't learn she had died until the following day, when they tracked down her body at a morgue in Chico. Jeanne would never hang another family photo on the wall, would never again tease her husband about his bad hearing, would never again visit her granddaughter in Oregon.

In the middle of the firestorm, a doctor and two nurses had taken the time to pray over Jeanne as morphine dripped through an IV. "Concentrate on your breathing," the surgical unit manager had coached Jeanne, resting his warm hands on her papery skin. He and his two colleagues stayed until her breathing stilled. It was a beautiful and awful way to bear witness.

Across the parking lot, David climbed back into his white sedan with Rachelle. They turned onto Pentz Road—one of the few opened routes—and raced downhill toward Chico. He was so excited to be moving that he nearly rammed a downed electrical pole. They passed silenced emergency vehicles, half-expecting the officers to direct them back to the hospital—but no one did. Paradise was unrecognizable, block after block leveled by flames. They drove over Pearson Road to the west, where two patrol vehicles—one from the Highway Patrol, the other from the Sheriff's Office—had crashed into the gully known as Dead Man's Hole and caught on fire. Rachelle thought it looked like something out of the Old Testament.

Pentz Road traced the town's eastern boundary, careening along the Feather River Canyon and past Kunkle Reservoir until, eight miles downhill, it bottomed out on the valley floor and connected to Highway 99. There David and Rachelle met a roadblock: Northbound lanes on the highway were closed. The officer stationed at the barricade wouldn't let them turn toward Chico. David explained that Rachelle needed to be taken to the Neonatal Intensive Care Unit at Enloe Medical Center—not the smaller facility in Oroville, which wasn't equipped for a newborn who had spent his first day of life breathing ash. The officer wouldn't budge. Rachelle grew hysterical. Desperate for her husband, she called Chris one more time—and finally got hold of him. After hours of trying to get in touch, their conversation was more practical than anything else. Chris had grown up in the county and knew every gravel road. He told Rachelle to meet him in the rice fields of Richvale, west of Oroville. They would find a way to Chico.

As David and Rachelle wended through the tilled farmland, a dozen other evacuees from the hospital were arriving in Oroville in a large sheriff's van. Tammy sat in the back, holding the hand of a nursing home patient. When the doors swung open, she saw familiar faces waiting in the parking lot. Her children, all five of them, had made it out of Paradise. Tammy sobbed as she ran toward her brood, thankful for this second chance. Her five children wrapped their arms around her. Her mother had already arrived in Chico, having driven down from Oregon to be with her daughter. Soon Tammy would reunite with her.

Rachelle, too, felt a surge of emotion as she and David parked behind her husband's white Suburban. Chris climbed out of their SUV, shook David's hand, and thanked him for being their angel. He and Rachelle didn't yet hug—Chris wanted to get her to the hospital. "Follow me, and I'll get us to Enloe," Chris said. Rachelle stayed in David's car, since he was better equipped to handle a medical emergency if something were to go wrong. But nothing did. As they arrived in Chico less than an hour later, parking near the leafy side entrance to the hospital, Rachelle finally allowed herself to breathe. They were safe. She sobbed as she held her flimsy pink gown together, black snot dripping from her nose.

It was the busiest day in the emergency room's 105-year history as Rachelle and forty-nine fellow patients from Feather River hospital were being rushed indoors. A nurse tucked Lincoln in the crook of her arm and settled Rachelle in a wheelchair. They had been expecting her. David waved goodbye as she was rolled indoors.

Rachelle never learned his last name. And never saw him again.

. . .

JAMIE PUFFED UPHILL, each step taking him farther from his daughter. He was lost and disoriented in the eddying heat. But he found his car within the grid of abandoned vehicles by listening for the reassuring chirp of its unlocking doors. He squeezed inside. He was blocked in nearly every direction. The blood-red wall of flames on the horizon blasted closer. Jamie jerked the Subaru off the Sky-way and into a ditch, barreling up the dirt slope. He edged around the yellow school bus that had gotten stuck after departing Paradise Alliance Church. Ahead was empty road.

As he prepared to hit the gas, a Cal Fire captain stopped him, pointing to a ninety-one-year-old woman in a wheelchair. She was being pushed down the mountain by her caretaker.

"Do you have room?" he asked. "They need a ride."

"Of course," Jamie replied.

He helped load the two women inside the Subaru, folding the wheelchair and stashing it in the hatchback. Now there was noth-ing left to stop him. He barreled down the road, screeching to a halt in front of Jill and Tezzrah. Jill, holding Tezzrah's hand tightly, had already told the girl that they might have to leave her father behind.

"If he isn't here within the last bus leaving, we are getting on and going," Jill explained, telling the seven-year-old that it might be their only way out.

But two minutes later, Jamie appeared: "See? I told you I'd come back," he said. Jamie picked Tezzrah up, twirling her in a hug. The baggy sleeves of his jacket draped from her skinny arms.

Once everyone was loaded in the Subaru, Jamie swerved through traffic downtown. Near the Dutch Bros coffee hut, south of the Walgreens, he turned onto the sidewalk, then blasted past town limits.

"I'm sorry if I smell bad. I was so scared I peed my pants," the caretaker whispered to Tezzrah.

"I don't smell anything, do you?" Jill said, winking at the girl.

Just when Jamie finally thought they were in the clear, they hit flames one more time. Beneath Lookout Point, which arched over

Butte Creek Canyon west of Paradise, both sides of the Skyway were being overtaken by the conflagration, which was roasting the wooden supports on the highway rails. Even the WELCOME sign, topped by an actual bandsaw blade fashioned into a metal halo, was now consumed. The token phrase—MAY YOU FIND PARADISE TO BE ALL ITS NAME IMPLIES—was illegible. After forty-six years of standing guard at the entrance of Paradise, the sign had fallen in an avalanche of flame. The heat even warped the bandsaw, ruining the halo. The town was mostly gone, and now its most famous icon was too.

Jamie continued downhill until the fire flattened and disappeared, the black sky breaking to blue. Behind him, Paradise was walled off by smoke.

"Where are we going to sleep tonight?" Tezzrah asked, piping up from the backseat. "I don't have my blanket."

Jamie steered through the entrance of the Silver Dollar Fairgrounds in Chico.

"I don't think we'll be going home tonight," he replied.

After dropping off the caretaker and her elderly charge at the fairgrounds, where they planned to reunite with the older woman's family, he headed across town to another convalescent home, where the Heritage staff was congregating. Pizza had been ordered, and the scent of melted cheese filled the home's hallways.

Jill's partner of seventeen years was checking patients in. The women could only look at each other, unable to hug for fear of breaking down. Nearby, Jamie found his friend—the nurse who had left to find her sons—hard at work. Her family was safe. Tezzrah grabbed a paper plate, heaped it with pizza, and sat down to eat.

"How are you doing?" Jill's partner asked.

"I don't know," Jamie said, blinking back tears. "I haven't heard from Erin."

To see so many others reunite was a reminder that his family was still incomplete. Now he thought only of his wife and their two younger daughters. Would they be a family again at the end of the day?

"Come here," Jill's partner said, pulling him into a hug. "I'm sorry. I'm sure it'll be okay."

A few hours later, after the patients had settled, Jamie and Tezz-rah stepped back into his Subaru. He had finally gotten a call from his wife. It had taken several hours for Erin to reach Chico, but she hadn't faced any difficulties, just traffic. It was late afternoon, and his younger daughters were starving. He drove to the In-N-Out Burger on Business Lane, across the street from the Chico Mall—and then he saw her. Erin was unbuckling Arrianah and Mariah from the back of her Buick Enclave. It was the most striking sight Jamie had ever seen—the woman he loved with their little girls. His family. He gazed at them, unable to look away. Erin's long hair blew in the wind as she tried to pull their daughters out of the car. The girls sat side by side in their car seats, miniatures of each other, looking around, dazed and confused, still sick.

Jamie hurried across the asphalt to reach Erin. For a second, he stared.

"You have to help me with these kids," Erin said, annoyed.

Their pitbull, Ginger, scratched at the windows; Tezzrah's pet rabbit, Cinnamon, cowered in her cage. Their cat was in the car somewhere, too. Jamie reached out and folded Erin into his arms, tears running down his cheeks. He wasn't going to let her go again—not for a second. Erin looked at her husband and started crying too. Their three girls looked on with puzzled sadness at see-ing their parents weep.

AN OFFICER BY A barricade blocked the yellow school bus from entering Chico, where the school district had opened a reunifica-tion center at the local Mormon temple, so Kevin drove the twenty-five miles south to Biggs. The hamlet of just over two thousand people was a blip off Highway 99, an island in the vast paddies. It was known as "the heart of rice country." Kevin pulled in to the parking lot of Pizza Roundup, an all-you-can-eat chain restaurant. The children badly needed to use the restroom. By the time they had gotten all the students through—Kevin guiding the boys' line, Mary helping with the girls—waiters had brought out hot pans of pepperoni pizza and pitchers of soda. All of it free.

Mary flirted with the farmers chewing on barbecued chicken wings and sipping from bottles of beer at a nearby table. She tossed her straight auburn hair, cheered by this moment of normalcy. One of the farmers mentioned that they had a friend who was on the board of the Biggs Unified School District, and soon she arrived with bottled water and candy.

"You guys got out!" the woman said, hugging Mary. "Paradise has been on the news all morning."

She offered up Biggs Elementary—where Mary's father had once taught—as a meeting point for parents to collect their children. It was only a few blocks away. As they arrived at the school, Mary nearly cried at the sight of the familiar brick building, still standing, even as so little of her childhood remained up the hill in Paradise.

Abbie and Mary led the children inside while Kevin cleared the bus. The ceiling of his "battle wagon" was black, encrusted with layers of soot and dust. Out the window, he watched as the students staggered across the concrete, coughing. No wonder—they had been breathing this grime for most of the ride. Kevin locked the bus door and followed them into the building, where a screening of *Moana* had been started and applesauce handed out.

Rowan Stovall sipped a can of cherry Fanta. Her mother, Nicole Alderman, thirty-two, was one of the first to arrive. She had spent the past few hours at the Mormon temple in Chico, waiting with a cluster of parents for news of the unaccounted-for students from Ponderosa Elementary. The principal at Ridgeview High had been charged with keeping them updated—but he knew little. He stood by and watched the parents clutch their younger children, still dressed in their jammies. Some of the adults wore work uniforms. After hours of waiting, his cellphone had pinged with a message. It contained a snapshot of the manifest, the names of the children scribbled in pen on a piece of paper. He read the list out loud and asked the parents to make the half-hour drive to Biggs if they could. Their children were alive.

Nicole hadn't been able to wait a second longer—road closures be damned. Rowan was her only child. She had led the convoy of parents down dirt backroads, past a water hole where she liked to take her "Rowboat" fishing, guiding them to Biggs Elementary. Now she burst into the classroom.

Rowan was coloring with crayons, her straight blond hair falling into her face. Looking up, she saw her mother. The ten-year-old ran to her and collapsed into her arms, cinching her hands around her waist. Kevin's eyes grew misty as he watched the reunion.

"I cried because I saw a deer burning," Rowan told her mother. "I tried to be brave. I'm so sorry, Mommy."

Nicole stroked her hair, tucking it behind her ears.

"It's okay, baby. It's okay."

AT THE OPTIMO LOUNGE, Captain Sean Norman had a mutiny on his hands. It was around 2 p.m., and while most of the town had escaped, his group was still trapped. Rescue was in sight, though: A bulldozer had cleared the fallen trees and flaming church bus off the mile-long stretch of the Skyway below the strip mall. Buses were about to arrive to carry the 150 stranded residents downhill to an evacuation center—but nobody wanted to abandon their vehicles. Norman cleared his throat, standing in front of the weary crowd and preparing to make a speech that he knew no one was going to like. For hours, they had complained of being hungry or needing to use the bathroom. He understood why they were upset, but he wished they would just *listen*.

"Okay, folks," Norman said sternly. "You have two options. If you have a full tank of gas to drive downhill, then go. If not, you are getting on the bus." From the back of the audience, a few hands shot up. "No questions until I'm done," he said, staring at them pointedly. "I am going to tell you the route that you are going to take to get out of here, and if you deviate from that route—to find your cat or check on your house—I don't have enough firefighters to come rescue you. Every single one of you that diverts off the course is one more person I have to go find, and one less engine to save your town."

As he finished speaking, the buses arrived. Officers directed the vehicles out of the lot, and Norman returned to his SUV. In the rearview mirror, he saw the roof of the Optimo catch fire. Within minutes, the newly renovated lounge would burn to the ground.

Norman drove south on the Skyway. The town was hushed, aside from the glug of fractured water pipes and crash of falling pines. Amid the desolation, Norman saw his colleague, a battalion chief, driving north. Norman tapped his horn and pulled over. The battalion chief had followed his father, a Cal Fire captain, into the service. He lived in Paradise with his wife and sixteen-year-old son, who had autism. The two men paused to talk, window to window. Norman explained what had happened at the strip mall.

"The same thing happened to me at the middle school," the battalion chief responded in his slow, deliberate drawl.

For a moment, there was a lull in their conversation.

"My house is gone," he added.

Norman had fought many terrible wildfires across the state, but the devastation had always been somewhere else, in someone else's community. None had ever felt this personal. Seeing the pain in his friend's eyes wrecked him. He didn't know the right thing to say, so he didn't say anything.

TRAVIS SAW THEIR HEADLIGHTS FIRST. They pierced the velvet darkness from South Libby Road, beckoning him forward. He sped the last few yards toward the fire engine, cutting the power on his four-wheeler. He introduced himself to the two firefighters from Nevada County. "They're unconscious, I think," he said, referring to Paul and Suzie. "Pretty much dead, or nearly there." The men crowded onto the back of Travis's all-terrain vehicle, wrapping their arms around one another. "Hold on tight!" one of the firefighters told his partner, clutching a duffel bag of medical supplies.

Thickets of gray pine, gutted by the firestorm and too weak to stand, crumpled around them as they sped through the forest. The four-wheeler juddered over volleyball-sized boulders. After about two miles, they arrived on the sloping lava cap. The men hiked downhill to the spot where Paul and Suzie lay. They hadn't moved, though their eyes blinked open as the firefighters approached.

Paul's skin was sloughing off, mottled red and black. Suzie's burn injuries were oozing a clear liquid, soaking what remained of her clothing. They needed to get them to a hospital, fast. It was clear that they had third-degree burns, meaning the flames had scorched their skin through the subcutaneous tissue, damaging the delicate nerve level. The firefighters worried that if their throats were too scorched to breathe, the swelling could suffocate them. One of them cradled Suzie as he carried her to the top of the hill. Her long hair swished against his arms. He placed her on the back of Travis's four-wheeler. Then he and his colleague returned for Paul. "This is going to hurt," they warned, reaching for his arms and legs.

"Just take me," he replied.

They carried him uphill, too.

The firefighters jury-rigged Paul's quad into a makeshift ambulance, sandwiching the man between them. The screwdriver had been lost in the darkness, so they used a pocketknife to get the engine running. Travis drove Suzie.

"Paul's medications are in the cooler," she murmured. "And the last tomatoes from our garden. They're the *last* two."

Travis knew it had to be a big deal if Suzie was pushing this hard. He hobbled down to the boulder for the untouched tomatoes. Maybe they would give them something to hope for, the seeds of a new garden.

As they gunned back for South Libby Road, Paul and Suzie shrieked. Their cooked skin slid and shed every time the wheels struck a boulder or rattled over a downed tree. Paul tried to apologize.

"Yell and scream as much as you want," the firefighter reassured him. "I'm sure it hurts."

Reaching South Libby Road, the firefighters lifted Paul and Suzie into their tall engine. They would be airlifted to UC Davis's burn center in Sacramento, where they would be treated for second- and third-degree burns. Travis didn't get a chance to say goodbye. The engine departed in a roar, leaving him alone, standing alongside his quad as the fire burned on.

. . .

AROUND DINNERTIME, the Camp Fire came for Station 36. Embers rocketed into the mature timber around it, setting the canopy aflame. Captain Matt McKenzie, who had spent the past ten hours trying to stop the fire from crossing Highway 70, had been told to abandon those plans and save his station.

"Do not let it burn," his chief had said.

The words echoed in his head. As he stood outside with his crew, McKenzie felt like he was on a motorcycle going 70 mph through a swarm of bugs without a helmet on—only it was the wind that was speeding and the bugs were firebrands. Across the highway, an Alameda County strike team was defending Scooters Café from certain destruction. Hidden behind a screen of tall pine, the Jarbo Gap station was about to catch fire too.

Inside, the crock pot remained on the counter, and the diced potatoes floated in their bowl of water. The French roast coffee grounds sat unbrewed in their pot—the failed beginning to their day. McKenzie thought of everything they had worked for as he looked up at his beloved station. Decades of protecting the land and the people, of being prepared for the worst, only to watch it burn. All of those years, and what was the point?

McKenzie and his crew fought to protect Station 36 as if it were any other building, pulling hose line out of their engines and hoping they didn't run through the thousand gallons of water too fast. The Vietnam War–era storage building out back—with its dusty lawn mowers, chainsaws, paint supplies, screws—combusted. The wind swelled and crashed. Embers plinked against McKenzie's goggles like hail, and detritus from the forest drummed against his helmet. The trees spat bark, cones, and sap at his face and singed small holes in his fire-resistant suit. It felt like an hour, though it was probably only five minutes.

The station had been saved.

Somewhere in the unending darkness, farther away than McKenzie could see, the remotely operated weather site tipped over and crashed behind its chain-link fence.

OBSERVATION: NIGHTFALL

Around 4 P.M., Sheriff Kory Honea gathered reporters for a press conference in the ill-lit basement of the main law enforcement office in Oroville. One journalist waited behind a blinking video camera and three more slouched in the front row. The rows of hard plastic chairs were otherwise empty. Most news outlets were focused on the Woolsey Fire, which was devouring mansions and mobile home parks in Malibu and threatening the homes of celebrities like Kim Kardashian and Chrissy Teigen.

Finally away from the front lines, officials collapsed into chairs or leaned against the wall at the front of the room. Among them were Paradise Police chief Eric Reinbold; Cal Fire unit chief Darren Read, who towered above the other speakers; Social Services director Shelby Boston; and county supervisor Doug Teeter. He lived on Rockford Lane, in a house built by his grandfather, and had just watched it burn before evacuating in a bulldozer. His face was coated in ash.

Honea stood at a podium in front of an American flag, staring out at the sparse audience. "Are we waiting on anyone else?" he asked the department spokeswoman. "We are only a minute out. Do you think we have anybody else?" The sheriff paused, then walked around the podium to lightly slap the back of a local reporter from his hometown news-paper, the *Redding Record Searchlight,* thanking him for coming. In the days to come, hundreds of journalists and TV news crews would stream into Butte County with big cameras and foam-wrapped microphones. They would pepper fire victims with questions about how they had survived and what their plans were. But for now, none of them had arrived yet.

"My name is Kory Honea, I am the sheriff of Butte County," Honea began, leaning into the microphone and gripping the sides of the lectern. "Uh, today we want to give you the information we have at this point, with regard to the Camp Fire and what has transpired. I want to first say, this was a very serious fire. It was a rapidly moving fire. A lot of the information we have is still preliminary. I want you to understand that."

He explained that the Camp Fire had left more than thirty-four thousand customers without power in Plumas and Butte counties and forced tens of thousands of people to evacuate Paradise. The blaze remained at 18,000 acres with no containment, he said, and it was making a run for Chico. There had been unverified deaths, he added. "We will get to those areas as soon as we can to check them out, but this is still a very active fire. Some of the areas where we are hearing the reports of deaths, we cannot get to safely. I can assure you, as soon as we are able to, we will. This is a significant crisis. It remains to be seen what the damage is."

He shuffled the printed papers on the lectern, then added: "This is the fire we always feared would come."

That day, Gavin Newsom was filling in for Governor Jerry Brown in Sacramento. Brown had left for Chicago that morning for the annual Bulletin of the Atomic Scientists board meeting—he had been named the organization's executive chair—after a stop in Austin the previous day for an event at the LBJ Library. While Honea was speaking, Newsom declared a state of emergency in Butte County. And from its headquarters on Beale Street in San Francisco, PG&E dispatched a tweet: "PG&E has determined that it will not proceed with plans today for a Public Safety Power Shutoff in portions of 8 Northern CA counties, as weather conditions did not warrant this safety measure. We want to thank our customers for their understanding."

KONKOW LEGEND

And these two good people ran and wandered for many moons, crying and nearly starving, until one day they halted near Anikato, which the white man calls the Trinity River. Wahnonopem had sent down the rains, the fire died out, the grasses were springing green again all over the land, the birds were singing everywhere, and Anikato was full with the fish shining and swimming in its limpid waters.

In a sheltered nook upon its banks they made a little home, but they built a kakanecome first. As the moons waned and came again, little children grew around them as plentiful as the grains of sand near the great water; and one day, long, long after, Peuchano and Umwanata, having grown very old, gathered their children and grandchildren around them and told them that the Black Spirit of Death was coming for them soon, but that before they went with him they wanted to sleep in their old welluda, their ancestral home, where they had first seen the wildflowers blooming and heard the glad songs of the birds singing among the pines.

And the women, the young maidens, and the little children waded into Anikato, and made themselves pure by ablutions, and knelt upon the banks, while the old men and the young men went down into the kakanecomes and purified themselves with the holy fire, and they all prayed that Wahnonopem, the Great Spirit, might lead them on their way to their far-off welluda.

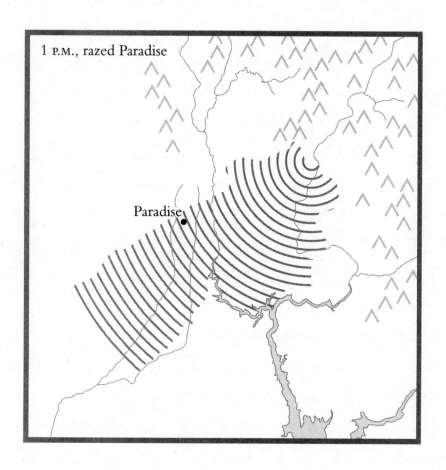

PART V

ASH

CHAPTER 16

UNCONFIRMED DEATHS

A few Chico police officers were slouched over their deli sandwiches at the command post at Butte Community College when Jeremy Carr noticed them. He approached, introduced himself, and asked what they had seen on that terrible morning. "It's not good," one of them replied, a respirator hanging loose around his neck. It wasn't the words that made an impression. Maybe it was the way the officer's face twitched or his voice cracked. Carr didn't know for sure. The thirty-one-year-old volunteer chaplain understood in that moment that the devastation that lay ahead was greater than he could have imagined.

Carr worked for the Butte County Sheriff's Office. His patrol uniform and badge allowed him into the closed circle of first responders: He could use their official radio and zip through California Highway Patrol checkpoints. But access meant only so much, and sometimes Carr felt extraneous. He didn't have the practical skills of a detective. Rather, his role was to represent comfort, even to the nonbelievers. To let firefighters and police officers know that if they wanted to talk, he was there. His specialty was not disaster but its aftermath.

On Friday, November 9, the day after the Camp Fire sparked, the command post was frenzied. More than five thousand firefighters had arrived from as far away as Missouri and South Dakota.

Trailers and industrial shipping containers of supplies filled the parking lots. The sky had taken on the hue of polished walnut, and the weather was chilly. Butte County was socked in by a noxious smog and the sun was obscured by ash, causing the temperature to drop and first responders to cough, their throats burning with every inhale. The Air Quality Index registered pollution above 500—ten times higher than the level considered healthy.

In San Francisco, the smoke was so oppressive that simply breathing was the equivalent of smoking eight cigarettes. Alcatraz Island and Muir Woods closed, and San Francisco's famed cable cars stopped running. Half-marathons were canceled in Napa, Monterey Bay, Berkeley. "You can't run in the smoke," a race director explained to a reporter. More than 180 public school districts closed, keeping more than a million children at home—about one in six of California's students. The state postponed its high school football playoffs. Paradise High had already forfeited. Hundreds of flights to San Francisco International Airport were delayed, and people wore N95 masks as they bustled about the Financial District.

A security guard manned the front entrance of Butte College, checking name badges and license plates. In-person classes had been canceled, as had final exams. Carr made the rounds, checking in with first responders. *Who has the worst job here?* Carr asked himself. *Who is most likely to be traumatized?* The full answer wouldn't come until after firefighters had tamped down the blaze within Paradise. They had recently ignited a backburn—an intentional fire to starve the conflagration of fuel—in an effort to save Chico and Durham. Flames were chewing through vegetation in distant canyons and gullies to the east. The National Weather Service had announced more Red Flag conditions—an indication of extreme fire danger—in the coming days.

The Camp Fire was already historic, having nearly wiped the town of Paradise off the map in only four hours. The buildings and bodies were still being tallied, but fatalities were expected to break the state record set eighty-five years earlier, when twenty-nine laborers were killed in Los Angeles by the 1933 Griffith Park Fire, known as the Park Holocaust. The men had been doing landscape maintenance, earning 40 cents an hour—a good wage during the Great Depression—when a small brush fire trapped them in a canyon. "I saw between 20 and 25 men burned to death today, screaming and fighting for life in a tornado of fire," a survivor had written in a firsthand newspaper account. "Terror-stricken, they leapt through the flames, continued on a few feet and went down in a welter of hot ashes and flames. . . . Some fought one another in blind terror and indecision. They had to fight something as the flames closed in."

The true extent of the damage, however, remained unknown. Hundreds of reporters were fanning into Butte County to report on the fire's aftermath; the turnout was so great that Sheriff Honea moved his ongoing press conferences to a smoky, low-slung building at the Silver Dollar Fairgrounds in Chico. He was doing interviews with journalists from as far away as China—becoming so well known that a clothing company would eventually sell $120,000 worth of T-shirts printed with a graphic of his face and the phrase HONEA IS MY HOMIE. (The money went to Camp Fire relief efforts.)

Soon Carr would join a search-and-recovery team tasked with looking for the dead. The team would comb the streets, hunt through ruined homes and cars, explore ditches and gullies. It was unrelentingly sad work, but it brought some fulfillment and meaning to the disaster. Tragedy, Carr knew, turned the human experience inside out and forced people to confront realities they never had to in times of prosperity. Trauma eventually came to everyone, in greater or lesser degrees. The trick was to notice the signs of distress: When a normally upbeat person shut down. When an introvert talked too much. When a leader drank to excess. How were first responders going to deal with this tragedy? What story were they going to tell themselves?

The chaplain's task was to make sure rescue workers had the space to engage with the hard questions. Carr believed God had called him to this job, though at first he hadn't understood why. He was physically commanding—big-boned and tall at six foot two—with a deep, reverberating voice. He had been raised in Monterey, and in high school his friends had elected him their campus chaplain—which surprised Carr at the time, since he didn't think he prayed or read the Bible as often as he should. But the role, Carr realized, drew him out of his shell and forced him to engage more fully with others. He went on to earn a degree in biblical studies at Azusa Pacific University in Southern California and then studied at Fuller Theological Seminary nearby. Feeling God's presence most keenly when he was helping others, Carr sought training as a chaplain in hospitals and hospices. At first, knocking on strangers' doors felt akin to a crime, as though he were intruding on their pain, but he came to see that his presence was important. He learned to exude peace and acceptance. Emotions needed to be felt, he would reiterate time and time again.

He had wanted to become a military chaplain, but that role required church experience, and a job opening at Foothill Community Church in Oroville had brought him to Butte County in 2015. Carr now worked as an associate pastor, volunteering with the Butte County Sheriff's Office to gain additional experience serving law enforcement officers. Carr had his own strategies to cope with the hopelessness that could rise like the tide when he bore witness to others' pain: journaling, crying, talking to his wife, Grace. Or listening to Adagio for Strings, the haunting orchestral work by Samuel Barber that is often performed at funerals and public observations of grief. He would have to rely on those mechanisms again when he and the others finished the search for the dead. Picking through the soot was a lonely job that forced first responders to swallow their emotions—but no one, Carr knew, could repress that part of themselves forever. Those in the field were already finding that out.

TEN MILES UP the mountain, every familiar place had been reduced to rubble and ruin. Much of the town was ash, from the downtown corridor to the country lanes. The Camp Fire had incinerated more than 18,800 structures and everything inside them, including the local history museum and its heirloom dolls, old mining certificates, antique shotguns, and a 54-pound plaster replica of the "Dogtown Nugget." Nor did the fire spare what was alleged to be one of the world's most valuable gemstones, the 500-pound, $280 million "Beleza Emerald," which a local couple said had been stashed in their home safe. In Paradise, 95 percent of the commercial buildings and 90 percent of the homes—about eleven thousand—were gone. Another three thousand houses had been leveled in Magalia and Concow.

Search and rescue workers discovered that flames had played hopscotch, destroying entire blocks while skipping over random structures. On one residential street, a smattering of lemon and persimmon trees had survived. They dropped their fruit to rot on the road's pocked macadam as the rest of the forest smoldered, releasing the musky scent of pine. The death toll—a number complicated by how little remained of the victims—rose and fell according to what workers could infer as they sifted through homes and vehicles. By the end of the search, though, it would stand at eighty-five people.

Only fire engines and news crew vans now rolled through Paradise. They barely slowed down at the toppled stop signs. Downhill, checkpoints blocked the four routes into town, barring residents from entering the burn zone. A few homeowners waited at the roadblock sealing off the Skyway, wanting to know what remained of their houses, their town. Many more evacuees languished—exhausted and shaken—at shelters in Chico, hoping for good news from up the mountain.

The police station on Black Olive Drive had somehow withstood the firestorm. Now officers gathered before their shifts in the station's back room, dressed in the yellow fire-resistant uniforms of firefighters. They drank lukewarm coffee and compared the filthy air filters in their patrol cars, then left to protect a ghost town.

Sheriff's deputy Tiffany Larson, thirty-two, couldn't pause to process the destruction while she had a job to do. The horror of the disaster was underscored by the desperate effort to find people who had not been heard from, a problem compounded by the downed cell-phone towers. An unknown number of victims remained missing—at one point, the count surpassed three thousand. Larson, an investigator, had been chosen to lead one of thirteen search-and-recovery teams. Carr might have seen her at base camp: dark hair stuffed under a black Adidas baseball cap, thick-rimmed eyeglasses.

Larson's job was to look. For what, exactly, she was not always sure. Officials called the work she and others were doing "welfare checks." The hope, fading day by day, was that the missing had escaped somehow and just hadn't made contact with their loved ones. In Chico, flyers printed with their faces plastered storefronts and coffee shops.

Larson led a team of three other sheriff's deputies and a chaplain whose house had burned down twice before: once in the 2008 Humboldt Fire, then again in the Camp Fire. She drove across Paradise, guiding the caravan past shopping carts toppled in the Kmart parking lot and over tree branches that crunched under her tires. They drove over a downed electrical line, barely pausing. Several days earlier, such a power line would have been a big deal, demanding the placement of orange cones and a call to PG&E; traffic would have snarled. But no longer. At Feather River hospital, IV poles, gurneys, and wheelchairs cluttered the helipad. Ash whirled in the puddled water. All that remained of the cardiology wing were four standing walls. The ceiling was gone, the interior a heap of rusted metal. Mangled pipes and electrical wires were scattered on the ground like a massive game of pick-up sticks. The remains of white cinderblock walls marked ruined offices: human resources, marketing, information technology. Another half dozen structures had sustained severe smoke and water damage, including the cancer center, the outpatient surgery center, and the Birth Day Place. In the surrounding neighborhoods, wind chimes tinkled on a few front lawns. The ordinary cheerfulness of their tune was jarring.

Larson parked at a mobile home park and got out of her SUV, instructing the men to follow. Their search, aided by a list of addresses, took them past compost bins set on the curb and campaign signs staked in the ditches. They picked their way through the first home, registering a mental list: Melted Rolling Rock and Heineken beer bottles stuck to the asphalt. Cans of vegetables so scorched and dehydrated that they sounded like maracas when shaken. The talismans of exurban life. And then, curved white shards on a box spring. "I think these are . . ." Larson trailed off, pointing at what looked like femur bones. "No, no, those are tack strips," said a deputy from the Alameda County Sheriff's Office, which was assisting in the wake of the disaster. He bent to look closer. This was his first wildfire, and he had never seen anything like it. "Bones have tiny air pockets in them."

Larson nodded and moved on. She was younger and more petite than the men she led, but she was from Paradise, and with her heritage came a certain authority. She had loved this community and known it long before it was reduced to an apocalyptic wasteland. When she spoke, they listened.

They hadn't found anything yet and didn't particularly want to, even though they knew that finding a body could help a family desperate for answers. Larson thought of her home on Newland Road, also lost in the fire, as she dug through the rubble. She and her husband, Bobby, had bought it six years earlier. They had met at Paradise High's homecoming dance her sophomore year, and that was that. They had been married for eleven years. Bobby worked for the Sheriff's Office too. He was nights; she was days. Since the Camp Fire ignited, they had seen each other for only about ten minutes at a colleague's home in Chico, where they were staying with their four dogs and two cats until they figured out what to do next. Larson's whole family—Bobby's side included—had lived in Paradise: three sisters, one brother, five brothers-in-law, and four sisters-in-law. Grandparents, too. Almost all of their homes had been destroyed.

On Walnut Street, a man leaned over a fence, wearing a crewneck sweatshirt that read ATTITUDE IS EVERYTHING. His was the only home on the road that had survived. The man hadn't been able to corral his cat, so he had defied evacuation orders and stayed, risking his own life. "Does he realize he's lucky he's not in a coroner's bag?" Larson grumbled. Nearby, she found a red rose, singed black around the edges. Fires were capricious in what they took and what they didn't, and somehow this bloom had survived. "Hey, Sarge, I got this for you," Larson said, smiling and turning to the deputy as she held out the flower. For a moment, he smiled too.

The work was grim, though, and sometimes precarious. An investigator on another team had recently stepped through the brittle concrete lid of a septic tank and fallen into raw sewage. Thousands of damaged tanks were hidden beneath the ash. A crew usually started in the garage, where the burnt frames of cars rested on their axels, aluminum pooled around them like congealed cake batter. Then they sought out metal box springs, an indication of a bedroom. Stoves and freezers implied a kitchen, washers and dryers the laundry room. There was little else to go on. Larson and her team looked for clues at the Pine Springs Mobile Home Park, nestled behind the Hope Christian Church on Clark Road. The park had held about sixty trailers, mostly housing for elderly residents. Only three of the units had survived. Two people were still unaccounted for. One of them, John Digby, seventy-eight, was a retired mail carrier and Air Force veteran who had been home with a sore throat when the fire broke out. His son, who lived in Minnesota, hadn't heard from him.

At a unit nestled under tall pines, Larson spotted a white wheelchair lift, still raised. The porch steps were gone; the railing hung in the air, leading to nowhere. Under it, an orange tabby cat with singed fur lay dead on its side. "Theoretically, if they had gotten out and into their car, this would be down," Larson said as she pointed to the lift.

"Unless there's another way out or someone took them," the sergeant said. "But the pet is still here. The wheelchair is still here. The lift's up. These are telltale signs."

"Yeah," Larson added, "somebody is not going to push it back up when they're trying to get the hell out."

The mobile home's aluminum roof had collapsed and buried everything underneath it. The search team couldn't go farther. Firefighter crews would have to return later with specialized machinery to pull the roof back, allowing forensic investigators with trained dogs to hunt for bone fragments, molars, dentures, surgical implants—anything to indicate the presence of a body. The specialized team in their gas masks, safety suits, and thick boots looked like astronauts on a moonwalk. Larson and her team didn't have such expertise or equipment, so she tagged the home with special tape, signaling an incomplete search. She hopped into her SUV and crossed the address off her list, then turned onto the next block. At the intersection, she paused to let a hearse headed to the morgue pass.

CARR SPENT SEVERAL HOURS with his team, also searching through houses. Now he needed to convince his colleagues—people like Larson—that it was going to be okay, that they would get through this tragedy. But he couldn't say it like that; it would sound trite. He needed to convey it somehow with his presence. After his team had returned to base camp, leaving the destruction in Paradise behind, Carr tried to make himself useful—but no one needed him quite yet. He found himself thinking about the dead. It was one thing to rob people of their possessions. But their lives? That was a whole other sort of senselessness. And in that moment, it came to him. *You are a pastor, you pray over the dead all the time,* Carr thought. *Go to them.*

The makeshift morgue was set up in Butte College's training facility, where students learned combat techniques on foam pads and took law enforcement classes. It had concrete floors and a big commercial bay with a roll-up metal door. A kitchen and a classroom jutted off the main room. A few officers were there, filling out paperwork that would remain with the bodies until they were driven to Sacramento for processing and identification. Brown paper bags, five-gallon buckets, and zippered body bags labeled with case numbers were arranged on the floor. Carr tried to imagine the lives of these people—the communities and the families they had left behind. Other chaplains were beginning the process of notifying next of kin. None had stayed behind with the victims in the cold training facility. "Would it weird you out if I prayed with the bodies?" Carr asked the clerk manning the front desk. The clerk shook his head.

Carr knelt on the hard concrete. He had been raised to value extemporaneous prayer. When he bowed his head and tried to pray, though, the words wouldn't come. He wasn't sure where to begin.

Over the next week, he would follow the same routine, spending a few minutes with each victim. Day after day, more bags and buckets would arrive, and Carr would feel horribly limited. Silence gripped him. He was reminded of the chasm between the world as it was and the world as it should be. He knew he was supposed to be a source of hope, but in moments like these it seemed impossible. There was no hope. He couldn't find a prayer within himself. *How am I going to show these dead humans love? Where are you, God?*

Then an answer emerged. Carr lowered his head. Under fluorescent lights, knees pressed into the unforgiving concrete, he prayed the Catholic requiem: "May everlasting light shine upon you," he said. "May the angels lead you into paradise."

OBSERVATION: IDENTIFICATION AND HOT SAUCE

After Carr finished his nightly prayer, the bodies—or more often just fragments of them—were driven ninety miles south to the state capital, where they were stored in a refrigerated semi truck in the parking lot of the county morgue. The building itself was full. The state's Office of Emergency Services had called in Sacramento County coroner Kim Gin to help with body identification efforts. She processed seven thousand bodies annually, including many from smaller counties, like Butte, that didn't have a dedicated facility. But even she was overwhelmed by the scale of this disaster. In the past nineteen years, she had processed, at most, five victims simultaneously—a fraction of the dozens coming from Paradise.

To deal with the unprecedented workload, Gin set up makeshift processing stations on the receiving dock and called in more people to help. She repurposed coffee tables and covered cardboard boxes with plastic sheets. The stench of smoke clung to the remains; investigators hacked dryly from the carbon particles. Any piece of evidence found near the victim helped investigators assign a name: a wedding ring, a purse, military dog tags, an antique muscle car, a partially burned Social Security card. Paired with missing person reports, these remnants provided context as to who the victim likely was.

Still, they didn't offer enough confirmation on their own, and false identification was worse than slow identification. Sometimes all Gin had to work with were a few molars or bones. Or, in two particular cases, the soft tissue on a victim's feet and a man's carbonized appendage.

Even matches that should have been straightforward were proving complicated. Of the eighteen dentists living in Paradise, ten had seen their offices burn down. Critical patient records had been lost. In the absence of dental information, Gin knew that any steel hardware with serial numbers—artificial hips, knees, shoulders—could prove helpful, since she could match the numbers to hospital records. The steel tended to be badly tarnished, so she relied on an old trick she had learned in school: soaking the devices in a natural corrosive—Taco Bell hot sauce. Her team grabbed handfuls of packets from the restaurant.

Along with the identifications came the task of assigning causes of death. That, too, was a challenge. Unless she could test for carbon monoxide in the blood to prove that the victim had perished from smoke inhalation, the death was simply marked "fire related." She taped large sheets of white paper to the walls of a conference room, one for each body—a floating cemetery of tombstone-like markers. On each sheet, doctors scrawled observations about a body in black marker. The cases were grouped by category: confirmed or unconfirmed.

Gin and Honea hoped to provide closure to families of the missing within three weeks, by Thanksgiving—but forensic pathology, especially at this scale, takes time. (More than eighteen years after the 9/11 terrorist attacks, about a thousand sets of remains still have not been identified.) To help the process, Gin accepted assistance from a company that offered its "Rapid DNA" technology for free—a first-of-its-kind collaboration following a mass tragedy. Richard Selden, the founder and chief scientific officer of ANDE Corp., flew in from Massachusetts. Many of the victims' organs, which had burned at temperatures upwards of 1,500 degrees, were crispy but intact. This meant that pathologists could gather blood and tissue samples, which Selden's team of technicians then placed in a microwave-sized instrument that parsed the DNA. Each scan took 104 minutes. They ran four of the machines simultaneously, the devices humming as they processed the samples. The soft tissue was so damaged that the technicians were not sure the specimens would register results—they planned to run tests on bones next—but, amazingly, the tissue delivered just the DNA answers Gin was hoping for.

In Butte County, relatives of the missing stepped inside mobile laboratories to have the insides of their cheeks swabbed. The ideal DNA for a match was from a parent or child; more distant relationships made for less certain results. From there, scans of the swabbed material took ninety-four minutes to process. When two scans matched, the computer screen would flash a probability, such as 100 million to 1, indicating the odds that the match might be false.

Identifications ticked up. The Department of Justice's Missing and Unidentified Persons Unit helped gather samples, as did local police units. In one case, the Royal Canadian Mounted Police even volunteered their services. Gin would name forty-one victims by the end of November, missing her goal but making good headway. In the conference room, she moved more sheets of paper to the "confirmed" wall.

CHAPTER 17

MAYOR OF NOWHERE

Less than a week after the Camp Fire roared through Paradise, the Town Council convened on a chilly Tuesday evening at the old Municipal Building in Chico. A white sticker had been smoothed onto the door of a first-floor office once occupied by the city's Chamber of Commerce. It read: PARADISE TOWN HALL.

The five councilmembers were in uncharted territory: without homes, without a workplace, without even email, because their server was down. In the interim, they had created a temporary account: townofparadisestaff@gmail.com. They couldn't return to Paradise until search teams declared an end to the body recovery efforts. As teams like Larson's sorted through the debris, linemen from PG&E were restringing electrical wires and Public Works crews were pushing toppled trees from the roads to make the town safe for habitation.

The meeting agenda for November 13 had been set the day before the worst wildfire in a century had razed their town. Now Paradise's leaders found themselves caught between two realities: the normality of the past and the uncertainty of the future. For the past few days, they had worked at the Emergency Operations Center, moved to the Chico Fire Training Center. A few of them had volunteered to help with a "windshield survey," driving through the carnage to get an estimate of the number of homes that had been destroyed so staff could start applying for disaster and recovery grants.

It looked more like a moonscape than a landscape. "It's emotionally hard," said Paradise Ridge Fire Safe Council chairman Phil John, who had once composed the "Wildfire Ready" rap, of the experience. "You look as far as you can see and there's nothing. Everybody you know doesn't have shit—not even their pets. Every time somebody asks, 'Can you check on my dog?' I'm like, 'I'm sorry, but you don't have a house. I imagine the dog is not there either.'"

Now, about forty people, some of them Paradise residents and others from nearby cities showing their support, filed into the chamber. An emergency session to pass a resolution for spending public money had been tacked on to the opening agenda for the meeting, which would then continue as usual.

Mayor Jones presided over the meeting, having recently won reelection with 4,417 votes. When the ballots were in, she had celebrated her victory at a fellow councilmember's home, serving grocery store cookies and chocolate sheet cake to about thirty friends, as well as a reporter from the small local newspaper, the *Paradise Post*. Jones had sipped red wine that night as she chatted with her supporters. With diamond rings on each hand and polished nails, she was the kind of woman who celebrated with a single glass of Zinfandel. An avid hunter, she had considered hosting a dinner of homemade venison stew at her remodeled farmhouse, which was decorated with the taxidermied heads of exotic animals she and her husband had shot on an African hunting safari. She had done so after her previous win, but this year decided against the hoopla.

And now her home was gone.

"My watch says we have to wait one minute before we start," Mayor Jones said, watching the clock. "There we go. Talk now!"

Paradise Fire chief David Hawks spoke first. He held the microphone close to his mouth and gestured at a printed map of the fire perimeter, explaining that the blaze was still raging in the evergreen forests near Bloomer Hill. At 130,000 acres, the wildfire was 35 percent contained—a figure determined by how much of it was confined behind fire lines and backburns. Hawks knew that at least forty-eight people had died. (By that point, only about half of the total victims had been located.) The flames had gone cold near

Chico, he added, and though a hotel on lower Clark Road had begun smoldering that afternoon, firefighters had quashed that threat within a half hour. "Obviously the fire swept through Paradise four days ago plus now, and there is very little to no fire activity in town," Hawks said. "Very little to no remaining fuel left." Fuel, in this case, meant houses.

Then there was a moment of silence. Everyone bowed their heads. Vice Mayor Greg Bolin broke the quiet to pray for families who had lost loved ones and needed solace, for first responders who had braved the fire, for those still fighting back the flames. "We love our town," he said.

They took roll and began. The first item on their pre-fire agenda: passing a proclamation to recognize November as National Runaway Prevention Month. "There is some town business that the council needs to handle," said Mayor Jones, "so I apologize for the mundane in the midst of tragedy."

The next item, said the town clerk, was an update on road projects—but no one wanted to take time to hear a now outdated report. There were bigger issues to discuss, like how to move forward when the entire town was mostly leveled and its people still evacuated. How would teachers keep up with their classes? Where would people live while they rebuilt? When would the debris be cleared? It had only been five days, and no one had any answers. Maybe they could have another community meeting on Thursday? "We are going to post news on the town Facebook page. Could you spread the word?" decided Town Manager Lauren Gill, turning to a smattering of TV journalists in the corner. "We are also trying to put together another page, maybe called Paradise Recovers? Paradise Rebuilds? It might take me twenty-four hours to get that up and running. We'll have all of our information there."

She wanted to host a community vigil at First Christian Church in Chico, too, in memory of those who had passed. "None of us have had any time to mourn or even look at the gravity of the situation and what it has done to us, so this will be a good opportunity for us to do that," Gill said of the vigil. "We're also planning on— on a fun note, because we are fun—we are planning on a community Thanksgiving. I'm getting a lot of excitement on that. I'll have more information soon. We'll all have a place to go and be together. Just have a Thanksgiving dinner and be together."

Seated behind the dais, Gill appeared stoic and calm, but inwardly she was struggling to process the disaster. The mayor, police chief, county supervisor, and five-member Town Council were homeless. So were sixty-five firefighters and eight police officers. Eight of the nine public schools in Paradise had sustained serious damage, as had two fire stations. Kmart, Safeway, McDonald's, and the newly remodeled Jack in the Box were ruined. The wooden flumes crisscrossing the West Branch Feather River had been destroyed. The beloved ponderosa pine forest had been hollowed out. Everyone knew someone who had been affected.

While the homes of her two sisters and her brother hadn't survived, Gill's own home near Butte Creek Canyon was unscathed. Her tube of mascara, she would discover, was still on the white quartz countertop, right where she had left it. She felt guilty that it was her house still standing and almost wished it had burned down too.

The rising death toll also filled her with dread. She questioned her role in the evacuation. Assistant Town Manager Marc Mattox, nagged by his conscience, worried that they would be taken to court for bungling the CodeRed alerts.

The Town Council didn't know any of this.

"Lauren, I just have to thank you for thinking so much about the community," Jones said. Her fellow councilmembers applauded.

Now several officials, each wearing a jacket emblazoned with the acronym of a different federal agency, spoke into the chamber's microphone. They assured the council that they were all on the same team. "This is a marathon, not a sprint," said Kevin Hannes, a Federal Emergency Management Agency official. "This is about Paradise, this is about Butte County, this is about California, supported by the federal government. We are not here to take over. We are here to directly support Governor Brown. We are committed to being here with you."

Everyone clapped again. Residents stepped up for public comment. They spoke because they needed someone to listen to what they had been through, even for three minutes. Like the man in the bright yellow jacket who oversaw the Paradise Community Guild. Or the woman in the baggy green hoodie who ran the Paradise Boys and Girls Club. She hadn't planned on speaking, but a hundred of the children had shown up for a meeting in Chico that afternoon, and she wanted to share how joyful that had been.

Then there was Ward Habriel, seventy-two, who had lived on Pentz Road with his wife, Cheryl, seventy-one. He wore a green flannel shirt and a cerulean LOVE PARADISE hat embroidered with a red heart. "Do you still love Paradise?" Ward said. "I have a real easy [answer] for that. Of course. Paradise is not several thousand acres of charcoal—it's about people. We have a certain spirit. I don't think tonight's a good night for complaints and gripes, moans and groans. . . . Maybe some of the buildings and roads you didn't like. But guess what? We have a great community." He reached for his wallet. He wanted to do something that was important to him, noting that the town had recently passed its Measure C sales tax, which now, obviously, wasn't generating much money. He stacked $6.26 on the podium. "I have spent a little bit of money in Chico. . . . I've added how much I spent . . . and I would like to pay that tax now."

"Woo, we'll take it!" Jones cheered.

Others weren't as supportive. Michael Orr, a local writer, shuffled to the podium, hands jammed into the pockets of his jeans. "I love this town," he began—then clarified the statement, realizing he was standing in Chico. "Not this town, our town. I love the community, the spirit that's still alive. The people are what our community is all about. . . . It's much more challenging when you're a town in exile, which is really what we are. We're spread all over the state." He thanked Gill for her leadership, along with the rest of the council, then directed his attention to the mayor. "I've been critical of you at times," Orr said to Jones. "For lack of vision and economic things and other things. My criticism today is the same thing. Where are you? The people need you. We are not hearing a word from you."

"I think I did twenty-five media interviews yesterday and at least the same amount today," Jones interrupted.

"Well, thousands of people don't own TVs," Michael responded.

"I know, but it's the job I've been given," Jones snapped. "There are only so many minutes in the day."

Michael continued: "You spent the last two months talking about how proud you were of your evacuation plan. Are you still as proud of it as you were then? Because most of the people coming down off that hill aren't very proud of it. A bunch of them have asked me to come here and ask you to resign. So that's what I'm doing today. Thank you." He folded his hands over his belly and edged away from the podium. The room fell silent.

"Thanks for your comments, and I'd like to answer that question," Jones said. "This is the same thing the media has asked me. I have said, 'Yes, I am.' Because we had an evacuation plan that was as robust as we did, because we had practiced it, because we had implemented it in the past." She sighed. "It wasn't perfect, but it actually worked. Without that plan, we would've had hundreds and hundreds more deaths in Paradise. Our people knew about the plan and they pretty much followed it. It was chaos, but it was sort of organized chaos, and I truly believe with all my heart that without that plan, many, many more people would have died."

No town or city had the means to evacuate every person at the same time, she added. "You can't build infrastructure big enough to do that, so I just want to say that again. . . . You can call the next speaker."

Less than an hour later, Jones adjourned the session. There wasn't much the council could do for a town that, at least for the moment, didn't exist.

EAST OF CHICO, Betsy Ann Cowley arrived home in Pulga that week with a suitcase full of sundresses and bathing suits. She had cut short her family cruise in the Dominican Republic after hearing of the Camp Fire. Cal Fire had stationed private security guards near the Caribou-Palermo Line, which crossed her property, to prevent anyone from disturbing the area around the ignition point. The agency's arson investigators had already climbed Camp Creek Road on November 8 to conduct a rigorous inspection. They had been concerned by the sight of a PG&E helicopter hovering above Tower 27/222. Now the investigators were having the tower dismantled. They had called in the help of an FBI Evidence Team to do metallurgical testing at its laboratory in Quantico, Virginia. Cal Fire and county officials hoped to prove that the equipment was the cause of the blaze.

A few Union Pacific Railroad workers who had befriended Cowley in recent years had noticed that her house and a scattering of cabins were among the wildfire's casualties. The rugged workers tried to anticipate what she might need upon her return to Pulga— her "Ladytown." They left containers of gasoline and potable water, along with several packets of cotton underwear they had carefully selected at Walmart. Though the granny panties were about two sizes too large, Cowley was touched by the gesture. She set her luggage down.

FIVE DAYS LATER, on Saturday, November 17, President Donald Trump landed at Chico Municipal Airport under a leaden sky. He was scheduled to meet with California governor Jerry Brown and governor-elect Gavin Newsom that morning. He stepped off Air Force One onto a runway littered with orange and white firefighting aircraft. Smoke obscured the morning light, and the air smelled of a wood fire. Trump wore a white shirt and a dark windbreaker with a camouflage baseball cap that read USA.

Mayor Jones had driven to the airport to greet him. She had received word of his visit the previous afternoon, while out at lunch with her husband in downtown Chico. The caller ID rang through as "White House." She had thought the caller might be a telemarketer but answered anyway. The official on the line asked if Jones wanted to meet the president. She inquired if the vice mayor might accompany her, but the official said no. If she told anyone, security concerns would force the White House to cancel the trip. Jones was on her own.

There was one problem: Her entire wardrobe had been consumed by flames. She had no clothing other than a handful of blue jeans and T-shirts she'd bought at a discount warehouse since evacuating. At the restaurant, her husband pointed out a boutique across the street. The owner outfitted Jones in twenty minutes, dressing her in a green cashmere sweater, black slacks, and a soft gray coat. Jones had twirled in the mirror. She looked good. As she left, she'd made sure to emphasize that the new outfit was for a big media interview, nothing more.

Now Jones waited in her silver Volvo. German shepherds had thoroughly sniffed the vehicle before she was allowed to drive onto the tarmac. She hoped the president would offer to let her ride with him. The official had mentioned Trump might propose such a thing.

Only a tragedy as big as the Camp Fire would have brought a sitting president to a town as small as Paradise. Even then, Trump struggled to remember the community's name. At several points throughout the day, he misspoke, calling Paradise "the town of Pleasure." Later that afternoon, at a press conference in Malibu, state officials would finally be forced to correct him. "And you're watching from New York, or you're watching from Washington, D.C., and you don't really see the gravity of it. I mean, as big as they look on the tube, you don't see what's going on until you come here. And what we saw at Pleasure—what a name, Pleasure—right now. What we just saw, we just left Pleasure . . ." Listening, Governor Brown trained his eyes on the ground, looking pained, until another official intervened and the president corrected himself: "Oh, Paradise. What we just saw at Paradise is just, you know, it's just not acceptable."

But before arriving in Malibu to survey the damage wrought by the Woolsey Fire, Trump would see Paradise for himself. Jones was the first to greet him on the tarmac, firmly shaking his hand and looking him straight in the eyes. She wasn't nervous. Trump had been briefed and knew who she was, calling her by her first name— not her title. He didn't offer a ride after all, so she drove in her Volvo near the back of his motorcade.

They wheeled across Chico toward the Skyway. Hundreds of people wearing N95 respirator masks lined the college town's streets, filming the president's passing on their cellphones. Some waved American flags or hefted campaign signs in the air to welcome him to their once bucolic corner of the Golden State. A smaller group held banners that read CLIMATE CHANGE!, MORON, WE ARE IN A DROUGHT!, and THE APOCALYPSE.

The smog thickened as the motorcade gained elevation, the city blocks of onlookers giving way to charred earth and matchstick trees so brittle that a puff of air could topple them. In yard after yard, homes had been leveled. Here and there a carved pumpkin rotted on concrete front steps, pardoned by the flames. A decorative skeleton lounged on one bench, arm resting on the seatback. An untouched ceramic nativity set perched atop the roof of a gutted hatchback. Chimneys stood guard over the devastation, as tall and even as gravestones. A local artist was planning to spray-paint portraits on some of them in honor of the victims. The motorcade passed a used car sales lot. One sedan had been whittled to its frame; the rest were untouched. Soot caked their windshields.

WEAR MASKS, an LED road sign cautioned. Construction crews and first responders were sifting through the wreckage, their faces hidden. Trump and the other elected officials did not wear masks.

Around 11:30 A.M., the motorcade pulled in to the Skyway Villa Mobile Home and RV Park. Trump began his walking tour at one of the burnt-out homes. A tattered American flag was draped over one ruined wall. Scraps of blue-and-white-starred fabric stippled the ground. The president navigated past a downed streetlamp, rusted ovens, and a metal shelf strewn with brittle terra-cotta pots. Jones and FEMA director Brock Long—who said he had never seen a worse disaster in his career—followed behind, along with Brown and Newsom. The pair of politicians stuck together, chatting minimally with the president. Jones, a registered Republican, was just happy that the three men appeared to be getting along. The president and the governor, who was about to pass the reins to Newsom, always seemed to be clashing with Trump over something. Now, though, they were friendly enough for Trump to slap Brown on the back.

Just two days after the Camp Fire, Trump had tweeted about the catastrophe, threatening to withhold disaster money from California. "There is no reason for these massive, deadly and costly forest fires in California except that forest management is so poor," he typed. "Billions of dollars are given each year, with so many lives lost, all because of gross mismanagement of the forests. Remedy now, or no more Fed payments!" Immediately, the head of the thirty-thousand-member California Professional Firefighters organization had issued a blistering response, calling the tweet "ill-informed, ill-timed and demeaning to victims and to our firefighters on the front lines." Newsom had added his bit: "Lives have been lost. Entire towns have been burned to the ground. Cars abandoned on the side of the road. People are being forced to flee their homes. This is not a time for partisanship. This is a time for coordinating relief and response and lifting those in need up."

Trump had in fact been incorrect. The federal government owned 57 percent of California's 33 million acres of forestland. Another 40 percent was owned by private landowners that included families, corporations, and Native American tribes. State and local agencies, including Cal Fire, were in charge of only about 3 percent. "Managing all the forests everywhere we can does not stop climate change—and those that deny that are definitely contributing to the tragedy," said Governor Brown to reporters at the state operations center in Sacramento after his quick return from Chicago. "The chickens are coming home to roost. This is real here."

Now, standing amid the wreckage, the president appeared visibly moved by the devastation. He kept commenting to Jones how unbelievable it all was, how it looked like bombs had gone off. He addressed Paradisians as "my people," which Jones took as a hopeful sign—it seemed they were deserving of federal aid after all. When Trump turned to face the press gaggle, he thanked the law enforcement officers and elected officials in attendance, including House Majority Leader Kevin McCarthy and Representative Doug LaMalfa, who represented Paradise and much of Northern California in the nation's capital. Neither congressman believed in climate change.

"We do have to do management and maintenance," Trump said. "I think everybody's seen the light. I don't think we'll have this again to this extent. Hopefully this is going to be the last of these because this was a really, really bad one." Then he brought up Finland, a "forest nation," mentioning how "they spend a lot of time raking and cleaning and doing things, and they don't have any problem. I know everybody's looking at that."

He had touched on an important topic. Many of California's forests were diseased and overgrown, weakened by the practice of putting out fires immediately. California's woodlands had once been so healthy that a horseback rider could trot through the forest with outstretched arms and never hit a single trunk or branch. Such a feat was no longer possible—the foothills had become too tangled. In truth, forests could benefit from controlled burns. Firefighters sometimes kindled these "good fires"—formally known as prescribed burns—when the weather was accommodating, ridding the forest of fuels that might later stoke a blaze into a runaway conflagration. By some estimates, nearly 20 million acres in California needed to burn in this way to overcome a century of misguided fire suppression. (These preventive burns were not without risk, though. In 2000, firefighters in New Mexico lost control of a prescribed burn and it ended up destroying 435 homes.)

Native Americans had long used these low-intensity burns to release nutrients back into the soil, control weeds, and improve wildlife habitat—and they worked. Trump was making reference to this practice, but many Californians scoffed at his insensitive delivery and limited understanding of forestry. There was rarely public support or funding for fire prevention efforts. Stickers reading MAKE AMERICA RAKE AGAIN and PROUD TO BE FROM PLEASURE soon proliferated on bumpers across Butte County.

Less than an hour after the president had arrived in Paradise, the motorcade left and returned to Chico to visit the incident command post. At a table made of plywood and two-by-fours, the president pored over a map with Cal Fire director Ken Pimlott, who explained that the Camp Fire had grown to 149,000 acres and was 55 percent contained. Trump walked among the weary firefighters and applauded them for "fighting like hell."

Afterward, a reporter inquired whether Trump thought climate change had been a factor in the Camp Fire. Trump said he thought it was more a matter of forest management. The reporter pressed. Had the Camp Fire changed his mind on climate change? "No," Trump said. "No. I want a great climate. I think we're going to have that."

FIVE DAYS LATER, it was Thanksgiving. And finally the seasonal rains arrived, rinsing the town of ash and helping to further contain the Camp Fire. Soot congealed like glue in the footprints of ruined houses, and rain droplets glistened in the pine boughs. For the first time in weeks, the sky turned from a murky gray to a light, chilled blue. And a familiar milestone had arrived in the midst of so much unfamiliarity: a holiday.

But there was no savory waft of sage and rosemary from roasting turkeys. No families gathered around the hearth as the ponderosa pines quivered outside the window, perhaps dusted in snow or studded with yellow warblers or long-eared owls. Life was in a holding pattern. The evacuation zone hadn't reopened. The meaning that Carr and others sought remained elusive.

More than fifty-two thousand people were still displaced. They were struggling, one by one, to find housing—the first step in regaining what was lost. Evacuees had scattered far and wide. At the state operations center in Sacramento, the director of the Office of Emergency Services felt hopeless. The Camp Fire wasn't Mark Ghilarducci's first disaster, or even his worst. The fifty-eight-year-old had served as the incident commander on the Oklahoma City bombing and had overseen the recovery of 169 bodies, nineteen of them children's. Their tiny appendages—fingers, toes—had been mixed in with the debris. The trauma had deterred Ghilarducci from taking government roles for the next twelve years, but he had returned to the work in 2012, when Brown had appointed him to his current post. Now the state's fire siege was testing his fortitude. "Everybody was punch-drunk coming off a series of these wildfires," Ghilarducci would later say of the 2018 fire season. "In all my years, I've not seen a fire like this. . . . This is everything in ash:

schools, hospitals, care centers, restaurants, churches, homes, roads. Absolutely everything."

Bus driver Kevin McKay and his girlfriend, Melanie, had moved into a second-floor room in a rundown hotel off Interstate 5 in Corning, about thirty miles northwest of Chico. Obtaining lodging any closer to home was impossible. Every hotel room was booked. Their room had scratchy blue carpeting and a low desk, where Kevin worked on his homework or compiled an inventory list for their insurance company. His knees bumped against the desk's underbelly as he racked his brain, noting every possession down to spatulas and dog dishes in a spreadsheet. His life, itemized. He and Melanie slept in the king-sized bed. His twelve-year-old son, Shaun, was relegated to the fold-out couch a few feet away. Kevin's mother, in decline from melanoma, had caught pneumonia and been admitted to a hospital in Red Bluff, where Kevin had once managed a Walgreens store. They planned to spend Thanksgiving with a former colleague who lived nearby.

Meanwhile, Jamie Mansanares and his family were crammed into a fifth-wheel trailer, which he had bought after the fire and hauled across the mountains from Nevada County after local dealerships had already sold out. They parked the trailer at the Butte County Fairgrounds—south of Chico in Gridley—until the site unexpectedly closed. Afterward, they settled outside a former colleague's house on East Avenue in Chico. It was a tight fit for six people—including Jamie's mother, who slept on the couch—and their pet cat and their dog, Ginger. The freezer was so small that Erin couldn't even cram in a frozen pizza. The oven didn't work; the water heater was broken. Every two days, she and Jamie carted Tezzrah and her sisters to a friend's house to bathe. They washed their clothes at the laundromat. This was all inconvenient, but it was better than being homeless. To free up space, they eventually took Ginger, who enjoyed eating from the cat's litter box, to a temporary shelter.

Jamie worried about their future. He and Erin had lost not only their house but also both their jobs. Heritage Paradise had burned to the ground. Now the entire staff needed to find new employment. Erin knew they could float for a few months on the money in their savings account and the cash in their safe, but not much longer than that. By Thanksgiving, she had secured a job at a rehabilitation center in Oroville. Jamie also took the first job he could find: an entry-level gig at Sierra Pacific Industries. He worked the graveyard shift at the timber company, sorting lumber and feeding it through the mill, his fingers stiffening in the cold. With their opposing schedules and long commutes, Jamie and Erin rarely saw each other.

On this holiday, though, they tried to forget about their dismal reality. They joined Erin's sister for a family dinner in the town of Bangor. The spread was more bountiful than anything Tezzrah had ever seen—enough food to fill several Tupperware containers of leftovers, though they wouldn't fit in their trailer's fridge. Erin missed the familiarity of her parents' worn dinner table and the comfort of their holiday traditions, like the carving of a turkey her brother had raised. Luckily, her father had made his signature "Martian salad," a dish of layered green Jell-O, canned fruit, and Cool Whip. It was Tezzrah's favorite and left a frothy white mustache on her upper lip.

For Thanksgiving, three public dinners had been organized for fire evacuees, one of them at Sierra Nevada Brewing Company. The owner of the brewery had decided to ferment a special "Resilience" beer to raise money for Camp Fire victims. On the day of the feast, Sheriff Honea tossed the first handful of hops into the tank and grinned bashfully for cameras. To celebrate the holiday, Travis and his wife, Carole, drove to the brewery from their hotel in Roseville, eighty miles to the south in the Sacramento suburbs, because they felt "too messed up" to be around their extended family or other "normal" people. They had been regulars at Sierra Nevada Brewing for years and wanted to get as close as possible to Paradise.

After the firefighters had left with Paul and Suzie Ernest, Travis Wright had stayed on the Ridge for a few more days. His Subaru, filled with his and Carole's most precious possessions and paperwork, had burned, leaving him with no means of transportation.

He'd waited for help to arrive, soaking in their unheated hot tub, scrubbing at his ash-caked skin. He'd napped on the couch. Late one evening, he'd watched as a line of police cars and vans inched down Edgewood Lane to collect the skeletons of his neighbors. Beverly Powers and her partner, Robert Duvall, trapped at death in separate trucks, unable to be together in their final moments. Ernest Foss, the musician, in his wheelchair in the driveway, his oxygen tank and a garden hose the only things left intact. His stepson and caretaker, Andrew Burt, outside their minivan with Bernice the dog. Joy Porter and her son Dennis Clark, less than a half mile from their home on Sunny Acres Way. Then the vehicles had left with their somber cargo, headlights slicing the darkness.

The Ridge had been strikingly quiet. Travis's thoughts looped endlessly. He'd wondered if he was going insane. Meanwhile, Carole hadn't known whether her husband was even alive. His cellphone battery had died, and the search and rescue crews hadn't checked their house. Travis had finally managed to hitch a ride to Chico with a firefighter two days after the fire. When he'd arrived at his in-laws' house in Rockland, he'd refused to unlace his work boots. He needed to keep them on, he thought, in case he needed to run. His hair and eyebrows were burned off in patches; it looked like he had mange. Carole had finally coaxed her husband out of his boots, then shaved his head. At a local mall, she'd helped him pick out a pair of checkered slip-on Vans as a replacement. Travis had wanted a pair since he was a child. After hearing his story, the teenage salesclerks pooled their money to pay for the new footwear. Travis and Carole were touched by the gift. Now they tried to blend in at dinner in the crowded brewery, which felt something like home.

Rachelle and Chris Sanders, meanwhile, opted to spend Thanksgiving at the home of friends in south Chico. They had moved into the couple's spare bedroom while they decided what to do next. Rachelle's grandparents still owned the family home on Pentz Road, and she didn't expect to receive anything from an insurance payout. Chris, meanwhile, fretted about how his landscaping business would fare in a town with no living vegetation. At least they had each other. Rachelle couldn't help but think they had it all: good health, friendship, love. For the holiday, all five children were present: Chris's two daughters; Rachelle's little ones, Aubrey and Vincent; and baby Lincoln, of course, who'd dropped one pound in the weeks after the Camp Fire and had been admitted to the neonatal intensive care unit at Enloe Medical Center. Luckily, he was home by the holiday.

Another of the public dinners took place at Cal State Chico, in the Bell Memorial Union. Fourteen long tables were covered in white cloth, and twinkle lights glittered along the rafters. Soft jazz pumped from speakers. Hundreds had volunteered to help. They couldn't replace what the victims had lost, but they could ladle mashed potatoes and gravy onto paper plates. It was better than adding to the warehouses of extraneous items that well-meaning people had pulled from the backs of their closets—old bikinis and prom dresses, even a shipment of white Crocs that wouldn't sell in stores—not understanding that what evacuees really needed was soap, toothbrushes, and underwear. (Eventually, officials had to rent an empty Toys"R"Us building, then a municipal auditorium, and finally two 20,000-square-foot warehouses to store the unwanted items.) The aisles at Target and Walmart had been stripped bare of necessities.

This university hall wasn't home, though, and it didn't feel like Thanksgiving, not to the people gathered. They sat in clusters: one couple here, then a few empty seats, then a family with their toddler. These survivors found their minds wandering back to the houses that no longer existed. There was no sense of when they might experience normality again. One evacuee mused about past holidays as she sat with her son in the auditorium. There had never been enough seats, not even at the children's table, she recalled. "It's strange for me," she commented to a volunteer clearing plates. "To tell you the truth, it's like you're dreaming but awake. That's the way I feel. To lose everything . . ." She trailed off, remembering the enchiladas that she usually rolled and slotted into the oven on this day.

The Camp Fire's devastation had provoked shock and horror

across the nation—even in Guy Fieri, who had shown up at 5 A.M. on Thanksgiving to do his part, setting up six smokers in the college's parking lot to help cook dinner. The famous restaurateur and bleached-blond TV personality basted more than seven thousand pounds of turkey. Firefighters, more accustomed to vanquishing active flames than chatting with a celebrity, huddled around him, sleep-deprived and awkward. Fieri clasped their shoulders, telling them they were superheroes without capes, that they were crazy—"remarkable!"—warriors. He praised their uniforms, which were clean and pressed. He called them "brother." He agreed to every request for a selfie. "You bet, brother!" Fieri said. "Whenever you're ready, I have turkey for you."

The grief had been overpowering, but light was seeping in. By the end of the day, containment of the 153,000-acre Camp Fire would hit 100 percent. Volunteers wearing plush turkey hats guided lines of people indoors to eat and give thanks.

ABOUT FOURTEEN HUNDRED PEOPLE were crammed into twelve temporary shelters that stretched to the mountains of Plumas County. Evacuees slept on inch-thick foam pads layered with pilled blankets and drank watered-down coffee or tea. An outbreak of the highly contagious norovirus—causing muscle pain, nausea, severe diarrhea, and vomiting—worsened conditions, sickening at least 145 people across four shelters. After the fire, the county had helped more than two thousand people find temporary housing. Usually, administrators expected to need to assist about 10 percent of the population—but the Ridge had brought challenging extremes: an influx of elderly people with special needs and families living below the poverty line. "That's who you see in the shelters," said Shelby Boston, the county director of social services. "You see the least-resourced individuals, the folks who don't have a support system."

The county had also set up emergency disaster shelters for animals in Oroville, Chico, and Gridley. Within twenty-four hours, all facilities were full. They housed more than thirty species, including a few wild turkeys that had been rounded up by mistake and a parrot prone to cursing. There were chickens, horses, cattle, sheep, goats, miniature donkeys, pigs, geese, llamas, fish, dogs, tame and feral cats, koi fish, and three hermit crabs (only one of which would survive). For the first time, the National Guard was mobilized to help with an animal operation. With four thousand animals in shelters and another six thousand animals trapped in the evacuation zone, county staff desperately needed their help. The shelters went through a ton of cat litter and twenty tons of feed every day. Meanwhile, the chickens continued laying eggs, which volunteers drove by the hundreds to UC Davis for toxicity testing. Even more eggs were thrown out because scientists couldn't say whether they were safe to consume. (They never got an answer. Funding for the study ran out before the eggs could be tested.)

At the Walmart in Chico, a refugee camp had sprung up in the parking lot and an adjacent field. Employees from the superstore set up portable toilets to accommodate the makeshift city. To stay off the damp ground, fire victims slept in their cars or in tents set up on shipping flats. In some ways, the encampment, which came to be known as Wallietown, was better than the Red Cross shelters; here, at least, the evacuees could keep their pets close. Residents of Chico volunteered their help, smoothing white tarps on the concrete and piecing together plastic shelving, which they filled with diapers, tampons and pads, sports bras, toilet paper, canned pet food, and children's toys. A row of canvas tents contained food. There were bulk bags of navel oranges, powdered pink lemonade, carafes of hot coffee, and boxes of frosted doughnuts.

Volunteer Mel Contant, dubbed Wallietown's "mayor," helped people get situated. Looking out over the encampment, she knew people's living situation would get ten times worse before it ever got better. Paradise had offered relative comfort to the everyman—to the teachers, the electricians, the laborers, who couldn't afford to live elsewhere. Even if fire victims were lucky enough to have insurance that paid for a hotel, they were still left without food and clothing. They showed up at the Walmart parking lot, hoping to gather supplies from the bins of donations. FEMA had opened a Di-

saster Recovery Center in the former Sears storefront at the Chico
Mall, but the lines there were long and the process so convoluted
that it put further strain on the fire victims. FEMA would distribute
$79 million in housing assistance over the next five months, a pit-
tance compared to the payouts from insurance policies. In Wallietown, a flyer posted on a bulletin board promised that a trailer of
laptops would arrive the next day in case anyone needed to reapply
for their driver's license or Social Security card.

On Thanksgiving Day, volunteers dropped off aluminum pans
of crispy turkey and fluffy mashed potatoes for everyone. Overhead, storm clouds churned, threatening to deluge the parking
lot—a known flood plain. The food steamed in the autumn chill.
Condensation dripped from plastic ceilings. In a few days, Walmart
would hire a private security company to roust the encampment
and seal off the field with a chain-link fence. But for now, on this
holiday, there was space enough to stretch out and eat.

Meanwhile, 475 people were still unaccounted for. Captain
Sean Norman was tasked with leading the field recovery operation.
He oversaw the search and rescue teams, which now included four
thousand people—even anthropologists, army infantry, and cadaver
dogs. It was the state's largest recovery mission. He was getting up
at 5 A.M. every day and reporting to his post, listening to his colleagues as they yelled at him and wept in anguish. At 10:30 P.M.,
he'd climb into bed, feeling broken, only to do the same thing the
next morning—pulling on his boots, knowing he had a job to do.
"My feelings didn't matter," he said later, "because there were a
whole bunch of people who needed an answer about their family
members."

Norman was unwittingly drawing the blueprint for body recovery efforts, developing a model for mass casualties that state officials
could one day use after an earthquake in San Francisco or Los Angeles. But the experience was crushing. "In the fire service, we
don't like giving up one inch of ground, one bedroom during a
structure fire, one extra acre," he would later say. "We don't like
losing, because loss, in our business, is people's properties—their
livelihoods and businesses. We exist to reduce loss. People would

tell me, 'You're helping people find closure.' I think we like to say that, but I don't think that's real. Do you ever find closure when you find you've lost someone?"

Despite the holiday, Butte County officials had already convened a housing task force. They had realized a new crisis was looming. The inferno had destroyed 14 percent of the county's housing, and now there wasn't enough to replace what had been lost. The rental vacancy rate countywide had been 1.5 percent before the Camp Fire. Now it hit zero. Like many places across the state, the local governments in Paradise and Chico had been slow to construct housing. In a good year, for example, the Town of Paradise had approved permits for no more than thirty-five new homes. Now Ed Mayer, the Butte County Housing Authority executive director, was grappling with the question of where people would live. "There is no availability in California," he said.

People were desperate. The day before the Camp Fire, 243 homes had been listed for sale in Chico. One week later, fewer than 75 remained on the market. Realtors saw offers come in at $50,000 above the asking price. In Chico, the median home price rose from $310,000 to $370,000. Rents skyrocketed. Even Paradise Police chief Reinbold had given up hope of finding something suitable, signing a lease on an apartment in a complex populated by college students. It was the only place with enough room for his three young children. His new neighbors openly smoked joints and partied on their balconies, unperturbed by his badge and uniform.

THE THANKSGIVING DINNER for evacuees at Chico State was not nearly as well attended as organizers had hoped. There had been issues with the buses hired to shuttle people in from the dozen evacuation shelters. Still, firefighters—who had been sleeping at the fairgrounds to free up hotel rooms—arrived to help with the meal, gamely trading their unwieldy firehoses and heavy axes for serving spoons. They snapped on plastic gloves and covered their heads with hairnets, some trying to stuff ponytails under the webbing. They unwrapped metal trays of rolls, lettuce, and turkey, vegan and otherwise, along with pumpkin pies with delicate whirls of whipped cream that, within a half hour, had collapsed into runny dollops. "Hot, green beans, hot!" yelled a volunteer, edging her way through the crowd.

Reinbold and other local figures and politicians stopped by the cloth-covered tables to eat with their fellow evacuees. Some brought their teenage children, who avoided the bright lights of television news crews—there to film a holiday segment—to hunch over their devices. A man in a Santa Claus costume hugged the smaller kids, who giggled and patted his white beard. Virginia Partain, sixty-four, bussed dirty plates from one of the tables. She'd lost her home on South Libby Road. She had photos of it saved on her phone, which she now took out to show a reporter. In one picture, the dogwood tree out front was festooned with red fall leaves. She took off her plastic gloves so she could scroll better. In the next photo, the garage door was crumpled on the driveway.

"That's my life now," Partain said, shrugging. "I just want to feel normal again." She had lived in a friend's home office for a bit, then moved to a rental in Chico, but the losses felt unspeakable. Partain, a Paradise High School English teacher—one of the students' favorites—knew she had to stop doing this, scrolling through photos of a home that no longer existed. She didn't want to ruminate. She was dressed in the same clothes she had worn during the evacuation: tie-dyed socks and a tie-dyed shirt. Her purple hair was pulled back in a loose ponytail. Turning away, she left to clear more tables. Anytime she spotted her students eating with their parents, she gave them a big hug and called them "my babies."

And then she filled her own plate and rested. The cranberry sauce ran into the turkey and into the side salad. There were things to be thankful for, things she had forgotten you should be thankful for. Clean air, the acrid odor of burning homes replaced by cool mountain breezes and the scent of gravy. Her life, her health. Rainstorms that fell and tempered the flames. A plate of hot food steaming on the table.

For a moment, she was full.

CHAPTER 18

SECONDARY BURNS

Sheriff Honea thought he knew how to prepare for the nightly press conferences at the Silver Dollar Fairgrounds. Now forty-eight, he had worked in law enforcement for more than two decades, handling murder cases as a homicide investigator and then a deputy sheriff before being elected to his current role in 2014. He had a law degree and experience in the district attorney's office, where he had worked as chief investigator. The previous winter, after the spillway on the Oroville Dam nearly failed and threatened to inundate the valley under a massive wall of water, Honea had called for a 188,000-person evacuation. Though controversial, the order had cemented his reputation as a decisive leader.

But the Camp Fire was unprecedented, and it had stunned his home county. Residents were in the beginning stages of filing what would amount to 27,784 insurance claims. One small insurance company, the Merced Property and Casualty Company, would be forced into insolvency. The staggering figures didn't include uninsured losses, which topped billions of dollars. Even those who had been lucky enough to have insurance worried about losing their coverage in the future. In Butte County, nearly six thousand homeowners had been denied coverage in 2018 because they lived in high fire risk areas. It was part of a trend across the state: Between 2014 and 2018, 340,000 customers in zip codes affected by past wildfires had been refused coverage. (In 2020, the California legislature would propose a law making this practice illegal.)

Journalists and residents alike wanted answers. Every morning that November and December, Honea woke to the knowledge that he would have to face their questions at 6 P.M. During the day, he compiled notes and decided what information to share, his anxiety building until his daily 4:30 P.M. meeting with his investigators. There was no easy way to make sense of the eighty-five fatalities. Of the victims, twelve had been participants in the county's In Home Support Services program, including Ernest Foss, the musician on Edgewood Lane. Despite the efforts of staff members to reach them, they had perished in their homes or while trying to flee, enduring horrific deaths. The official tally included only people who had died in the fire or directly from fire-related injuries, so those like Jeanne Williams, who died of a brain bleed on the helipad at Feather River hospital, weren't counted among the victims. (Wrongful death suits against PG&E eventually included an additional fifty people who had died of interrupted medical care, smoke inhalation, or stress.) Honea worried about the long-term emotional impact of the tragedy, even on himself. "I wake up at 3 A.M. and lie awake, my mind running," he later said. "There's this general sense of impending doom."

Inevitably, the biggest questions always came at the tail end of each day's press conference. Did authorities know yet why the Camp Fire had ignited? Who was responsible for causing such death and destruction? Honea didn't have answers. Rumors were swirling that PG&E was at fault, and in early December, when the company submitted two electric incident reports to the California Public Utilities Commission, speculation reached a deafening pitch. The reports acknowledged outages on the Caribou-Palermo Line at 6:15 A.M. and the Big Bend Distribution Line at 6:45 A.M. on November 8—suspiciously close timing to the Camp Fire.

Honea understood that the answer wasn't so simple. The utility might have sparked the fire, but other factors had already transformed the Ridge into a tinderbox.

Climate change was one of the biggest drivers. As temperatures rose and periods of drought increased, the state's forests withered. Rainfall became erratic, and snowpack melted weeks early. Under these conditions, even the smallest blazes could mutate into a destructive megafire, exploding in size and intensity as they ripped across the parched landscape. Global warming had managed to change even the rhythm of lightning. Researchers found that strikes could double if greenhouse gas emissions aren't reduced—significantly increasing the threat of uncontrollable fire. Overall, the amount of land burned in the West is expected to rise sixfold.

But the state's flawed forestry management had also played a role. By imposing a century's worth of colonial fire suppression policies, foresters had allowed the woods to become diseased and overgrown. Sickened by insects—mostly bark beetles—and packed too close together, more than 129 million trees in California died during the state's recent drought. Rather than allowing low-intensity burns that might have cleared out this kindling—as Indigenous people have long practiced—nearly all fire was stamped out. Attempts at "prescribed" burning, or so-called "good" fire, failed at scale, slowed by the bureaucracy of permitting and complaints about smoky air. The risk compounded.

And Paradise's haphazard growth over the last century was also at fault—an act of willful disregard in view of its treacherous location. When George Bille, one of the town's first prominent settlers, arrived in Paradise in 1909, the community was loosely parceled into generous tracts; the only structure on his 160-acre homestead was a wooden miner's cabin. Bille cleared the old-growth oak and pine from his lot, seeded a barley field, and dug a well. But by 1960, after his death, his property had been parceled into tracts for seventy-five homes. Across the Ridge, builders carved subdivision after subdivision into the red dirt. The building regulations were minimal, applying only to properties that consisted of five or more parcels, at which point state zoning laws mandated the addition of a sewage system, sidewalks, and gutters. To skirt these rules, developers divided properties into just four parcels, selling the land to their friends and business partners, who quartered and sold them again, and so on, in an approach known as "four by fouring." The

Butte County Board of Supervisors, more interested in encouraging development in the city of Chico than in policing new construction uphill in Paradise, had allowed the town to expand without proper planning or infrastructure.

"We just kept growing and growing, creeping and creeping," remembered Dona Gavagan Dausey, who served as vice mayor on Paradise's first Town Council in 1979. "It was like spider vein growth. Every time I drove down the street, I would find another road that I had never heard of before. . . . They were selling all of these quarter-acre lots for three thousand dollars. People were buying them up like pancakes."

The notion of limiting how people could build on their land was a relatively new one. In much of California, growth had always been a matter of manifest destiny. "The future of Paradise is absolutely assured," read an editorial in the *Chico Record* in July 1908. "It is no longer a question as to whether it will survive. The only possible question is, how fast will it grow?" Development cut through forestland everywhere in the Golden State. Rural counties gratefully collected the property taxes for these new homes. By the time the California legislature voted in the 1970s for long-term planning, zoning, and environmental reviews to inform land use decisions, Paradise was already locked into place as a town as vulnerable as it was beautiful.

For the most part, the population had been happy with the town's indiscriminate sprawl. There were no rules, no requirements. No developer's fees, no zoning. It had taken four attempts for Paradise to incorporate as a town before it finally wrested control from the Butte County Board of Supervisors in 1979. Its residents loathed government intervention and didn't want the county making decisions for them. In 1991, voters had recalled the Town Council after it proposed establishing a downtown sewer district. In 2015, when the water district tried to increase rates by 30 percent—adding about $6 per household bill—two board members had been recalled, and a third was ousted in the next election. The district manager had been forced to take a new job in Chico. The architects of Paradise's original General Plan bemoaned the lost potential of their town. They kept trying to compensate for past failures, only to face a community unwilling to commit to any sort of change, whether large or small.

The hodgepodge growth of communities like Paradise was enabled, of course, by the spread of electricity. Following World War II, electrical lines ferried energy from hydroelectric plants in the Sierra Nevada to places where development hadn't historically been possible. By the 2010s, one in four Californians—about 2.7 million people—lived on flammable land, a number that continues to skyrocket despite the inherent risk. ("There's something that is truly Californian about the wilderness and the wild and pioneering spirit," Newsom would later tell reporters as he explained why he wouldn't block building in fire-prone areas.) California's relentless hunger for energy had allowed PG&E to cannibalize other utilities until it became a monster conglomerate. The state grew too reliant on PG&E to be able to halt its expansion.

And then PG&E became too powerful to halt at all.

ON TUESDAY, DECEMBER 11, 2018, Meredith E. Allen, the senior director of regulator relations for PG&E, submitted a supplemental incident report to the California Public Utilities Commission. In the report, Allen explained that PG&E had done an aerial patrol on the afternoon that the Camp Fire ignited and noted a flash mark—essentially a black scar left by the fire—on Tower 27/222. She declined to say when the tower had last been climbed for a detailed inspection—a fact important for deciding the company's liability, and its fate. Legislators were considering whether the utility should be broken up or restructured. Newsom, preparing to assume the role of governor in January 2019, was threatening a state takeover—though in truth, few lawmakers wanted to deal with the headache of safely providing power to 16 million Californians.

PG&E wasn't ready to take official blame for the Camp Fire. Allen finished December's supplemental incident report to the Public Utilities Commission by saying: "These incidents remain under investigation, and this information is preliminary. The cause of these incidents has not been determined and may not be fully understood until additional information becomes available."

The company's attention was focused on a more pressing battle that was about to take place in a San Francisco courthouse—near the utility's headquarters on Beale Street—over how victims of the Camp Fire would be compensated for their losses. Those impacted by the 2017 Wine Country wildfires were also still waiting to be compensated; they had filed about $30 billion in claims, sending PG&E to the brink of bankruptcy. In the days after the Camp Fire, the company's stock had lost half its value. (Hedge funds had bought much of the downgraded stock, becoming the largest shareholders.)

Already the public was expressing fury—not just with PG&E but also with the system that protected it. In late November, dozens of protesters from Butte County had driven to the Bay Area to speak during public comment at a California Public Utilities Commission meeting. They were disillusioned with a system that they thought protected PG&E. "This is the second straight year that we've had to choke on the carbonized remains of our neighbors," a Chico resident remarked. "It's evil, but it's not illogical, because if you're going to bail out PG&E, to provide public money to socialize their losses, what is their incentive to be safe? To act safely? There is none."

As PG&E was scrambling to save face, and damages, in court, Butte County district attorney Mike Ramsey had joined forces with Cal Fire to launch a joint criminal investigation into the utility. It was a tough challenge: Ramsey would need to impanel a special grand jury and present enough evidence to prove that the company had been "criminally reckless and grossly negligent." Resulting charges could range from manslaughter to environmental crimes. If the utility pleaded not guilty, then the case would go to court—at a great cost to taxpayers, particularly if PG&E's lawyers asked for the trial to be moved to San Francisco, where there was a greater chance of an unbiased jury.

But Ramsey, seventy, wasn't intimidated. Gregarious and grandfatherly, the owner of the meth lab tank turned aquarium had light blue eyes and bushy white eyebrows, as well as an acerbic sense of humor that could turn on a dime—a peculiarity that unnerved opposing attorneys. They could never tell when he was making a joke. As a young man, Ramsey had studied history, math, and physics at the University of California, Berkeley, before returning to his hometown of Oroville to work as a substitute teacher, then as a staff writer at the local newspaper. After a stint as a court reporter, he had gone on to earn a law degree. He had been elected and reelected eight times in Butte County—the longest-serving district attorney in California history.

For Ramsey, the Camp Fire had been personal. His youngest daughter, pregnant with her fifth child, had almost died while evacuating with her other four children from Paradise; her husband, a Cal Fire captain, had been helping out with evacuations in Concow. (Their infant son—delivered at home—was the first baby born in Paradise after the Camp Fire.) Paul and Suzie Ernest's daughter, Arielle, worked in Ramsey's office, and ten of his staff had lost their homes. Like every prosecutor, Ramsey believed in holding people responsible for their actions, and under California law, a corporation could be criminally charged just like an individual. (This is why PG&E had faced criminal charges for the San Bruno explosion in 2010, when its failure to turn off the gas in its broken pipelines eradicated a neighborhood and killed eight people.)

Still, PG&E would be a tough plaintiff to hold accountable, because of its power and resources. Ramsey knew this from firsthand experience. He had spent much of the previous year mired in talks with the company's attorneys about its role in igniting the 2017 Honey Fire, which had burned 150 acres near Honey Run Road and scared "the hell out of Paradise," as Ramsey put it. The blaze had sparked when an untrimmed tree branch snagged a PG&E distribution line. On November 5—three days before the Camp Fire—PG&E had finally agreed to a $1.5 million settlement that would fund a four-year power line inspection program overseen by Cal Fire. Though Ramsey considered it a victory, PG&E had won too: It had avoided adding another criminal charge to its record and escaped the scrutiny of the feds, who were still monitoring the utility after the San Bruno explosion.

PG&E'S TAINTED TRACK RECORD was alarming not just to the general public. It also had a demoralizing impact on its twenty-three thousand employees, many of whom lived and worked in the very communities that had been devastated by its mistakes—some had lost their homes, too. Even so, many of the linemen, who could be seen driving through Chico and Paradise in blue company trucks, became lightning rods for people's anger. "The people on the front lines don't make policy," explained Bob Dean, fifty-seven, who had lived in Paradise for fifteen years and was a representative at IBEW Local 1245, which represented about twelve thousand of the utility's employees. "We just know that when something bad happens, we have to run and fix it. It's crushing. PG&E probably caused the fire, and our guys are taking the blame."

By Christmastime, the simmering rage toward PG&E had reached a boiling point. Across Butte County, residents hurled trash at the utility's workers. They spat and yelled. They slashed their tires and painted "85"—the number of the dead—in red paint on their vehicle doors. Hotels wouldn't rent them rooms; restaurants refused them service. One employee said he'd had a gun pulled on him.

Sitting in his blue truck at a gas station, Luke Bellefeuille held his breath as a stranger approached, hoping the other customer would pass. He found himself struggling to justify working for the utility. PG&E had been a good employer to the thirty-eight-year-old—it offered benefits, health insurance, and a retirement plan, a rarity in his hometown of Paradise. It had taken him many years to earn a spot at the company, scoring a job first as a painter, then in hydropower maintenance. Bellefeuille, who stood six foot seven, had been proud of his extra-tall uniform, his blue company-issued truck, and his civic-mindedness: roadwork, welding, flume repairs. But that was before he'd become one of 89 employees who had lost their homes in the Camp Fire, and one of 122 whose homes had been destroyed by PG&E-caused wildfires since 2017. The only trace of Bellefeuille's property was the in-ground swimming pool. He and his girlfriend, along with her nine-year-old son, had only recently moved into their newly constructed dream home on the lower Skyway. Now, when they should have been celebrating the anniversary of their move, they found themselves homeless.

Bellefeuille's co-workers had reported being flipped off or sworn at by local residents. "We're out here busting our butts every day, even though we lost everything," Bellefeuille said. "And people treat us like this? We're victims too." He knew he had become a scapegoat for his employer's failings—but he didn't think he could quit. Paradise was all he knew. He had learned to fish in the Feather River and had ridden his first dirt bike through these foothills.

And every December, Bellefeuille had sipped hot cocoa at Mountain View Tree Farm, then helped his father, a general contractor, chop down their prize pine and lug it home. But this year, he knew, would be different. Ash was drifting like snow across Paradise. The tree farm lay in ruin, the pines scabbed with charcoal. When a friend offered to let him join his company clearing destroyed lots, Bellefeuille stopped hesitating. He turned in his resignation at PG&E. He figured the career change was a way to give back to his hometown.

The business manager of PG&E's union wasn't surprised. Over the past twenty years, Tom Dalzell had seen stress like Bellefeuille's accumulate in many employees as "one crisis after another" eroded their faith in PG&E. "One San Bruno," he said, "goes a long way toward undermining public confidence."

AFTER TWENTY-NINE DAYS, it was time for the people of Paradise to return to their town. It was mid-December, and the Butte County Sheriff's Office had sent a CodeRed alert to announce the news—the end of the evacuation organized by zones, just as it had begun. Only this time, the alerts dispatched properly. The text read: "The Butte County Sheriff's Office, the Paradise Police Department and other agencies will lift evacuation orders at 10 A.M. for

'Zones 3, 8 and 14' along the Pentz Road corridor." This had been the first area hit when the Camp Fire stormed through.

Rachelle didn't know how to feel. She had fled a neighborhood but was returning to a "zone scheduled for repopulation." She tucked Lincoln into his carrier, pulling the blue visor over his face so the sun wouldn't shine in his eyes. She wasn't prepared to see her home again—or rather, what was left of it. As Chris started the Suburban—the same one he had driven the morning of the fire— she slipped into the backseat next to Lincoln and checked her phone. Chris backed out of the driveway of their friends' house. Reindeer decorations stood on the front lawn, the grass limp and yellow. Christmas wreaths dangled in the windows, and lights looped the flagpole.

A half hour later, after trying first to turn onto Pentz Road, which was still closed, Chris reached the town limits on the Sky- way. He flicked off the radio. He and Rachelle noted a series of neon orange signs: DETOUR AHEAD! and ROAD WORK AHEAD! and ROAD CLOSED AHEAD! They joined a line of vehicles. California Highway Patrol officers herded them through a checkpoint in a church parking lot, making sure their driver's licenses had a Paradise address, then handed them white hazardous materials suits in case they wanted to sift through the toxic rubble. An officer swiped or- ange paint on their windshield: proof they belonged. Storm clouds churned in the sky. Chris and Rachelle sat in silence. Lincoln was asleep, fists tucked under his chin.

Near Pentz Road, the landscape became unfamiliar. Ribbons tied on mailboxes signified that the property had been checked for possible victims. Behind the new chain-link fence around Feather River hospital, orange and red leaves littered the parking lot. The berries spangling the bushes had been baked like raisins. The hos- pital was closed indefinitely; its thirteen hundred employees had been laid off. Chainsaws shrieked as crews felled trees along the streets, their limbs frosted with ash. Only a few blocks on, Ra- chelle recognized some markers of their old life: Trash bins lining the road. The Ace Hardware, its windows advertising a Christmas tree special that had already expired. On the left side of the road, the owners of one destroyed house had strung up a new American flag in the front yard. The flag was the brightest color for miles. Rachelle could see into the yards of houses that had once been cloaked by forest. In every direction, tiny figures in tiny white suits hunched over the remnants of their homes. In the vast openness, their pain seemed raw and glaring.

And then they saw their house—or where it should have been. Rachelle knew it would be gone; she had already seen it burn on the day of the fire. The crepe myrtle saplings that Chris had planted that fall were now blackened twigs. So were the purple oleander, their thorns so sharp that they made rosebush thorns feel like butter knives. A mangle of bikes and bedframes lay scattered around the garage. Even the bathtub was unrecognizable. Chris parked and got out first, walking over to the house's footprint and toeing the ruins. His boots crunched over broken glass and rusted screws. A pitcher's mound, painted in white on the driveway, was still visible on the concrete, marking the spot where he and Vincent had once tossed the ball.

"The whole wall fell in?" Rachelle asked, staring aghast at the garage as she stepped from the car. She left Lincoln to sleep in the SUV. Underfoot, the soot was as viscous as paste. The air reeked of smoke and chemicals. "Is this the window frame?" Chris didn't respond. "I was okay until I saw the kids' bikes," Rachelle told him, laughing and then crying, dabbing at her eyes with a shirtsleeve, trying not to smear her mascara. "It doesn't matter how much they tell you it's gone. It's different when you see it. I don't feel like there's anything left to search for."

They hugged for a moment. Rachelle knew her wedding bands were in the wreckage somewhere. She had taken them off a month before and stored them in her bedroom nightstand, after her fingers had become too swollen from pregnancy to wear them. But she knew she wouldn't bother to sift through the ash, desperate to find what had once been so precious to her, only to find warped rings she couldn't wear. Mourning her losses again seemed masochistic, like a fire's secondary burns. What was the point?

Rachelle spotted her elderly neighbors in the adjacent lot. They plucked decorative deer from their lawn and stacked them near the front steps, salvaging the few items they could locate. She returned to the Suburban and unbuckled Lincoln from his car seat. The baby's forehead furrowed as she lifted him into the morning brightness. This was the home he was supposed to have returned to, the home where he was supposed to have taken his first wobbly steps under the dogwood in the backyard and learned to throw a baseball in the driveway. His siblings would have played board games with him in the "Motel 6" back bedroom. Outside in the summertime, they would have pointed out constellations to him, backs pressed against the grass. Grief catching in her throat, Rachelle turned her back on that now impossible future.

CHAPTER 19

REBIRTH

On a Wednesday in early May, people waited in line at Paradise Town Hall. Filing past contaminated water fountains covered with hazard signs, they made their way to the window of the Development Services office, which was adorned with yellow streamers. They came loaded with questions and paperwork, seeking approval to install trailers on their burnt-out lots, or permission to rebuild, though the Town Council was still figuring out the requirements for new construction. "Can we make our town more fire-resistant and still keep it affordable?" Mayor Jones often asked herself in the months after the Camp Fire. She didn't know the answer. She and her husband had purchased a house in Chico, beating out four other bidders, while they waited to rebuild themselves. New housing would at the very least have to comply with the stricter state building codes enacted in 2008.

The members of the Town Council, burned out of their homes like everyone else, weren't prepared for these existential challenges. They had recently voted to allow residents to live in travel trailers on their contaminated properties despite the health hazards, but FEMA had threatened to cut emergency funding and they had rescinded their decision, infuriating the community. "The amount of metals in that ash from asbestos could be dangerous," said Bob Fenton, the FEMA administrator overseeing the wildfire recovery efforts. "I simply asked the question: Is it hazardous? Or is it not? If it's not, then I no longer have the authority to remove debris from private property. If it is, I do."

Reminders of the tragedy were still easy to find. Residents talked of suffering from "fire brain," and the landscape mirrored their ravaged emotional state. Shuttered convenience stations still advertised gas prices frozen at an appealing $3.27—while the current prices, now topping $4.00, presented an added hardship for those displaced and facing long commutes, like Jamie's family. The roads in Paradise were rutted where vehicles had burned, and millions of tons of debris littered the gullies and ridges: rusted cars, liquefied play sets, mangled rebar, shattered glass. The state agency leading the $2 billion clean-up had cleared only 17 percent of the town's lots. Paradise's drinking water had been contaminated with cancer-causing benzene after the district's pipes had melted; officials estimated it could take up to two years and $50 million to restore water service, a sobering testament to the wildfire's reach—it had destroyed even the underground infrastructure.

The fire had consumed the town's tax base, and with customers gone, business owners didn't know whether they would reopen. Thousands of residents had filed lawsuits against PG&E, hoping to receive some kind of settlement to help them regain what they had lost. A handful of litigators, who called themselves the Northern California Fire Lawyers, plastered billboards along the Skyway, and attorneys from as far away as Texas opened offices in Chico. They brought in Erin Brockovich, of the Hinkley contamination case, to speak with potential clients.

Downhill from Paradise, cities on the valley floor struggled to absorb the overflow. The number of prescriptions filled at the Walgreens in Chico doubled, as did the number of driver's licenses issued by the Department of Motor Vehicles office and the amount of sewage handled by Chico's water district. Compassion for fire victims had given way to resentment on the part of those now inconvenienced by their presence. When Chico's mayor and a councilman failed to support a state bill that would have accelerated the construction of new housing, bitter residents organized a recall effort, saying they had lost faith in the leaders' ability "to guide the community" in the aftermath of the Camp Fire. Many more residents of Paradise left California altogether, as a housing official had predicted before Thanksgiving. They settled in places like Hawaii and Florida, or Washington, Oregon, and Idaho, even Micronesia, where they could leave the tragedy far behind. They would be counted among the 1.2 million Americans displaced by natural disasters in 2018.

The ones who stayed—and who were standing in line for permits that day in May—faced the growing skepticism of urban planners. There was a sense that the usual calculus had changed, that fire-prone communities must discuss not only how to rebuild, as the Town Council was currently doing, but whether they should rebuild at all. Between 1970 and 1999, 94 percent of the roughly three thousand houses destroyed by wildfires in California had been rebuilt in the same spot—and often burned down a second or third time. Rebuilding was a difficult and daunting process, compounded by the high cost of construction materials and labor as well as the rush to meet the two-year deadline for payouts from insurance companies. One year after the 2017 Tubbs Fire, which had destroyed three thousand homes in Santa Rosa, only fifty-five had been rebuilt—less than 2 percent of those lost.

In Sacramento, legislators had met at the state capitol earlier in May to discuss the fate of places like Paradise as part of Wildfire Awareness Week. It no longer seemed possible to ignore the destruction wrought in the past five years by drought, overgrown forests, a warming climate, and a vulnerable electrical grid. "We need to have that tough discussion," said Ghilarducci, the former Oklahoma City bombing incident commander who now worked as the director of the state's Office of Emergency Services, to his colleagues. "There are too many risks. It's not just here. You're looking across the country at climate-related disaster impacts—they are severe. There are communities getting wiped out time and time again, with lives lost, when they could have been prevented by not building in certain areas."

James Gallagher, thirty-eight, a Republican whose legislative district in the State Assembly included Paradise, wasn't ready to abandon the town. Taking the microphone, he argued before his fellow lawmakers that barring development in certain places seemed rash. "If we say, 'Here are the places in California we can build and not build,' then I think we are missing a major point," Gallagher said. "Should we deprioritize places that have earthquake risk? Should we deprioritize places that have flood risk? Because we are going to start running out of places where we can build."

ON MAY 22, 2019, the Paradise Town Council held a special meeting to present its building plan to the public—and to hear from a PG&E executive. Thanks to donations from a local nonprofit, the council had enlisted the help of a firm, Urban Design Associates of Pittsburgh, that had assisted New Orleans after Hurricane Katrina. Before an audience of about a hundred in the Performing Arts Center, representatives from the firm flipped through a slideshow of neat neighborhoods whose electrical lines were buried underground— a feature that could stave off future wildfires. The images were computer-generated and pristine, the vision extravagant. The downtown would be walkable, and the business district would have an $85 million underground sewage system. Residents listened, wide-eyed. Though the state legislature had announced plans to funnel $31.8 million to fire-affected communities like Paradise, to restore lost property tax revenue, and to provide another $10 million specifically for Camp Fire recovery, the funds would not be nearly enough to cover this futuristic plan. "The town cannot afford this," one man said bluntly.

Then Jones announced the next speaker: "At this time, we'll move on to item 1C, which is a presentation by Aaron Johnson from Pacific Gas & Electric Company on their future energy plans for the town of Paradise."

Johnson, dressed in khakis and a collared blue shirt, timidly approached the podium. The previous week, Cal Fire had released an official report that held PG&E responsible for the Camp Fire. The state agency had shared its findings with Butte County district attorney Mike Ramsey, who had already formed a Criminal Grand Jury as part of his investigation. Sworn in on March 25, the nineteen members—the number would eventually drop to sixteen after three people couldn't make the time commitment—had been meeting every week in secret to review the evidence. Not even their friends and family knew of their task.

Publicly at least, PG&E could no longer dodge culpability. Johnson, a former energy adviser to the California Public Utilities Commission and now PG&E's vice president for electric operations, was the one chosen to face Paradise. He didn't anticipate a positive reception. At a November meeting at Chico State University, he had been booed off the stage by angry Butte County residents—a searing experience. His cheeks had flushed redder than his hair.

Meanwhile, at the company's headquarters in San Francisco, a fresh cadre of executives had arrived at the glass offices. Ten new directors had been appointed to the utility's board, along with a new CEO, Bill Johnson, sixty-five. Johnson, who had taken the reins earlier that month, had previously led the Tennessee Valley Authority for six years. But he was best known for his tenure as the CEO of Duke Energy Corp.—a position he held for just a few hours before stepping down, just long enough to collect more than $40 million in exit payments. (When Duke Energy merged with Progress Energy, where Johnson was CEO, in 2012, he had signed a three-year contract to head up the new company. On the day the merger went into effect, he was forced to resign at midnight.)

"Um, thank you for having me," Aaron Johnson began, staring out at the packed auditorium. "I'm going to admit to being a little bit nervous here tonight." Chuckles rippled through the audience. "I do want to start on a personal note and just say, I've been up here representing the company since the fire. . . . As you all know, I'm sure, Cal Fire announced last week that it was determined that our electrical transmission lines near Pulga were the cause of the Camp Fire. While we have not seen the full report yet, we accept this determination. This news is a weight that we carry, and all PG&E employees feel that burden." Johnson gripped the microphone. "It wasn't unexpected. From the outset, we have tried to be as transparent as possible that our lines may have caused this fire. . . . So, as I stand in front of you here tonight, on behalf of my company, I want to apologize for the role our equipment had in this tragedy. Nothing I can say in front of you is going to undo that, and I know that. We also understand it creates an obligation for us to do the right thing for this community."

Johnson thanked the first responders who had risked their lives during the Camp Fire and announced PG&E's new Wildfire Assistance Fund, a $105 million program intended to help those who had been displaced by wildfires in 2017 and 2018. The utility, he announced, also planned to spend the next five years installing a new underground electrical system in Paradise—a small piece of the plan that an observer had earlier deemed too expensive. At this news, the Town Council stood in a show of support. "We don't deserve that, but I'll take it," Johnson joked. "Our new CEO, Bill Johnson—no relation, unlucky for me—came up here during his first full week on the job. One of the things he did was have lunch with a group of our employees who lost their homes in the fire." That group, the vice president said, had only one request for the CEO: that Paradise needed to be rebuilt in a way that made it safer and more robust. PG&E was going to try to make that happen, he finished.

Residents lined up for public comment. For Warren Harvey, the utility's newfound contrition did little to temper his bitterness. "I've been a resident here since 1947, so if you don't agree with me, that's okay, but I got here first," Harvey began, addressing Johnson. "Here's my problem with you. You're going to underground the power lines, but you've already cut down seven trees on my property that were over a hundred years old. You didn't need to kill the value of my property. Now that you have decided that you're involved, just make me whole, and I'll be happy. Get your checkbook out—my name's Warren Harvey—and I'd love to see a check from you."

As a chorus of boos arose, directed at him, Johnson looked overwhelmed. "I hear you," he replied.

. . .

JUDGE WILLIAM ALSUP had his own plan for making PG&E confront its negligence. After Aaron Johnson's visit to Paradise, the judge, who was responsible for overseeing PG&E's probation for the San Bruno explosion, ordered all of the company's executives and board members to tour the Camp Fire burn scar. Their guide, District Attorney Ramsey, organized the stops. Privately, the two men called it the Tour de Ashes. Ramsey hoped company officials would be swayed into doing the right thing for the people of Butte County after witnessing the damage caused by PG&E's neglected infrastructure.

Earlier that year, Judge Alsup had learned about the settlement the company had brokered with Ramsey over the Honey Fire. Alsup was livid that the company had evaded responsibility once again, and so in late January 2019, he had summoned PG&E's attorneys to his chambers at the Northern District of California court, near San Francisco's Civic Center Plaza, for a dressing down. The utility had recently appeared in the same courthouse to fine-tune the details of its bankruptcy protection, but on that midwinter morning, its attorneys instead headed to Courtroom 12 on the nineteenth floor. The room had high ceilings and wood-paneled walls. The air conditioning blasted. Alsup, seventy-three, liked to keep his chambers cold and his audience alert.

"Every time I ask you to admit something, you say you're 'still under investigation,'" Alsup said to PG&E's legal team. "Usually a criminal on probation is forthcoming and admits what they need to admit. You haven't admitted much." He reminded the lawyers that the company was still on probation until at least 2022. That probation, he ruled, had been violated when PG&E sparked the Honey Fire in Butte County and didn't properly alert its probation officer. As punishment, Alsup was considering whether to force the company to implement an expensive and sweeping new wildfire prevention proposal. "Now, we need to say another thing," Alsup said, raising his eyebrows pointedly. "Climate change and global warming make it a lot worse—or at least worse—and so [does] the drought . . . but the drought did not start the fire. Global warming did not start the fire. According to Cal Fire, PG&E started them, all of them. So that raises the question: What do we do? Does the judge just turn a blind eye and say, 'PG&E, continue your business as usual. Kill more people by starting more fires'? I know it's not quite that simple, because we've got to have electricity in this state, but can't we have electricity that is delivered safely?"

Kevin Orsini, a $1,500-an-hour attorney from a prestigious New York City law firm, stepped to the front of the chambers. At forty, he had dark, wavy hair that had begun to retreat. He leaned in to the microphone, talking quickly, trying to make his points before Alsup could interrupt, as he tended to do. "We have an inherently dangerous product, is the fact," Orsini acknowledged. "We have electric running through high-power lines in areas that are incredibly susceptible to wildfire conditions. . . . PG&E understands and accepts that it has a credibility problem. Which is why I couldn't stand up here and say to Your Honor, 'Trust us. We've got it.'"

Alsup looked pained. "What troubles me," he replied, "is . . . come June 21—I'm unofficially saying that's the start of the wildfire season and it will run to the first big rain in October, November, or December . . . six months, and it will be dry as can be and the fire season will be on us and the emergency will be on us, and will we be seeing headlines 'PG&E has done it again, started another fire' and some other town has burned down?"

"We share that concern, Your Honor," Orsini said.

Five months later, Alsup—not so convinced by PG&E's "commitment" to change—arranged for the Tour de Ashes. Now the group boarded a nondescript white bus and ascended into the foothills. The excursion was conducted under such secrecy that not even Paradise's councilmembers were initially aware it was taking place. Alsup didn't want any distractions. No media, no cameras. And certainly no self-congratulatory press release from the PG&E public relations team.

First stop: Jarbo Gap. The group of about thirty-five got off the bus at Scooters Café, where they heard from several Concow residents. The wildfire was more destructive than anything they had ever seen, said resident Peggy Moak. After PG&E electrical lines fell and blocked the main evacuation route out of Concow, she and her husband, Pete, had chopped them apart with an ax, not knowing whether they were electrified, to help their neighbors escape. Now she was worried that her beautifully rugged community would scatter and disappear completely. FOR SALE signs flecked the residential streets around the reservoir. Moak saw more of the signs with each passing week.

Next, the group headed to the Paradise Performing Arts Center, and on the way, Ramsey played video footage from first responders and residents as they fled for their lives, synchronizing the bus's route with the depicted locations. The driver stopped in front of a ruined hillside home on Norwood Drive. Ramsey flipped to a ten-minute 911 recording from three generations of women: an immobile grandmother, her daughter, and her disabled granddaughter. From the Emergency Command Center in Oroville, a Cal Fire captain listened to the Hefferns—Matilde, sixty-eight, Christina, forty, and Ishka, twenty—die in the exact spot the PG&E executives were now looking at. "Ma'am, can you get out of the house?" the captain asked Matilde on the recording. It was 9:19 A.M. "We can't breathe," she said. "Please help us." He told her to try kicking through the wall or going to the front porch to wait for first responders. Help was on the way, he promised. Matilde's daughter and granddaughter moaned and screamed in the background. The matriarch, breathing heavily, explained that every door was on fire and her skin was burning. "Don't leave me," she said. "Please, please, sir, please. Are we going to die?" The line crackled. "Are you in the house?" the dispatcher asked. "Are you out?"

He never received an answer.

The bus began moving again. At a mobile home park near Feather River hospital, Sheriff Honea boarded. He showed the executives a photo of what remained of the three women: a small, clear bag filled with bone shards. Ramsey shared the story of another victim, a woman who had refused to evacuate her trailer—located in the park they were sitting in—without her cats. One of his investigators had finally succeeded in persuading her to leave, only for her to slip out of his hands and run back into her home as it burst into flames. Weeks later, the same investigator had had to return to collect her remains.

What the group wasn't told, but might have read in the papers, was that twelve more burn victims—among them Paul and Suzie Ernest—remained in serious condition at a specialized burn center in Davis, that dozens of people had sought refuge in a lake, and that one man who had died remained unidentified, and would be for years, if not forever, his remains unable to be returned to his family. And lately, a rash of suicides and heart attacks had taken out community leaders. The fire had gone cold, but the disaster continued.

At the Performing Arts Center, the board ate sandwiches from a local food truck, Camp Fire BBQ. A group of residents lined up to talk, including former mayor, and self-described "former town drunk," Steve "Woody" Culleton, schools superintendent Michelle John, and the general manager of the irrigation district, Kevin Phillips. John described the trauma the students had endured, how their lives were forever altered. The teachers were the glue holding the children together. "How do you build a town without water?" Phillips, of the water district, beseeched them. Butte County Sheriff's Office investigator Tiffany Larson, who had spent weeks picking through the rubble for bodies, told them her home was gone—and how could that be, when she had always felt safe in the middle of Paradise? She and her husband likely weren't coming back, she added, because the cost of rebuilding was too high.

Cellphones jingled. The first Public Safety Power Shutoff of the summer—an electrical blackout like the one in Calistoga in 2017—was about to begin. Seeing the executives glance down at their phones, Steve Bertagna, a Paradise Police sergeant, lost his temper. "What the hell are we?" Bertagna asked them. "Some third-world country that when the breeze blows, we have to turn off the power? We're the fifth-largest economy in the world!" Bertagna had joined the force seven years earlier after serving as the mayor of Chico and running a small business selling stereos. At fifty-four, he had invested himself in the community, determined to serve others in any way he could. He waved at every child from his patrol car and promised to give "every little old lady a hug if she needed it." Now Bertagna was tired of hearing chipper people say Paradise was "surviving and rebuilding." His department had been gutted, and it wasn't okay. "Can't you people fucking deliver power safely?" he yelled. His audience was quiet, their cellphones silenced.

Later that afternoon, the bus dipped down the Skyway and descended to the valley floor, heading south to San Francisco. In the spongy soil along the roadside, the executives caught their last glimpse of Paradise—eighty-five memorial crosses, staked in a row.

SPRING, THOUGH, had brought new hope. Volunteers planted thousands of hyacinth and daffodil bulbs at the entrances to Paradise. The pine sapling that had been donated for the town's tree lighting ceremony in December grew taller. The post office reopened, along with the Holiday Market, Starbucks, and Ace Hardware. Three dozen physicians from Feather River hospital joined the staff of a small health clinic downtown. The VFW hall, in the process of being reroofed, had plans to resume Wednesday night bingo.

Nature regenerated, healthier for the burn. In the ashes, a medley of wildflowers thrived: poppies and scrambled eggs, wild oat and St. John's wort, scarlet Indian paintbrush and fluted lupine. They ran up the hillsides and bunched in bright splotches around the remains of front steps and along picket fences, growing in places where they never had before. Trees pushed new green shoots through the earth. One couple had planted a new pine sapling in the hollowed trunk of its predecessor. A black bear cub lumbered through the forest and nosed for blackberries in the brush. Mosquitoes multiplied in puddles; they had never been so populous. The red-legged frogs were uproarious in their evening chorus.

And the seniors of Paradise High were about to celebrate their graduation. Faith Brown, seventeen, could hardly wait. Before the wildfire, she had found Paradise stifling. Sometimes, when she took a date to a restaurant, the waitress would text her father that she was out with a boy. "Give him a hard time," he would reply, because he didn't want his daughter to date until she was older—when she had a college degree and he was in a nursing home. She had grown up on June Way, near the Mountain View Tree Farm. While many of Faith's classmates had planned to go to Butte College, she had spent the previous year eyeing Oregon or Colorado, where no one knew her and there were bigger universities.

The fall semester had seemed as if it would end on a good note. At five foot nine inches, Faith was petite for an outside hitter, but she could jump and helped the varsity volleyball team go to the state playoffs. She was smart—she had skipped kindergarten—and received mostly A's and B's. After school, she worked at the Boys and Girls Club assisting younger students with homework, and on weekends she and her friends roamed Paradise freely. They knew every hiking spot and swimming hole on the Ridge. They walked on the railroad tracks near the flumes and watched sunsets from Lake Oroville. When there was nothing else to do, they piled into Faith's green Subaru Outback and drove around looking for unexplored roads.

But then came the Camp Fire, and all that changed. Faith was living with her parents at the DoubleTree hotel in Sacramento; her mother didn't want to stay in Chico because of the traffic congestion and the people with glazed-over eyes who wandered around like zombies, traumatized by the wildfire. Faith slept on an orange fold-out couch with springs that dug into her back. Her parents slept in the only proper bed. There was a dresser with three drawers; they each got one for their few belongings. They washed the dog dishes in the bathroom sink and ate at Denny's most nights. Disagreements got blown out of proportion. Faith wished she had a door to slam and a place to be alone. She longed to be with her friends, who were staying in trailers and hotels in Chico, but her mother wanted the family to be together.

On Instagram, some of Faith's classmates posted photos from San Francisco, Davis, and San Jose. They had moved without saying goodbye. Faith found solace in music, compiling playlists of country music with songs like "What Happens in a Small Town" by Brantley Gilbert and "You're Gonna Miss This" by Trace Adkins.

She named one playlist "Paradise," another "Home." When she drove around Chico, Faith could almost pretend it was still possible to turn onto the Skyway and navigate to her old house. On social media, she posted photos of the Paradise she had known: the historic Honey Run Covered Bridge, which had burned down, and the duck pond where she had posed for prom photos the previous year in a form-fitting black gown, her arms around two friends, both in red.

Faith's final semester of high school felt endless, robbed of normal milestones. In the spring, it was tradition for seniors to pick a night to watch the sunset, then have a massive sleepover together in the parking lot. There was also prom, graduation. Would her diploma even say Paradise High School? She didn't know. After the wildfire, school officials had scrambled to lease vacant warehouses and office buildings in Chico. Classes had first been held in a former LensCrafters store at the Chico Mall, where canned holiday music echoed in the tiled hallways. The students called it Mall School. Then, in January 2019, Paradise High had moved into a cavernous office building near the Chico Municipal Airport, where Faith now reported every day. Students nicknamed their new quarters the Fortress because it was on Fortress Street.

Some days, it felt like she was moving at 65 mph, unable to take in the passing landscape. Her parents offered to make an appointment with a therapist, but she wasn't ready. "Are you OK?" friends kept texting her. "Yeah, I'm OK," she would reply. Then it would hit her—how she wasn't okay at all. How she had driven alone through six hours of flames and seen the McDonald's where she had ordered McFlurries after school catch flame, along with the church she had attended as a child and the Dutch Bros hut where she had worked. How she felt older. And how she desperately missed the little town she had once hoped to leave far behind.

BY MAY, AT LEAST one of Faith's wishes had come true. Her diploma would carry the name of her longtime high school. In early June 2019, 220 seniors came together to graduate on Paradise

High's football field, part of a treasured tradition. Volunteers arranged white chairs on Om Wraith Field for the students, and parents sat on the metal bleachers, clutching programs that listed their children's names. They had anticipated this moment for so long. The seniors clustered on the turf, a sea of green and white gowns, the boys in creased khakis and the girls with lipsticked mouths and curled hair stiffened with hairspray. Students always seemed older on graduation day, but this group had grown up particularly fast, matured by disaster.

That the ceremony was even taking place was a small miracle. Schools superintendent Michelle John had pulled off the impossible. The very features of Paradise that residents had cherished—the isolation, hulking trees, sloping hills—had greatly complicated the clearing of burned lots, and Paradise High had been closed for the past 210 days. Undeterred, John had fought her way through the bureaucratic red tape to make graduation at the school happen. (Her success would feel deeply bittersweet three days later, when her husband, Phil John, the chairman of the local Fire Safe Council, would die of complications from a heart attack—likely compounded by the stress of the Camp Fire—sustained during a bike race. He was a half mile from the finish line.)

The evening was perfect. Golden light filtered through the charred pines. Principal Loren Lighthall, forty-six, paced near the stage in his signature black Crocs. The students arrived in two uneven lines, converging with fist bumps, elaborate handshakes, explosions of confetti, and long hugs, before filing into the folding chairs. They wore leis made of folded dollar bills. Lighthall moved to the podium, testing the microphone and shuffling papers. The audience broke out in cheers: "Lighthall! Lighthall! Lighthall!" "They're talking about you, Lance," he joked, scanning the rows of seniors for the face of his seventeen-year-old son. "Welcome and good evening, graduates, faculty, and friends. I am Loren Lighthall, the proud principal of Paradise High School."

He paused. It would be his last time saying these words. Over the past two years, Lighthall and his wife, along with their seven children, had become integrated into a community they adored.

But after their 2,200-square-foot home had burned down, they realized they couldn't afford, emotionally or financially, to help rebuild what was lost. Lighthall cherished his students: He had attended every sports game and stocked his office with candy and granola bars so that teenagers would drop by and chat. He was proud to count rodeo champions, beauty queens, and the only National Merit Scholar in the region among his senior class. But that spring, his job had begun to feel increasingly untenable. The daily trauma and his family's cramped apartment in Chico were too much to deal with.

Paradise High School hadn't burned down, but some of its storage buildings for sports equipment and uniforms had, and the main building had sustained smoke damage. About four hundred pine trees on campus were at risk of toppling at any moment. Enrollment had dropped by half, and normal classes weren't expected to resume until August 2019. More than 90 percent of the district's thirty-four hundred students had lost their homes, as had dozens of teachers and several school board members. After school, most students drove to temporary homes as much as an hour away, like Faith, or crammed into thirty-foot travel trailers. Lighthall knew that many families were waiting out the semester before making the decision to move permanently. They had one foot in each world, straddling their old life and their new one. Some of Lighthall's students hadn't even known what address to type on their college application forms. When Lighthall announced his resignation, it made the Sacramento evening news.

That morning, eating some celebratory egg burritos with seniors in the cafeteria, he had started to feel pangs of regret over the decision. Students he'd known for two years flooded into the cafeteria. The building felt different, smaller and quieter, after their long absence, and smelled of dust and citrus cleaning solution. Outside, crews had painted the bleachers and watered the flowers, which had been planted in the shape of a P, on the lawn. Dead pines had been felled, sawdust hanging in the air. Volunteers had set up the stage where Lighthall now stood, looking out on the students he loved.

The crowd quieted, and after Lighthall, other speakers climbed onto the stage. There was Bob Wilson, a San Diego businessman who had written every student, teacher, administrator, bus driver, and custodian a $1,000 check after reading about the Camp Fire in his local newspaper, and Sheriff Honea, of course. A few students spoke too. They remembered Mall School and the Fortress and the weirdness of seeing airplanes from their English classroom. The sky drained of color as the graduates snaked across the field, Faith among them. She had decided to attend college near Paradise after all. Airhorns blared from the audience. Lighthall handed diplomas to his students, and then to his son—a moment he would watch on a video recording a half dozen times in the following weeks.

"Graduates, please stand and face the audience," the principal said. "It is with great honor that we present the Paradise graduating class of 2019!" Mortarboards flew through the air. The seniors rushed off the field, scattering confetti in their wake. The Camp Fire had ended life as they had known it—and now graduation would do the same. Lighthall watched as they folded into the backseats of family minivans and friends' sedans to leave, wishing that graduation wasn't ending, wishing that he could stay. He waited an hour longer, until he was the last person on the field.

OBSERVATION: ANNIVERSARY

On the anniversary of the Camp Fire, more than fifteen hundred people crowded the parking lot of the old Bank of America branch in Paradise. Only 2,034 residents had returned—a fraction of its population of 26,500. Governor Newsom had reclassified the town as a "rural area." There were two grocery stores, three gas stations, and five restaurants. Most establishments closed by 7 P.M. The ruins of scattered houses puckered the ground.

Around 10:30 A.M., the microphone clicked. Mayor Jones announced that the ceremony was beginning. Officials like Representative Doug LaMalfa planned to speak. Though LaMalfa's congressional district had been repeatedly struck by climate disasters, he maintained that global warming wasn't real. "The climate of the globe has been fluctuating since God created it," he had said during a candidate forum in Redding earlier that year. But LaMalfa, often wearing his signature white cowboy hat, his graying goatee neatly trimmed, was never one to miss a press opportunity.

As Jones finished speaking, there was a ruckus from the Skyway.

"We can't hear you," a resident shouted at her.

"That's such a Paradise thing to happen," a man said,

chuckling at the dysfunction. He was standing next to Steve "Woody" Culleton, the former mayor who had made a point of walking inside the home he and his wife were rebuilding on Forest Service Road at 8:30 A.M., exactly the time they had evacuated the previous November. Now his car alarm screeched, interrupting the ceremony. Culleton fumbled with his key fob to mute it.

The day was already hot, the sun toasting bald heads pink. A toddler laughed, reaching for his mother's nose. LaMalfa talked about the strength of Paradise. Other people spoke of hope, resiliency, and perseverance—the buzzwords of disaster recovery. Councilman Mike Zuccolillo took the microphone as 11:08 A.M.—chosen to mirror the date, 11/08— neared. The town had planned eighty-five seconds of silence for the eighty-five people who had lost their lives. Across Butte County, everyone was uniting for this moment to remember, even if they couldn't physically be together.

"You've heard the word 'resilient' time and time again," Zuccolillo said. "This is the true definition of it."

An unnatural stillness descended outside the old Bank of America branch. There was a subdued shuffling of flip-flops, sneakers, work boots, and sequined ballet flats. Tissues were pressed to damp eyes. The people gathered thought about many things in that minute and a half. The eighty-five victims. The keys that no longer opened front doors. The neighbors who had moved away. The feeling of being asked, "Should you *really* rebuild?" when the answer wasn't so simple, because where else would they go? Home was here, and their families were here, and this town was all they had ever known.

They thought about the few homes that had inexplicably survived, an odd kind of luck, the kind that spared some people and not others for no apparent reason. The rumble of debris trucks weighted down with the ashen remains of their lives. The rumble of construction trucks loaded up with fresh-milled lumber and new dreams.

They thought about all the things they had lost: the Gold Nugget Museum, the Elks Lodge, Mendon's Nursery. The old Paradise sign topped by the bandsaw halo. The one million crippled ponderosa pines. The sound of children on playgrounds, the bustle of downtown. Eighty-five seconds of silence for how much it had hurt, and still hurt. Silence, too, for the things that they had held on to: Johnny Appleseed Days and Gold Nugget Days; trick-or-treating for full-size candy bars on Lancaster Loop; balmy summer evenings at the drive-in movie theater, a mattress thrown in the truck bed; the red dirt that stained their clothing; how every phone number began with 877 and directions were simply "upper," "middle," or "lower" Paradise—no explanation needed; the cool waters of the Feather River; the air that smelled like heaven after the first winter rain or the first warm day of summer. Somehow it all felt holy.

If a town was only houses and buildings, would they still be gathered here?

CHAPTER 20

RECKONING

I n March 2020, as the coronavirus pandemic raged and the state
of California entered lockdown, PG&E's lawyers reached an
agreement with Butte County district attorney Mike Ramsey.
He had concluded his report, though it wouldn't be released for a
few more months. The company announced that it would plead
guilty to one count of unlawfully causing a fire and eighty-four
counts of involuntary manslaughter. (An eighty-fifth death was
proven to be only indirectly related to the wildfire; a Magalia man
had committed suicide rather than burn to death in his home,
which in a cruel twist of fate had then survived the firestorm.) It
was a remarkable step—and one of the greatest admissions of corporate criminal liability in United States history. Eight years earlier,
the petroleum company BP had made news by pleading guilty to
manslaughter for the 2010 Deepwater Horizon explosion, which
killed eleven people and polluted the Gulf of Mexico with millions
of gallons of crude oil. But those convictions paled in comparison
to PG&E's eighty-four counts. The utility had become one of the
country's most prolific criminals.

A senior inspector helping out in Ramsey's office had spotted
immediate parallels between the Camp Fire and the San Bruno
explosion in 2010. He pointed to issues with PG&E's record-
keeping and inspection processes. The oversights were so egregious

that if PG&E had been an actual person, a district judge said, it would have faced the maximum sentence of ninety years in state prison. But because a corporation couldn't be sent to prison, the most the court could do was levy the maximum possible fine for the felonies: $3.48 million. (This amounted to about 20 cents per PG&E customer; in 2017, the company netted $4.6 million in revenue *every day*.) An additional $500,000 would go to Ramsey's office as reimbursement for its investigation, plus $15 million to provide water to residents. In all, for the deaths alone—excluding fines—Ramsey could only seek $10,000 per person killed in the Camp Fire. "We decided to take responsibility for the role our equipment played in this fire," CEO Bill Johnson told a reporter. "We didn't make this decision lightly, but in the end, I think this is the best course forward, particularly for the victims." He said he hoped the plea would "move along the process of rebuilding."

For the families of the victims, however, this hardly seemed like enough. "I don't believe justice is served by a $3.5 million fine," said Joseph Downer, whose brother, Andrew, had died in the blaze. "This is my big brother. This is the guy who taught me so much in life. PG&E took all of this from me, all of it."

Laurie Teague said she had expected more than a wrist slap for PG&E, particularly in view of the terror she knew her seventy-nine-year-old stepfather, Herb Alderman, had experienced before perishing. "He didn't drive, but he was sharp as a tack," she recounted. "He knew everything that was going on around him. [A physical therapy receptionist] had been on the phone with him on the morning of the fire. She said he called at least five times, each time sounding more panicky and desperate. He was pleading for help, begging for someone to come and get him. His last words were: 'The fire is two houses away, help me.'"

Paul Ernest had not survived the conflagration either, despite Travis's best efforts and multiple skin grafts. Arielle Funk, his daughter, was both grief-stricken and livid. "He spent nine months in intensive care with burns on thirty percent of his body," she said. "He spent the last year unable to walk and breathe on his own.

November 8, 2018, has become that day for our family—the defining mark. Before the Camp Fire, or after the Camp Fire." She continued: "He won't be present at our annual family beach bash. We won't get to eat his homegrown produce, which he was so proud of. We won't get to honor him on Veterans Day—the only day he allowed us to treat him—or celebrate his favorite holiday, Thanksgiving. He's not here to help my mom through the biggest hardship of her life."

David Shores, whose brother Don and his wife, Kathy, had died, was more pointed: "Throughout history, there have been many instances where one individual caused deaths and faced years of imprisonment. When a corporation is involved, they only face a hefty fine, which . . . cuts into their profit margin for one year, and [then they] go back to business as usual. Mass murderers are not tolerated in our society. A corporation should not be a shield to mass deaths."

Ramsey had looked into the possibility of charging specific individuals for the Camp Fire, but he knew he couldn't meet the burden of proof that California law required to pursue charges against company executives, past or present. There simply wasn't enough evidence to indict them. Many of the decisions that had led to the Camp Fire, his investigation found, had been made in the 1980s, 1990s, and 2000s. It was almost impossible to prove that a person who had made a decision then had understood that they were laying the groundwork for a catastrophic wildfire decades in the future and had willfully chosen to ignore that risk.

A 696-page report released by the California Public Utilities Commission in December 2019 had already documented some of PG&E's missteps on a grand scale. That month, the state agency imposed a nearly $2 billion penalty on the company for causing fires in 2017 and 2018. Across the utility's entire transmission system, it had found a pattern of inadequate inspection and botched maintenance. On the Caribou-Palermo Line alone, twenty-nine high-priority problems had been discovered. The potential consequences were so dire that PG&E permanently shut the line down.

But the biggest finding in the government agency's report involved Tower 27/222, whose hook had broken and sparked the Camp Fire. The CPUC report revealed catastrophic negligence. Company rules mandated that each transmission tower be climbed once every five years for a detailed inspection. Yet this one had not been scaled since 2001—seventeen years before the historic wildfire. "Timely replacement [of the hook] could have prevented the Camp Fire," the report concluded.

THAT MONTH, as PG&E was admitting its guilt, Geisha Williams—who had been the CEO of PG&E at the time of the Camp Fire—listed her luxury Marin County home for sale. Located in Tiburon, the house had sweeping views of the bay and the Richmond–San Rafael Bridge. Williams, who had abruptly resigned in January 2019, banking $2.5 million worth of severance payments in cash, had bought the house in 2008, one year after joining PG&E. She stood to earn back upwards of $4.7 million on the sale—more than the price PG&E had had to pay for the eighty-four lives lost on her watch.

THE CONCLUSIONS of the ninety-two-page report compiled by Ramsey's office were damning, detailing how a company had reduced inspections and overlooked failing equipment on a century-old system. The Caribou-Palermo Line had been erected after World War I by the Great Western Power Company, which PG&E had acquired in 1930, nearly ten years after the line was built. The inherited equipment was never fully cataloged, but many of the original pieces appeared to have remained in place, outliving the laborers who had installed them.

Investigators found extensive deterioration on parts of Tower 27/222, as well as bolted-on replacement plates, which indicated that PG&E was "aware that the hooks and holes were rubbing on each other causing wear." An engineer with expertise in failure analysis and a meteorologist hired by government officials concluded that the weakening of the broken hook was consistent with about "97 years of rotational body-on-body wear." The ancient hook had been whittled down to the barest thread of metal—less than one eighth of an inch, about the width of the tip of a ballpoint pen—when it finally snapped. Wind and weather had catalyzed its slow decay. It had been bought for 22 cents in 1919; a replacement hook in 2018 would have cost $19.

The hook's deterioration would have been apparent in any reasonably thorough inspection, but over the decades, such checks had occurred less and less frequently. Per company policy in 1987, crews were required to patrol the transmission line three times a year: once from the ground, twice from the air. They were tasked with climbing 5 percent of the towers annually. By 2005, the line was being patrolled only once a year, with a closer inspection once every five years. As the number of inspections dropped, so did their thoroughness—as did PG&E's expenses. From 2001 to 2008, PG&E inspected the Caribou-Palermo Line only by helicopter. In 2001, more than one and a half days had been spent on these aerial inspections. By 2018, the inspections concluded within a few hours. Though PG&E employees maintained that they strictly followed company procedures for the inspections, Ramsey's office found that they did not, largely because of confusion over what those procedures actually were. Even PG&E couldn't fully explain its process. "This inability to determine who made decisions and upon what those decisions were based frustrated efforts to identify individuals potentially personally liable for policies that led to the conditions which caused the Camp Fire," read the report.

The departments within PG&E had also been extremely siloed, Ramsey found. The tower division oversaw maintenance of the transmission towers; the line division oversaw maintenance of the electrical lines. The hooks fell somewhere in between, succumbing to "wear and wear and wear with no one looking at them, until obviously disaster struck." He compared it to buying a used car and failing to invest in its maintenance, driving it until it fell apart.

"It was the hook that took the lives, the hopes, the dreams, the health, the sanity, the wealth, the happiness of a community," Ramsey said later in court. "But etched into the very soul of this community is a concern: What will happen next? Will this happen again?"

AT THE NORTH BUTTE County Courthouse in Chico, the main hearing room was bathed in the dim blue light of a projector. It was June 16, 2020, the day of PG&E's sentencing. The meeting had been rescheduled multiple times as the pandemic known as Covid-19 spread across the nation, infecting and killing hundreds of thousands of people. Now, with the proper health and safety precautions in place, the legal fallout of the state's deadliest and most destructive wildfire was finally reaching its long-awaited conclusion. The families of the dead were hoping for closure.

That week, PG&E's $57.65 billion bankruptcy restructuring plan had been approved. Approximately $13.5 billion would be put into a trust to pay wildfire victims; another $25 billion would go to settlements with local governments, owners of insurance claims, and fire victims who had filed civil suits. The hedge funds, which owned not only most of the company's stock but also investments in its debt and insurance claims, now stood to profit handsomely. (They had bought these claims from insurers who had paid for homes destroyed in the fire, some for as little as 30 cents on the dollar.)

There had been no state takeover, as Governor Newsom had threatened, and no vastly reimagined utility. Instead, PG&E had opted to split its operations into regional divisions. The board of directors would be overhauled again, and another new CEO installed. Plans were announced to sell the company's century-old Beale Street headquarters; a move to nearby Oakland would offer a reduction in costs and a fresh start. Meanwhile, the California Public Utilities Commission had carved out a new safety enforcement process. If PG&E bungled it at any point during the coming wildfire season, it stood to lose its operating license—at which point the state could opt to take over the company's electrical grid. (It was an empty threat. Preventing the company from operating at all would leave millions across the state powerless.)

At 9 A.M., court entered session. District Attorney Ramsey and his staff, as well as PG&E's attorneys, wore homemade or surgical masks. Seating in the courtroom was limited, and chairs were mostly empty. Because of the virus, proceedings were being broadcast live online. Bill Johnson, the outgoing CEO of PG&E, stood behind a pale wood lectern. On this Tuesday morning, his interim replacement—Bill Smith, a PG&E board member and former AT&T executive—hadn't joined him in court. Johnson's mask hung around his neck so the audience could see his face. His hands were clasped in front of him and he shifted from foot to foot. Butte County Superior Court judge Michael Deems slowly read aloud the names of the victims, who had suffered gruesome, preventable deaths. The court projected their portraits on the wall. "How do you plead?" Judge Deems asked.

Johnson paused to take in the photo of each victim. Shirlee Teays, ninety, found inside her home on the Skyway, holding a framed photograph—of what, investigators couldn't tell. Cheryl Brown, seventy-five, and her husband, Larry Brown, seventy-two, seated in their recliners. Don Shores, seventy, and his wife, Kathy Shores, sixty-five, seated in theirs. Teresa Ammons, eighty-two, outside her trailer with her purse. Ethel Riggs, ninety-six, unable to reach the manual release of her garage door. Lolene Rios, fifty-six, in her basement with her four dogs. Richard Brown, seventy-four, the father of nurse Chardonnay Telly, splayed under his truck. TK Huff, seventy-one, facedown in his garden, having crawled ten feet from his wheelchair in an attempt to escape the flames.

"Guilty, Your Honor."

David Young, sixty-nine, who crashed his minivan into a tree while evacuating, then burned alive along with two pets in its cargo area. Rose Farrell, ninety-nine, three months shy of her hundredth birthday, next to her wheelchair on her front porch. Evva Holt, eighty-five, in the backseat of a burning truck on Pearson Road. Marie Wehe, seventy-eight, in a burning truck on Windermere Lane. Julian Binstock, eighty-eight, in the shower with his border collie mix, Jack. Sara Magnuson, seventy-five, under a wet carpet in her bathtub, while on the phone with her neighbor, Cal Fire dispatcher Beth Bowersox. Three generations of Heffern women—Ishka, twenty, Christina, forty, and Matilde, sixty-eight—in their bathtub. Travis and Carole's neighbor Paul Ernest, seventy-two, who had spent most of the past year in a hospital burn unit, holding on long enough to give his wife, Suzie, the will to live.

"Guilty, Your Honor."

Herb Alderman, seventy-nine, home with a sprained ankle. Judith Sipher, sixty-eight, home with the flu. Andrew Downer, fifty-four, home in his wheelchair. Anna Hastings, sixty-seven, home with severe scoliosis. Rafaela Andrade, eighty-four, in the house her husband had built by hand for them. Gordon Dise, sixty-six, who returned home for something and never made it back out. Joseph Rabetoy, thirty-nine, without a car on Angel Drive. Barbara Carlson, seventy-one, and her sister Shirley Haley, sixty-seven, their arms wrapped around each other at their home on Heavenly Place.

"Guilty, Your Honor."

A half hour later, after acknowledging PG&E's guilt eighty-four times, Johnson read from a printed statement. He could never take away the pain felt by the many thousands of people who had been harmed by the Camp Fire, he said. PG&E was "deeply sorry" for the fire and its "tragic consequences." To some, the words sounded awfully similar to those uttered by a former CEO after the 2010 San Bruno explosion. "On behalf of PG&E, I apologize, and I apologize personally for the pain that was caused here," Johnson said. "I make this plea with great sadness and regret, and with eyes wide open to what happened and to what must never happen again."

Later that afternoon in his Oroville office, Ramsey became emotional as he ran through a list of all that had been lost in the Camp Fire. "PG&E destroyed not just these people, but an entire community," he said. "It killed a town." Before him, pieces of Tower 27/222 had been dismembered and arranged on a long table for public viewing. "We hope that, as Mr. Johnson said today, this [verdict] will live on. Safety must come first—not a matrix of risk analysis versus profit. PG&E must do this or, like these victims, die."

THE FIRE: LIGHTNING SIEGE OF 2020

Storm clouds churned. Electrons rubbed and fizzed as the purple summer storm gathered strength above the Pacific Ocean, then whirled toward California and kissed the shoreline. Soon the energy exploded. Thunderclaps reverberated, though no rain followed. More than twelve thousand lightning strikes—at 50,000 degrees Fahrenheit, about five times hotter than the surface of the sun—slapped the parched ground. There had been no appreciable rainfall for months, no reprieve from the changing of the climate, no answer to the man-made—and man-ignored—warming. As lightning struck in California, hurricanes were boiling in the Caribbean Sea, so many on the horizon that researchers turned to the Greek alphabet for new names. The hottest temperature on earth was recorded at Furnace Creek in Death Valley: 130 degrees Fahrenheit. A respiratory pandemic raged on and on and on, and record heat waves caused rolling blackouts.

From the historic lightning siege, more than 560 uncontrollable wildfires ignited. During the late summer of 2020, they leaped and roared across California, devouring everything in their paths. More than 3 million acres scorched, hundreds of thousands of fire refugees fleeing from their homes, fire tornadoes pummeling the ground. Birds fell from the jaundiced sky, their feathers singed charcoal black. The Joshua trees burned, never to repopulate that stretch of Southern California desert. The redwood trees—many older than the first European settlers to colonize North America, some older than Christ—burned too, along with homes and neighborhoods tucked deep within the forests. Licking and spitting sparks, the flames zipped through the isolated gullies and ravines and coughed into the sky, regaining their proper place on the land. Fire would always prevail.

But in the first early hours of a mid-August day—when all was still calm and the future was inconceivable, when the warming of the globe seemed deniable, when a wildfire was a worthy and beatable foe, forever the "new normal"—there was only darkness and potential.

Above Butte County, lightning roiled and struck ground.

And then: the first crackle of red flame.

KONKOW LEGEND

With the next sun they started; the young men first to clear the way and frighten the wild beasts, and the women, the young maidens, and the little children, with Peuchano and Umwanata in their midst, in the middle of a long line, with the old men bringing up the rear. For many days they journeyed thus over the mountains and across streams, always making the kakanecomes first before they slept at night—until, one evening, they saw away off in the distance a green valley, with the setting sun shining upon it. They halted, and Peuchano and Umwanata were brought to the front.

Shading their old eyes with their feeble hands, anxious and silent, they gazed long and trembling upon it, until one by one the tears chased each other down their old wrinkled faces, and falling upon their knees they looked upward, and with clasped hands and sobbing voices, cried, "Welluda, welluda, once more!" And the young men took up the glad cry, stronger and stronger, as it went: "Welluda, welluda, our home!" And above it all, rising sweet and solemn above the grand old pines, the songs of praise of the young maidens to Wahnonopem, the Great Spirit, who had brought them safely through so many dangers to welluda, their old home. And in the long, long years—as many as the stars above—around the campfires of the tribe at night the story was told by the old to the young; and I tell it to you, as it came down to me.

REBURN

Nearly two years after the Camp Fire, a lightning-sparked wildfire in Plumas County raced into Butte County, pushed on by the Jarbo Winds. The North Complex Fire, as it was called, broke the ranks of the state's largest wildfires, and Butte County found itself atop the leaderboard of misery once again. The firestorm of September 2020 killed sixteen people and leveled the hamlets of Berry Creek and Feather Falls. An hour's drive from Paradise, the communities were tucked in the foothills near Lake Oroville, directly on the other side of the Feather River Canyon. As the fire leaped toward Paradise, evacuation warnings again went out for the town.

Captain Matt McKenzie, now forty-four, propelled himself into the middle of the firestorm. Many things had changed for him since the Camp Fire—most notably, his home station. He had departed Jarbo Gap the previous year for a new post in Paradise. Working at Station 36 had left him with a lingering sense of grief, particularly on the first anniversary of the Camp Fire. He had imagined that November as a fresh start. "I don't know why I thought it was going to be some huge weight lifted off my shoulders," McKenzie had said at the time. "It's just time. It's three hundred sixty-five days—not some miracle number. I had it set in my mind that after a year, we would be able to move on." Back then, at least, he had been grateful that the county had made it another year without a major wildfire, though he knew conditions hadn't changed. This year, they hadn't been so lucky. But McKenzie had ruptured his ACL earlier that summer and was on light duty until he could get surgery. Still, ever the softhearted one, he was intent on helping a friend evacuate his animals. McKenzie couldn't stand to think they might burn alive. He wedged his truck through the flames and sped into another disaster.

At the same time, Sean Norman, who had been promoted to battalion chief, was trapped with a handful of residents at the local fire station in Berry Creek. One person was so badly burned that in order to start an intravenous line, a paramedic had to drill the needle into his bone marrow. Norman, now forty-nine, felt a weary sense of déjà vu. "At some point next week, a bunch of people are going to come home," he said later. "A wife like my wife and kids like my kids. They're going to sift to find two or three things that didn't melt. I could have never imagined the things and the places where this job has taken me. . . . I'm just tired. I'm tired of seeing dead animals. I'm tired of seeing dead people. I'm tired of seeing destroyed communities. I just don't want to see it anymore." Yet one image offered him hope: the look on his daughter's face when she heard a whinny from the barn. She had finally persuaded her parents to buy her a horse, whom she named Sparkles.

As fires continued to ravage California, the top ranks of the Butte County unit shifted again and again. Darren Read retired, and David Hawks took his spot. When Hawks retired in 2019, John Messina, now forty-seven, was propelled to the top of the pecking order. He had taken over supervision of a county that had seen more than 30 percent of its land burn in the last decade—and would probably see even more soon. "I gotta be honest with you, it's tough to watch another part of the county be destroyed by fire," Messina said. "It's tough to watch another rural community devastated, because we know how hard that recovery process is. It hangs over your head." But Messina had kept one memento from the Camp Fire. On his new desk, he displayed the soup-can-sized rock that he had used to anchor his map. Someday he planned to get it engraved.

Meanwhile, dispatcher Beth Bowersox was struggling to move on. In the past two years, she had battled post-traumatic stress disorder, brought on by the 911 calls she had answered on that fateful morning. In 2019, Bowersox had finally worked up the courage to attend a work-sponsored retreat for the disorder in Boise—but while she was there, a freak tornado had ripped across campus and torn off the roof of the building she was in. Still, Bowersox had found the workshop helpful, even posting publicly about the experience on Facebook: "Going to the retreat didn't fix me . . . it doesn't work that way. But it has given me hope (something I was in drastically short supply of) and tools to keep moving forward to heal and become my true self." After taking a three-month medical leave of absence earlier that summer, Bowersox found herself at her workstation again as the North Complex Fire roared. She fielded a potential evacuation order for herself, along with more frantic 911 calls—and Bowersox wasn't sure how much longer she could hold it together.

Up the Skyway in Paradise, the recovery process remained slow and arduous. The town was rebuilding without any major changes to its transportation system—despite the evacuation issues made apparent during the Camp Fire. Only 298 homes had been rebuilt, while another 1,149 permits had been issued. Community leaders, including the schools superintendent, the high school principal, the keeper of Paradise Lake, and the fire chief, had retired or moved away, further destabilizing the community that remained. Others took their places: Steve "Woody" Culleton was running for the Town Council again, as was Luke Bellefeuille, the PG&E worker who had left the company to help rebuild his hometown and now channeled his civic-mindedness in a new direction. Marc Mattox, the former assistant town manager, had recently returned to Town Hall to head up the Public Works Department. He had relocated to Chico the previous year for a job as an engineer with the city—but he returned to work in Paradise a few months later. He knew his heart would always be on the Ridge.

One key person, though, wasn't sticking around. As the North Complex Fire threatened to sweep into Paradise, Town Manager

Lauren Gill, now sixty-two, was packing up her office. Her five-year contract with the town had ended. Though more than ninety people had applied for her job, Kevin Phillips, who had overseen the Paradise Irrigation District, had beaten out the competition. Gill was relieved to be leaving. She had gotten married in May and wasn't sure whether she and her new husband would stay in Paradise. "Some days, I think the only way to clear my mind is to go somewhere else and change this scenery," she said. "But you also can't just run away from something or leave it behind. You have to deal with it. I'm not going to let this fire defeat me." In January 2021, though, the town would release an after-action report calling out the "lack of coordination" at Town Hall on the morning of the Camp Fire, which had led to a delay in alerts that "could have caused major confusion and further compromised the safety of Paradise residents."

As Gill said her socially distanced goodbyes at Town Hall, Travis Wright was preparing to defend his bungalow on Edgewood Lane from flames yet again. This time Carole was with him; she had asked for the day off from work. Not knowing whether her husband had survived the Camp Fire had left an indelible imprint on her psyche. She wasn't going to leave him this time. The couple was still living with Travis's sister in Chico until they could finish repairing the damage to their home: The Camp Fire had burned holes in their roof and scorched some walls. Finding a contractor was difficult; there was more money to be made in new construction. Travis and Carole were also waiting for a settlement from the class action lawsuit they had joined against PG&E in 2018. Travis knew, though, that the cash would amount to little compared to what he had lost.

He still thought about Paul every day, tormented by the thought of what he might have done differently to help save his friend's life. Paul's four-wheeler was still parked in Travis's garage. Sometimes Travis could still pretend his neighbor was alive, about to walk over and hop back on the ATV for an afternoon ride. In their kitchen, Travis and Carole had crafted a small shrine for him: a candleholder Paul had carved from knotted wood, a program from his funeral service, and one of the packets of tomato seeds that Suzie had handed to every guest. "Please plant and enjoy," the packet read, "in honor of Paul Ernest."

Meanwhile, Jamie and Erin Mansanares were preparing to evacuate Palermo, the unincorporated town south of Oroville where they had moved in 2019. Their home in Magalia wasn't rebuilt yet. The relocation had been hardest on their middle daughter, Mariah, now five. Her best friend had moved to West Virginia, part of a mass exodus that sent thousands streaming out of Butte County to more affordable places. Mariah was inconsolable. She missed her friend and hated their new house. "Why did the fire pick us?" she beseeched her parents. Tezzrah, nine, and Arrianah, four, seemed to be handling things okay.

Still, Jamie hoped to get his family home by Thanksgiving. This time, they had constructed the house with fire-resistant cinderblock, even building a tiny playhouse out back with the leftover materials. Everyone was excited to return, though they knew that their timeline would likely get pushed back, as had happened frequently in the past two years. (They wouldn't end up returning home until 2021.) Every six months, Erin wrote the $77 check to renew their P.O. box in Oroville, hoping it would be the last. She liked to work on their lot, clearing brush and hacking at the weeds, channeling her anger and grief into something productive. It helped her focus on the future. Erin planned to attend occupational therapy school in Sacramento. Jamie had been promoted to maintenance director of a Chico nursing home. They had made a contingency plan, too. If their home burned down a second time, they would move to Washington State, where they had friends.

The wildfires in California had again attracted the attention of President Trump. As he spoke with the press in August 2020, he reiterated sentiments that were uncannily similar to those of 2018. "And I see again, the forest fires are starting," Trump told reporters. "They're starting again in California. And I said, you've got to clean your floors. You've got to clean your floors. They have many, many years of leaves and broken trees. And they're like, like so flammable. You touch them and it goes up. I've been telling them this now for three years, but they don't want to listen. The environment. The environment. But they have massive fires again in California. Maybe we're just going to have to make them pay for it." In October, he did in fact try to make the state pay for it by denying aid to communities that had been affected by six major wildfires—the North Complex Fire in Butte County was not among them—then quickly flipped his decision after the outcry that ensued.

Farther from the North Complex Fire, ash fell from the skies above Northern California. The soot fell on Tammy Ferguson's home near Sacramento, where she now lived with her fiancé and worked as a nurse in a hospital near the state capitol. And it fell on Kevin McKay's new house in Chico, where he and his girlfriend, Melanie, and his son, Shaun, had decided to move for good. As always, Kevin busied himself with work, digging an in-ground pool in the backyard and keeping the oak canopy neatly trimmed. He had kept his ruined lot in Paradise, just in case he tired of the valley heat and wanted to move back uphill in retirement, but it felt like that chapter of his life had closed. His parents' cabin in Magalia, which had survived the fire, had been sold after his mother died. It was time to move on.

Kevin was on track to graduate from Chico State University in May 2021. He had scaled back on classes after the fire, pushing his graduation timeline back. But as of the fall of 2020, he was registered for twenty-one credits—all online because of the pandemic. Though he wasn't transporting students to school these days, he regularly drove uphill to Paradise to service the buses and make sure they ran smoothly. On the drive, Kevin listened to a series of audiobooks by the historian Shelby Foote, who had written extensively on Kevin's latest topic of fascination: the Civil War. An avid history buff, Kevin was curious to learn more about the genesis of that year's protests over racial injustice. This aptitude for learning revealed itself in other ways, too. Recently, Kevin had made the dean's list. And after adopting a rescue dog, he named the pooch after a historical figure he greatly admired: former British prime minister Winston Churchill.

Across Chico, the firestorm terrified Rachelle Sanders's children. As the sky turned a granulated orange outside her apartment in the city, her nine-year-old daughter, Aubrey, hid under her covers, incapacitated by fear and shaking uncontrollably. Vincent, now eleven and starting middle school, was testy. In these moments, Rachelle was left to juggle her older children—along with Lincoln, not yet two years old—on her own. Chris had died of acute myeloid leukemia—a rare type of cancer in the bone marrow—in September 2019. He had been diagnosed two months after the Camp Fire. Though Chris had dreamed of seeing Lincoln nosedive into his first birthday cake, he hadn't lived long enough to attend his son's celebration. Rachelle missed her husband, who had loved her and their family unwaveringly. The thought of their baby growing up without his father filled her with sadness.

To make things worse, in April 2020, Rachelle's ex-husband, Mike, had been apprehended for sending sexually explicit text messages to a person he believed was a sixteen-year-old girl but was in fact a sheriff's deputy investigating him. Mike was charged with sending harmful material to a minor, engaging in lewd or lascivious behavior, and communicating with a minor for the purpose of engaging in sexual conduct. As the court hearing neared, Rachelle gained full custody of their two children. She remained in permanent "full mom mode," driving Aubrey to her competitive cheer practices nearly every afternoon and helping Vincent learn to use a planner for his homework assignments. Most days, Rachelle felt she was barely holding it all together. But there were moments of levity. Earlier that year, she had opened a storefront for Premium Landscaping in Chico. (After Chris had died, she had taken over their business.) As Rachelle stared at their new suite, her children ringed around her, she felt an upswell of pride. Chris would have been so proud.

Rachelle still received CodeRed alerts from Paradise, and as the fire neared her former town, an evacuation warning pinged on her cellphone. Officials had heeded the hard-learned lessons of the Camp Fire. Rather than having local authorities, like a town manager or the Police Department, issue evacuation alerts, only the Sheriff's Office could send them, preventing confusion and redundancies. Though Rachelle didn't think about the disaster as much these days, she did have moments. On a recent trip to pick up laptops from her children's school, she had gotten stuck in traffic—always a trigger for her. The red wash of brake lights reminded Rachelle of being stuck in David's car, unable to move and facing certain death, clutching her newborn son in her arms. Now, re-reading the alert on her phone, Rachelle knew she likely wouldn't return to Paradise. When the town had burned down, it had taken all of her memories—good and bad—with it. Her new life in Chico felt like a clean slate.

Maybe someday the town she had known would rise strong and whole again under the tall pines. A new wooden sign topped by another bandsaw halo was in the process of being staked off the Skyway, welcoming visitors to the Ridge. Perhaps someday the sign would again ring true, letting the weary and the curious in on a secret that everyone in town already knew: that this really and truly *was* paradise.

ACKNOWLEDGMENTS

On an early morning in October 2017, I drove north from San Francisco to Santa Rosa under an ashen sky. Overnight, ferocious winds had fanned more than a dozen wildfires across Northern California's wine country. The evacuations were hectic; forty-one people had died. My editor at the *San Francisco Chronicle* had dispatched me to the working-class neighborhood of Coffey Park—the epicenter of the damage. He wanted to know what remained of it.

I arrived and parked my Corolla. It was eerily quiet. A gutted sedan rested on its side in the middle of the neighborhood park—I wondered what had happened to its driver. Even the mailboxes and road signs had melted. Residents got lost on familiar streets as they tried to locate what remained of their homes. To more easily find them the second time around, they chalked their last names in the driveways.

In the following year, I drove the ruined streets of that fire zone weekly, passing those same driveways. I listened carefully to the stories of fire victims and wrote about their traumas and triumphs for the newspaper. I sat in evacuation shelters and homeless encampments, outside hospital burn units and at new construction sites where more than fifty-six hundred homes were slowly being rebuilt, one by one. I accompanied a couple as they brought their firstborn home from the hospital, though "home" was a guest bedroom in a relative's house, which didn't really feel like home at all. I attended a high school prom where seniors grappled with the idea of embarking for college. Should they defer admission? They had already lost so much and couldn't bear the thought of leaving their friends and family behind too. I watched a first grader try—and fail—to ride a donated bike without training wheels. This bike was bigger than her old one, which had been a present for her sixth birthday, just a few days before the fire. She still cried over the destroyed gift.

After witnessing such devastation, I was convinced that a fire season couldn't get worse—but I was wrong. In 2018, the Camp Fire shattered more than records; it destroyed a state's collective sense of safety. But even that wasn't the end. In 2019 and 2020, fires again burned into the collective psyche. The "new normal" was here to stay.

I have spent much of my journalism career bearing witness to the human cost of climate change. Parents left without children, children left without parents. Tens of thousands of ruined homes. Hundreds of thousands of people calcified in trauma that will never heal. But through it all, I have hoped that my reporting would force others to heed the wildfire crisis unfolding in California. I have hoped that it would deeply honor fire victims and their communities. For every person who has trusted me with their precious story: Thank you. For Santa Rosa and Redwood Valley. For Mariposa, Ventura, Malibu, Redding. For Berry Creek, Feather Falls. For Concow, Magalia, Butte Creek.

For Paradise.

To the people of Butte County—more names than I have space to list—thank you for welcoming me into your lives. I don't have sufficient words to convey my gratitude. I will carry your stories with me for the rest of my life. Cal Fire's Matt McKenzie, David Hawks, Sean Norman, and John Messina. (See, John? I told you I wouldn't forget your name.) Paradise Police Chief Eric Reinbold. Former Town Manager Lauren Gill, and Assistant Town Manager Marc Mattox. County Supervisor Doug Teeter. District Attorney

Mike Ramsey. The kindest gatekeepers: Jill Kinney of Feather River hospital, Sonya Meyer of Heritage Paradise, Megan McMann of the Butte County Sheriff's Office, Scotty McLean of Cal Fire. Thank you for always getting back to me promptly, even on Friday afternoons. At Paradise Town Hall, Dina Volenski and Colette Curtis always greeted me with a hug. Away from Town Hall, Chris Smith and Steve "Woody" Culleton kept me apprised of local gossip and goings-on. *Paradise Post* editor Rick Silva and former *Chico Enterprise-Record* reporter Camille von Kaenel are the kind of local journalists any community would be lucky to have; thank you for folding me into your pack. Jim Broshears and Calli-Jane Deanda have never stopped trying to make Butte County more fire safe.

Then there are the people who became like family. Phil and Michelle John opened up the spare bedroom of their home on Acorn Ridge Drive and copied me a housekey printed with sunflowers. They made Paradise feel like home. Later, after Phil died and their beautiful house overlooking the canyon was sold, Manuel and Lourdes Sanchez-Palacios took me into their home. Thank you for the homemade pesto, for the brown paper bags heavy with walnuts from your trees, and for rescheduling your weekly date night during my trips so we could share dinner. I'm also thankful to Xan Parker and Lizz Morhaim, members of Ron Howard's documentary film crew, who let me crash in their hotel room and made me belly-laugh so hard that I almost peed my pants. I'm eternally grateful that our paths crossed. (And if you haven't watched *Rebuilding Paradise,* you must.)

This book is the product of more than five hundred interviews and years of full-time wildfire coverage. I even enrolled in a professional firefighting academy to better understand fire. (Thank you to Marin County Fire Department Chief Jason Weber and Battalion Chief Jeremy Pierce for taking me on; I'm ready to hop on an engine as soon as you'll have me.) It's the product of coming to love a community that I embedded in: spending countless hours strolling across Paradise on my evening walks, buying ice cream sandwiches from the Holiday Market, eating green curry at Sophia's Thai.

The people whose lives I've chronicled in this book offered me unfettered access to their day-to-day lives without any expectation. To maintain journalistic integrity and impartiality, they were not compensated for their time, which makes their loyalty to this project all the more humbling. To know them is one of the greatest gifts.

Rachelle Sanders swiftly answered every one of my last-minute text queries with novel-length answers. She has always been unflinchingly honest about her past, as was her late husband, Chris—may he rest in peace. Their children, baby Lincoln and the big kids, Aubrey and Vincent, exhibited incredible patience as I rode along in their family's Suburban to cheerleading and baseball practices, or just to grab takeout for dinner, peppering their mother with questions the entire time.

Erin and Jamie Mansanares received me with warmth from the get-go. I've never met parents so willing to sacrifice for their children. They work incredibly hard—and they give a damn, particularly for the elderly folks in their care, a demographic that society tends to leave behind. Their daughters, Tezzrah, Arrianah, and Mariah, are incredible, loving little girls who are destined for great things.

Kevin McKay drove me and a bunch of high schoolers around on Bus 991 and was quick to extend dinner invites with his brood at their new house in Chico. Abbie Davis and Mary Ludwig spoke at length about their experiences, even though the retelling brought a jagged shadow of pain. (They also persuaded me to order my own pair of clogs.) Travis and Carole Wright, you remind me of my own parents—generous, compassionate, warmhearted. I hope that after this book, the Italian Cottage gives you a lifetime of free sandwiches.

Away from Butte County, my never-ending gratitude goes out to all those who loved this book into being, particularly my editor at Crown, Amanda Cook, whose compassion and care have been unparalleled and whose exacting edits made every draft infinitely better than the last, though the process could sometimes be painful. Still, I wish I could work with her forever. My literary agents,

Larry Weissman and Sascha Alper, believed in this book from its inception and provided constant support and enthusiasm. I was only twenty-five at the time—but not once did they ever doubt me. And the rest of the team at Crown: Katie Berry, David Drake, Gillian Blake, Annsley Rosner, and Zach Phillips, along with the marketing and publicity teams, who championed this book behind the scenes and helped get it into readers' hands. Katie, in particular, was a delightfully ruthless line editor. Julie Tate was an ace fact checker; I definitely owe her a taco for cleaning up my mistakes. Any errors that remain are my own. Thank you to Kate Hedges and Eric Josephson, who graciously shared the Konkow legend with me. And as my sensitivity reader, Adrian Jawort read through the Konkow legend, along with information about Indigenous burning practices, with a fine-toothed comb.

Thank you to former *San Francisco Chronicle* managing editor Kristen Go, who hired me, and former editor in chief Audrey Cooper, who celebrated my book with too much champagne and frozen pizza. Managing Editor Demian Bulwa is one of the best narrative editors I've had; I wouldn't be the reporter and storyteller I am without him. Trapper Byrne, the *Chronicle's* former politics editor, promised a younger me that one day I would write my own ticket—keeping me in journalism when I was prepared to quit and try my hand at another career. Thank you for that, and also for all the coffee.

I am lucky to have been able to call the *Chronicle* newsroom my first professional home, and even luckier to have started my career in California, where I've worked alongside some of the best journalists in the business. Storytelling doesn't exist in a vacuum, and this book would not have been possible without the incredible investigative and narrative reporting on the Camp Fire from my colleagues at the *Chronicle*, along with the *Los Angeles Times, The Sacramento Bee*, the *Redding Record Searchlight*, the *Chico Enterprise-Record*, the *Chico News & Review*, the *Paradise Post*, and others. Biggest of thanks to one of those journalists, Jason Fagone, for always answering my frantic phone calls and reassuring me that my fears were all "part of the process." WWJD—"What Would Jason Do?"—has become a common household saying in my little flat. I could write a second book on just how amazing he is, though I'm sure that's the last thing he would ever want. Still, be warned, Jason! It might happen . . .

The kindest group of people volunteered to read early versions of this manuscript and, in the process, made it so much sharper. Thank you to Jason Fagone (again), Heather Knight, Taylor McNair, Kate Rodemann, and Trisha Thadani. And to Eli Saslow, for answering a cold email in late 2018 to help me make sense of the book writing process. My mentors—Jacqui Banaszynski, Liz Brixey, and Michael Merschel—were instrumental in helping me get my start in journalism. I wouldn't be here if it weren't for them.

Chronicle staff photographer Gabrielle Lurie is the bridge between our fire world and "normal" life. I never would've been able to manage the transitions without her. Thank you for weathering more firestorms and flat tires, murder hotels and dirt roads than either one of us should ever have had to endure. (Maybe someday we'll meet the Kardashians.)

J. D. Morris was first my interpreter when it came to the intricacies of PG&E, and then he was just my friend. Thank you for so patiently explaining the difference between a distribution line and a transmission line. Trisha Thadani was my best friend, period. I'm not sure what I would have done without our "bitch and bike" sessions or our dreamy summer road trip through Oregon with "Van Man." Though our relationship ultimately didn't survive this book, I remain grateful to Alex Campbell, who cooked me shrimp scampi when I forgot to eat and consoled me through every bout of depression over the dying Joshua trees. I only wish I could have been home with you more often. And it was my dearest and oldest friend, John Handel, who convinced me on a warm summer afternoon at Indian Rock Park in Berkeley—smoke from the Pawnee Fire unfurling in the summer sky—that this book was worth writing. Thank you for believing in me.

To the rest of my dear, neglected friends: Thank you for your warmth and understanding. In particular: Jordan Hendricks and Ulyssa Mello; Bre Herrmann and Brendan Foley; Noah Berger; Marie-Louise Brunet; Tania Ermak; Kevin Fagan; King Kaufman;

Erin Mansur; Shayanne Martin; Cahner Olson; Stuart Palley; Sofi Pechner; and Otis R. Taylor, Jr. Thank you also to the Brick House in Yuba City—the halfway point on my long drives from San Francisco to Paradise—for the coffee and homemade blueberry scones. And to my therapist, Renee, who gave me back my mind. She has taught me that even my compassion has its limits—and that resiliency is the secret to healing. (She also always kept my favorite kind of tea in stock; thank you, Renee.)

Most of all, to my family: my aunt/second mom Margaret, my godparents, Bill and Gretchen, my parents, Steve and Rose, and my brother, Daniel (who once came to visit me in San Francisco and got dragged all the way to Oroville for a story). They are my blue skies after weeks of smothering gray. My sweet father always told me that my words are worth hearing, that people want to read them, and that if I work hard enough, I can make anything happen: I love you more than anything else on this earth, Daddy.

KONKOW LEGEND

The Konkow legend that is scattered through the book is drawn from A. G. Tassin, "The Con-Cow Indians," *Overland Monthly,* vol. 4, no. 19 (1884), pp. 7–9, quod.lib.umich.edu/m/moajrnl/ahj1472.2-04.019/13.

CHAPTER 1: DAWN AT JARBO GAP

Interviews: Meteorologist Jan Null; Cal Fire captains Matt McKenzie and Mike Kennefic; Cal Fire firefighters Jared Mack and Dakota McGinnis; Butte County district attorney Mike Ramsey; Scooters Café owners Dan and Bonnie Salmon.

9   **only 0.88 inches of precipitation:** Jan Null, National Weather Service meteorologist, in discussion with the author, April 18, 2019.

10   **Only 2 to 3 percent of the wildfires:** Victoria Christiansen, Forest Service interim chief, "Opportunities to Improve the Wildland Fire System" (Keynote address, Fire Continuum Conference, University of Montana, Missoula, May 21, 2018), fs.usda.gov/speeches/opportunities -improve-wildland-fire-system.

10   **when high fire risk was forecast:** At the time, PG&E was de-energizing some distribution lines when humidity fell below 20 percent, with sustained winds at 25 mph and gusts above 45 mph. "Public Safety Power Shutoff Event," Emergency Preparedness, Pacific Gas & Electric Company, accessed May 22, 2020, pge.com/en_US/safety/emergency -preparedness/natural-disaster/wildfires/public-safety-event.page.

11   **It registered winds blowing at 32 mph:** RAWS USA Climate Archive, Jarbo Gap Remoted Automated Weather Station Data, Monthly Summary for November 2018, accessed May 22, 2020, raws.dri.edu/cgi -bin/rawMAIN.pl?caCJAR.

11   **humidity plummeted to 23 percent:** RAWS USA Climate Archive.

11   **It was forecast to hit 5 percent:** According to the Storm Prediction Center, relative humidity was expected to drop as low as 5 to 15 percent

in the Sacramento Valley. By comparison, the Sahara has an average relative humidity of 25 percent.

11   **Covering 1,636 square miles:** U.S. Census Bureau, *Quick Facts,* 2010, census.gov/quickfacts/fact/table/buttecountycalifornia,chicocitycalifornia ,US/LND110210#LND110210.

11   **flames had ravaged the foothills 103 times:** "Butte County Office of Emergency Management Local Hazard Mitigation Plan," Appendix H, Table H-2, Oroville, CA, October 2019, accessed May 26, 2020, butte county.net/Portals/19/LHMP/2019/LHMPUpdateAppendicesOnly.pdf? ver=2019-11-13-130507-400.

11   **the Poe Fire, in 2001:** California Department of Forestry and Fire Protection, Butte Unit, "Butte County Community Wildfire Protection Plan 2015–2020," updated Nov. 3, 2015, accessed May 26, 2020, p. 7, butte county.net/Portals/14/Evac%20Maps/2015_Countywide_CWPP_FINAL .pdf.

11   **the Butte Lightning Complex and Humboldt fires:** Ibid.

11   **His outpost was perched on a knob of land:** "Station 36: Jarbo Gap," California Department of Forestry and Fire Protection, last modified 2013, accessed May 22, 2020, buttecounty.net/fire/FireFacilities/ FireStations/Station36.

11   **Two captains—he was one of them:** The other captain was Mike Kennefic. Mike Kennefic in discussion with the author, January 15, 2019.

12   **trying to impress McKenzie:** The men described McKenzie as approachable, someone who clearly knew what he was doing but didn't carry himself with arrogance. He was widely described as having "the demeanor of a cowboy." Jared Mack, Cal Fire firefighter at Station 36, in discussion with the author, July 2, 2019; Dakota McGinnis, Cal Fire firefighter at Station 36, in discussion with the author, July 3, 2019.

12   **Scooters Café, a family-owned restaurant:** Dan Salmon opened Scooters Café in 1998 after purchasing and renovating an abandoned minimart on 41 acres of land near the Feather River. His wife, Bonnie, joined the business a few years later. "It's the wild, wild West," she said of the area. "We have bears and mountain lions and outlaws and in-laws. When I moved here from Florida, it wasn't the California I thought it would be." In 2019, Dan and Bonnie retired and sold the restaurant to an Oroville couple. Scooters Café is now Jake's Burgers. Dan and Bonnie Salmon, co-owners of Scooters Café, in discussion with the author, August 16, 2019.

12   **a stone lodge turned into a diner:** The stone lodge is the historic Rock House Diner, which was built in 1937 with locally sourced rock as well as an apple-green mineral known as vesuvianite jade. In 2018, the Camp Fire gutted the diner. Even its iconic cowboy sign was destroyed.

12 **a market with two gas pumps:** Canyon Lakes Market, known as the Dome Store for its distinctive shape.

13 **80 percent of the county's crime:** Butte County Sheriff's Office, "Meth Facts," accessed May 22, 2020, buttecounty.net/sheriffcoroner/methfacts.

15 **The fridge was stocked with fresh groceries:** Contrary to popular belief, the food firefighters eat at their stations isn't purchased with taxpayer dollars. They pool their money to buy groceries every week. Captain Matt McKenzie wanted me to make sure everyone knew that.

THE FIRE: PREVAILING WINDS

Interviews: Meteorologists Neil Lareau, Jan Null, Rob Elvington, and Alex Hoon; climate scientists Daniel Swain and Daniel McEvoy.

17 **It had been more than seven months:** Jan Null, 2019.

17 **the state's historic drought:** National Integrated Drought Information System, "California Is No Stranger to Dry Conditions, but the Drought from 2011–2017 Was Exceptional," 2017, drought.gov/drought/california -no-stranger-dry-conditions-drought-2011-2017-was-exceptional.

17 **the fourth warmest in California:** NOAA National Centers for Environmental Information, "Climate at a Glance," January 2018 to October 2018, accessed May 26, 2020, ncdc.noaa.gov/cag/statewide/time-series/ 4/tavg/10/1/2018-2018?base_prd=true&begbaseyear=1901&endbaseyear =2000.

17 **five years of chart-topping heat:** NOAA National Centers for Environmental Information, "State of the Climate: National Climate Report for Annual 2018, January 2019," ncdc.noaa.gov/sotc/national/201813.

17 **July was five degrees warmer:** NOAA National Centers for Environmental Information, "Climate at a Glance: Statewide Time Series," published March 2019, accessed March 30, 2019, ncdc.noaa.gov/cag/ statewide/rankings/4/tavg/201807#1. Jason Samenow, "On Fire: July Was California's Hottest Month Ever Recorded," *Washington Post,* August 9, 2018, washingtonpost.com/news/capital-weather-gang/wp/2018/08/ 09/on-fire-july-was-californias-hottest-month-ever-recorded/.

17 **was at 74 percent for a common evergreen shrub:** "Informational Summary Report of Serious or Near Serious CAL FIRE Injuries, Illnesses and Accidents," California Department of Forestry and Fire Protection, Oroville, CA, November 8, 2018, p. 4. This information is available by public records request. See fire.ca.gov/programs/fire-protection/ reports/.

17 **The historical average during November was 93 percent:** "Informational Summary Report," p. 4.

17 **broke records all summer:** Daniel McEvoy, assistant research professor,

climatology, Western Regional Climate Center, in discussion with the author, March 12, 2019.

CHAPTER 2: ALL ITS NAME IMPLIES

Interviews: Rachelle and Chris Sanders; Mayor Jody Jones; former mayors Steve "Woody" Culleton and Alan White; Town Manager Lauren Gill; former town managers Chuck Rough and Donna Mattheis; Town Council members Michael Zuccolillo, Melissa Schuster, Scott Lotter, and Steve Crowder; former Town Council member Dona Gavagan Dausey; Chamber of Commerce director Monica Nolan; Paradise Irrigation District manager Kevin Phillips; Paradise Recreation and Park District manager Dan Efseaff; assistant town manager Marc Mattox; Town Clerk Dina Volenski; Butte County supervisor Doug Teeter; former Butte County supervisor Jane Dolan; historian Don Criswell; *Paradise Post* editor Rick Silva; Paradise Unified School District superintendent Michelle John; Paradise Police chief Eric Reinbold and lieutenant Anthony Borgman; Butte County Fire Safe Council director Calli-Jane Deanda; Paradise Ridge Fire Safe Council chairman Phil John; Northern California Ballet owner Trudi Angel; Kevin McKay.

18   **The average Paradise resident was fifty years old:** This was the median age. U.S. Census Bureau, *Quick Facts,* 2018.

18   **roughly 26,500 residents:** U.S. Census Bureau, *Quick Facts.*

18   **a fifteen-bed sanatorium in 1950:** "History: Our Heritage," Adventist Health Feather River, accessed November 9, 2018. adventisthealth.org/feather-river/about-us/history/.

18   **only thirty-one thousand infants had been born:** Leslie Gail Gillham was the first infant born at Feather River hospital, on November 12, 1968. She received gifts from the hospital administrator and chaplain, and all personal expenses were waived for her family. As of 2 P.M. on October 19, 2018, 31,807 babies had been born in the Birth Day Place. Breanna Bork, executive director at Feather River Health Foundation, email message to the author, April 11, 2019.

20   **daffodils bloomed along the roadsides:** The Paradise Garden Club started the "Daffodils Across the Ridge" project after the 2008 wildfires as a way to liven up the dreary landscape. Daffodils are considered "the flower of hope." Volunteers planted more than five thousand bulbs at the Paradise library, Bille Park, Aquatic Park, Terry Ashe Recreation Center, and public schools and churches that first year—and many more since then. As of November 2018, 162,000 daffodils had been planted in Paradise. Because the remaining bulbs were lost in the Camp Fire, the Garden Club planted twelve hundred white hyacinths in 2019. They're arranged in the shape of arrows pointing into Paradise on three thoroughfares: the Skyway and Pentz and Clark roads.

22   **losing by 250 votes:** Trevor Warner, "Council Members Speak on Zuc-

colillo Removal," *Paradise Post,* March 13, 2015. paradisepost.com/2015/03/13/council-members-speak-on-zuccolillo-removal/.

22   **It had been there for forty-six years:** "Chamber in Action," *Paradise Rising Resource Guide,* 2019–2020, accessed May 26, 2020, p. 94, issuu.com/paradiseridgebusinessjournal/docs/final_e6b19dbbc20267.

22   **May you find Paradise to be:** The iconic sign was located on the Skyway below Paradise. Contractor Charles M. Todd designed and built it in June 1969 at a cost of $3,635.52 for materials and labor. Diamond International donated the five train trestle timbers used for the wooden sign face. The fir trestles came from a historic section of railroad built in 1914 near Ramsey Bar along the West Branch Feather River. The 12-foot halo was made from two metal saw blades salvaged for $40. Construction was finished in 1972. The sign was moved a decade later when the Skyway was widened to four lanes. "The town has changed," Todd said of the September 1982 move. "But like the sign, Paradise still lives up to its name."

24   **This time he won:** Rick Silva, "Bolin, Zuccolillo and Schuster Win Seats," *Paradise Post,* November 9, 2016. paradisepost.com/2016/11/09/update-bolin-zuccolillo-and-schuster-win-seats/.

24   **and opened his own landscaping company:** Chris named his company Premium Landscape Services.

25   **Paradise spread across a wide ridge:** Town of Paradise, California, "About Paradise," accessed November 9, 2018, townofparadise.com/index.php/visitors/about-paradise.

25   **an unincorporated village of 11,500 people:** U.S. Census Bureau, *Quick Facts,* 2018.

25   **After the Gold Rush thrust California into statehood:** The historian Kevin Starr writes that the Gold Rush established the "founding patterns, the DNA code, of American California." This is key to understanding how the Town of Paradise came to be. Kevin Starr, "Striking It Rich: The Establishment of an American State," in *California: A History* (New York: Modern Library, 2007), pp. 80–100.

26   **In 1863, the 461 surviving Konkows were marched:** The California cavalry marched the Konkows from Chico. Tragically, hundreds of people were abandoned on the road. Round Valley Reservation supervisor James Short wrote: "About 150 sick Indians were scattered along the trail for 50 miles . . . dying at a rate of 2 or 3 per day. They had nothing to eat . . . and the wild hogs were eating them up either before or after they were dead." In 1996, members of the Round Valley tribes began re-creating the walk, turning a sordid moment in history into a story of resilience and healing. Benjamin Madley, "The Civil War in California and Its Aftermath," in *An American Genocide: The United States and the California Indian Catastrophe* (New Haven: Yale University Press, 2016), pp. 319–20.

26 **a man named William Leonard:** Robert Colby, "Introduction," in *Images of America: Paradise* (Charleston, S.C.: Arcadia Publishing, 2006), p. 7.

26 **more than $150 million worth of gold was unearthed:** Lois McDonald, "A Wealth of Minerals and Water: Mining on the Ridge," in *The Golden Ridge: A History of Paradise and Beyond* (Paradise: Paradise Fact and Folklore, 1981), p. 12.

26 **The most valuable find was a 54-pound nugget:** Ibid., p. 12.

26 **remembered a visiting miner:** The miner was Granville Stuart. He arrived from Iowa as a youth and settled near Magalia in 1852. Clyde A. Milner II and Carol A. O'Connor, "Moving West," in *As Big as the West: The Pioneer Life of Granville Stuart* (New York: Oxford University Press, 2008), p. 22.

27 **as far east as Denver:** Colby, "Introduction," p. 10.

27 **residents passed a $350,000 bond:** "Urban Water Management Plan 2015," Paradise Irrigation District, accessed November 20, 2018, pidwater .com/docs/about-your-water/water-supply/357-pid-uwmp-2015/file.

27 **a home sold for $205:** The local sawmill cut 9,000 to 10,000 board feet daily for new homes. The demand was so high that at Crego Sawmill, the sawdust heap towered three stories high. Colleen Sharp Muto, "The Crego Mill of Optimo," *Tales of the Paradise Ridge,* vol. 22, no. 1, June 1981, PAM 22020, California Historical Society, p. 5.

27 **a 1940s Chamber of Commerce brochure:** The full excerpt from the brochure reads: "All roads in Paradise lead to some point of scenic beauty; some inspiring mountain picture; some beautiful blossoming or bearing orchard; some entrancing trout stream; some attractive home and garden; some picnic spot where native birds and little animals give a continuous opera of Nature's song and story. Come lose yourself in Paradise a while and learn what living really is. . . . Presenting you an opportunity for a more healthful, more complete and more enjoyable life in an ideally natural environment," Paradise Chamber of Commerce, ca. 1940s, PAM 5342, California Historical Society.

27 **the population more than doubled:** Paradise became popular among retirees, with real estate brokers and agents advertising the community to an older demographic. In 1962, the population was 11,000. Two decades later, it was 22,000. James E. Alley, "The Paradise Irrigation District," *Tales of the Paradise Ridge,* vol. 3, no. 1, June 1962, PAM 22020, California Historical Society, p. 11.

27 **applied for a permit to open the first ballet studio:** Trudi Angel, owner of Northern California Ballet, in discussion with the author, January 8, 2019.

27 **280 miles of additional private roads:** Paradise has eighty-five miles of paved public roads. There were few sidewalks, the town's General Plan notes, because of "the desire of the community to maintain the rural char-

acter of the town." This also meant that construction of private roads was scattershot. For example, the minimum allowable street width standard is 20 feet, but in Paradise, it was 16 feet—despite an urgent need for fire access. Town of Paradise, *1994 General Plan,* vol. 1, pp. 130–36.

28   **from 3.3 million in 1919 to 39.4 million in 2018:** Ibid.

28   **The median property value in Paradise was $205,500:** Ibid.

28   **more than a thousand full-sized American flags:** "Parade of Flags," Town of Paradise, California, accessed November 9, 2018, townofparadise .com/index.php/visitors/community-events/9-uncategorised/101-parade -of-flags.

28   **allowed a ninety-nine-year-old to compete:** Mary Nugent, "Ridge Woman Celebrates 100 years, and a Town She Loves," *Paradise Post,* April 21, 2018, chicoer.com/2009/01/10/ridge-woman-celebrates-100-years -and-a-town-she-loves/.

28   **volunteers baked a thousand pies:** Monica Nolan, director of the Paradise Chamber of Commerce, in discussion with the author, September 13, 2019.

28   **In 2001, nearly 62 percent of the Ridge's population:** Thomas Curwen, "In a Town of Unanswered Questions, Paradise Tries to Imagine Its Future," *Los Angeles Times,* December 24, 2018, latimes.com/local/ california/la-me-paradise-rebuilds-20181224-story.html.

28   **from among thirty churches:** Butte County telephone book, 2018.

29   **a majority of the county's vote went to the Republican:** Election Summary Report, Butte County Clerk-Recorder Registrar of Voters, November 23, 2016, accessed May 26, 2020, clerk-recorder.buttecounty .net/elections/archives/eln35/35_statement_of_vote.pdf.

29   **"Paradise Rants and Raves!":** The group can be found at facebook .com/groups/268820559989723/.

29   **it cost the owners $80,000 to install a septic system:** Jody Jones, mayor of Paradise, in discussion with the author, January 16, 2019.

29   **known as Poverty Ridge:** McDonald, "The Mystery of the Past and the Tracking of the New," in *Golden Ridge,* p. 3.

29   **"darn nice place to starve":** Ibid., p. 4.

29   **a $300 monthly stipend:** "2018 Candidate Election Information," Town of Paradise, California, accessed February 7, 2019, townofparadise.com/ index.php/forms-and-documents/town-clerk/elections/1769-candidate -broshure/file.

30   **the fresh coat of paint on Town Hall:** Linda Watkins-Bennett, "Paradise Town Hall Gets Makeover: A Council Member's Donation Helped Make It Happen," Action News Now, November 5, 2018, actionnews now.com/content/news/Paradise-Town-Hall-Gets-Makeover-499737831 .html.

30   **would generate $1.4 million annually:** Lauren Gill, Paradise town manager, in discussion with the author, July 10, 2019.

30   **thirty-seven mobile home parks:** Butte County telephone book, 2018.

30   **Adverse Childhood Experiences:** ACEs are traumatic experiences that impact a child's developing brain. The three categories are abuse, neglect, and household dysfunction. In Butte County—which had the highest rate of ACEs of California's fifty-eight counties—76.5 percent of adults had experienced at least one ACE. Center for Youth Wellness, *A Hidden Crisis: Findings on Adverse Childhood Experiences in California,* San Francisco, 2014, centerforyouthwellness.org/wp-content/themes/cyw/build/img/building-a-movement/hidden-crisis.pdf.

30   **qualified for free or reduced-price lunch:** Michelle John, Paradise Unified School District superintendent, in discussion with the author, May 1, 2019.

31   **siphoning 21 percent of the town's students:** Ibid.

31   **A new high school:** The school was Achieve Charter High School. It opened after a lengthy battle with the public school district. Amanda Hovik, "Achieve Charter High School hosts ceremony for school's opening," *Paradise Post,* August 10, 2018, paradisepost.com/2018/08/10/achieve-charter-high-school-hosts-ceremony-for-schools-opening/.

31   **voters had approved a $61 million bond:** Steve Schoonover, "Five Local School Districts Have Bonds on Nov. 6 Ballot," *Chico Enterprise-Record,* October 21, 2018, chicoer.com/2018/10/21/five-local-school-district-have-bonds-on-nov-6-ballot/.

31   **Adventist Health, whose $5,000 donation:** Dan Efseaff, Paradise Recreation and Park District manager, in discussion with the author, October 3, 2019.

CHAPTER 3: RED FLAG OVER PARADISE

Interviews: Paradise Fire chief David Hawks; Butte County district attorney Mike Ramsey; meteorologist Alex Hoon; Jamie, Erin, and Tezzrah Mansanares; Stanford senior research scholar Michael Wara; attorneys Mike Danko and Frank Pitre; California Department of Water Resources guide Jana Frazier; Pulga owner Betsy Ann Cowley; Paradise mayor Jody Jones; Paradise town manager Lauren Gill; Butte County supervisor Doug Teeter; pyrogeographer Zeke Lunder; Travis and Carole Wright; Suzie Ernest; Mike Ranney.

41   **The National Weather Service issued a Red Flag Warning:** Red Flag Warnings are issued when a combination of strong wind, low relative humidity, and low fuel moisture are present. Experts agree that the Camp Fire wasn't the strongest offshore wind event of the season, but the gales came at the driest time, and fuels were extraordinarily parched. In November 2018, conditions were more similar to what climatologists would

expect to see in August. U.S. Department of Commerce, National Oceanic and Atmospheric Administration, *Service Assessment: November 2018 Camp Fire,* National Weather Service Western Region Headquarters: Salt Lake City, January 2020, accessed February 2, 2020, p. 14, weather.gov/media/publications/assessments/sa1162SignedReport.pdf.

41   **which staffed three additional engines:** David Hawks, text message to the author, July 12, 2019.

42   **The earthquake registered a 6.9 magnitude:** "The 1989 Loma Prieta Earthquake," California Department of Conservation, accessed March 1, 2019, conservation.ca.gov/cgs/earthquakes/loma-prieta.

43   **They advertised their programs:** Jay Matthews, "Hard Work, Low Pay, Miserable Conditions Popular in Calif. Corps," *The Washington Post,* March 26, 1982, washingtonpost.com/archive/politics/1982/03/26/hard-work-low-pay-miserable-conditions-popular-in-calif-corps/a37498db-18ed-46b1-be74-5d7056846f79/.

43   **"Learn skills, earn scholarships":** California Conservation Corps, accessed April 3, 2019, ccc.ca.gov/.

46   **More than 125,000 miles:** PG&E has 106,681 circuit miles of electric distribution lines and 18,466 circuit miles of interconnected transmission lines, which serve 5.4 million customers—defined as homes or businesses, so the actual number of people was much higher. Pacific Gas & Electric Company, "Company Profile," accessed May 26, 2020, pge.com/en_US/about-pge/company-information/profile/profile.page.

46   **five times the circumference of the earth:** The circumference of the earth is just under 25,000 miles.

46   **to more than 16 million people:** Pacific Gas & Electric Company, "Company Profile."

46   **most of California—equivalent to one in twenty Americans:** The United States has a population of about 328.2 million people.

46   **Nearly half of its grid crossed land:** This amounts to about 30,000 square miles of overhead electrical circuitry. Sumeet Singh, "Big Data," in *Fire-Threat Maps & the High Fire-Threat District (HFTD),* California Public Utilities Commission, accessed May 20, 2019, cpuc.ca.gov/FireThreatMaps.

47   **As millions of trees across the Sierra Nevada died:** Data collected by the U.S. Forest Service show that a record 129 million trees have died across 8.9 million acres in California. Bark beetles and drought are the main culprits. U.S. Forest Service Pacific Southwest Region, *2017 Tree Mortality Aerial Detection Survey Results,* Washington, D.C.: 2017, accessed May 26, 2020, fs.usda.gov/Internet/FSE_DOCUMENTS/fseprd566199.pdf.

47   **one of the largest direct action lawsuit settlements:** Studies have shown that hexavalent chromium causes serious illness, including cancer. PG&E released the heavy metal from its natural gas compressor station

over a fourteen-year span (from 1952 until 1966), and the chemical lingered in the groundwater for decades. In 2012, water regulators discovered that the contaminated water plume near PG&E's gas compressor station had spread. The Lahontan Regional Water Quality Control Board slapped the company with a $3.6 million fine. Since then, the population in Hinkley has dropped along with the property values. By 2015, Hinkley Elementary School, the Hinkley post office, and the only market had closed. Between 2010 and October 2014, PG&E bought about three hundred properties belonging to residents whose wells had been contaminated. Unable to do anything with the buildings, the utility—which now owns about two-thirds of Hinkley—had them bulldozed. It could take as long as thirty to sixty years to clean up the groundwater, PG&E spokespeople have said. "PG&E Hinkley Chromium Cleanup," California Water Boards, accessed May 26, 2020, waterboards.ca.gov/lahontan/ water_issues/projects/pge/. Jeremy P. Jacobs, "Another Pollution Battle Looms in Erin Brockovich's Town," *New York Times,* August 18, 2011, archive.nytimes.com/www.nytimes.com/gwire/2011/08/18/18greenwire -another-pollution-battle-looms-in-erin-brockov-58590.html.

47   **The ensuing Trauner Fire:** Public resources code says letting a tree grow too close to power lines is a misdemeanor. Tom Nadeau, *Showdown at the Bouzy Rouge: People v. PG&E* (Grass Valley, Calif.: Comstock Bonanza Press, 1998), p. 8.

47   **and a historic schoolhouse:** The Rough and Ready School was rebuilt. On September 23, 1995, it was dedicated as a Nevada County Historical Landmark. Nadeau, *Showdown,* p. 105.

47   **diverted more than $77.6 million from its tree trimming budget:** Kenneth Howe and Rebecca Smith, "Tree Trimming Pact Lowers PG&E Fine to $29 Million," *San Francisco Chronicle,* April 3, 1999, sfgate.com/ news/article/Tree-Trimming-Pact-Lowers-PG-E-Fine-to-29-Million-29 38340.php.

47   **a local reporter who covered the case:** Nadeau's book on *People v. PG&E* is delightful and reads like a true crime thriller. The hearing took place in a cabaret nightclub called the Bouzy Rouge—the site of a former bull-and-bear-fighting arena—which Nevada County rented for the trial for $1,700 a month. The stuffy room featured velvet drapes, flounced lace curtains, primrose wallpaper, and dour portraits of former miners. The foreman of the jury that declared PG&E guilty was named—of all things— Dave Fickle. Nadeau, *Showdown,* p. x.

47   **a natural gas pipeline owned by PG&E:** National Transportation Safety Board, *Pipeline Accident Report: Pacific Gas and Electric Company Natural Gas Transmission Pipeline Rupture and Fire,* NTSB/PAR-11/01 PB2011-916501, Washington, D.C.: August 30, 2011, ntsb.gov/ investigations/AccidentReports/Reports/PAR1101.pdf.

48   **first responders initially thought a jetliner had crashed:** "Initially we

thought a jet airplane went down from San Francisco airport," San Bruno Fire captain Bill Forester told a local reporter. "The sound, the noise of it was deafening. It sounded like a jet engine could still be running." Others reported what they thought was a gas station explosion to 911. "First Responders Recall San Bruno Explosion," CBSN Bay Area, September 14, 2010, sanfrancisco.cbslocal.com/2010/09/14/first-responders-recall-san-bruno-explosion/.

48 **registered as a 1.1 magnitude earthquake:** Paul Rogers, "San Bruno Blast: PG&E Settles Nearly All Remaining Lawsuits for a $565 Million Total," *San Jose Mercury News,* September 9, 2013, mercurynews.com/2013/09/09/san-bruno-blast-pge-settles-nearly-all-remaining-lawsuits-for-a-565-million-total/.

48 **gouging a 72-foot-long crater:** National Transportation Safety Board, *Pipeline Accident Report,* p. 18.

48 **to shut off the gas:** PG&E finally shut off the gas after 95 minutes. A federal investigation found that this response time was "excessively long" and "contributed to the extent and severity of property damage." National Transportation Safety Board, *Pipeline Accident Report,* pp. x, 89.

48 **"It was beyond words":** The resident was Susan Bullis. She lost her seventeen-year-old son, William, her husband, Gregory, her mother-in-law, Lavonne, and her dog, Lucky. Bullis was at a nurses' meeting in Sunnyvale when the explosion occurred in her neighborhood. Her testimony can be found on the CPUC's website. California Public Utilities Commission, *Declarations of Susan Bullis, Betti Magoolaghan and Robert Pelligrini,* Burlingame, CA: 2012, pp. 1–2, cpuc.ca.gov/uploadedFiles/CPUC_Public_Website/Content/Safety/Natural_Gas_Pipeline/News/DeclarationsofSusanBullisandBettiMagoolaghanandRobertPellegrini.pdf.

48 **crews had welded the thirty-inch natural gas pipeline incorrectly:** National Transportation Safety Board, *Pipeline Accident Report,* p. x.

48 **found to contain fabricated data:** According to the federal investigation, PG&E made up numbers or provided insufficient data. The utility also "pre-interviewed" witnesses before trial. National Transportation Safety Board, *Pipeline Accident Report,* p. 1.

48 **or to be printed in erasable ink:** Jaxon Van Derbeken, "PG&E Worked on Some Documents with Erasable Ink," *San Francisco Chronicle,* April 28, 2011, sfgate.com/news/article/PG-E-worked-on-some-documents-with-erasable-ink-2373432.php.

48 **had already fined PG&E $1.6 billion:** Previously, the highest penalty levied against an American utility was in New Mexico in 2000, when El Paso Natural Gas Co. paid $101 million for a fatal explosion that killed ten people, five of them children. The California Public Utilities Commission broke that record with its $1.6 billion penalty for PG&E following the San Bruno explosion. California Public Utilities Commission, "San Bruno

Incident Report," accessed May 27, 2020, cpuc.ca.gov/General.aspx?id= 5476.

48   **slapped with a second, $3 million fine:** California Public Utilities Commission, "San Bruno Incident Report."

49   **shelling out $5.3 million:** Douglas MacMillan and Neena Satija, "PG&E Helped Fund the Careers of Calif. Governor and His Wife. Now He Accuses the Utility of 'Corporate Greed,'" *Washington Post,* November 11, 2019, washingtonpost.com/business/2019/11/11/pge-helped-fund -careers-calif-governor-his-wife-now-he-accuses-utility-corporate-greed/.

49   **paid nearly $100 million in fines:** George Avalos, "PG&E Fined $97.5 Million for Improper Back-Channel PUC Talks," *San Jose Mercury News,* April 26, 2018, mercurynews.com/2018/04/26/pge-fined-97-5-million -for-improper-back-channel-puc-talks/.

49   **a couple of bottles of "good Pinot":** Peevey was the president of Southern California Edison before moving to the CPUC, which regulates the Southern California utility. Other emails show Peevey dining with top PG&E officials at his Sea Ranch home on the Sonoma County coast. Emails between PG&E and the CPUC were released after a federal probe into the relationship between the two entities. The investigation found that lax oversight by the CPUC contributed to the San Bruno explosion. Brian K. Cherry, PG&E vice president of regulation and rates, email to Thomas E. Bottorff, PG&E senior vice president of regulatory affairs, May 31, 2010, pgecorp.com/sfg14/PGE_PaulClanon Letter.pdf.

49   **a 44-foot gray pine:** California Public Utilities Commission, *Citation Issued Pursuant to Decision 16-09-055,* D.16-09-055 E.17-04-001, San Francisco: 2017, cpuc.ca.gov/uploadedFiles/CPUC_Public_Website/ Content/News_Room/E1 704001E2015091601Citation20170425.pdf.

49   **the worst wildfires in modern state history:** Kevin Fagan et al., "A Fire's First, Fatal Hours," *San Francisco Chronicle,* October 2017, sfchronicle .com/bayarea/article/Wine-Country-fires-first-fatal-hours-12278092.php.

49   **A fourteen-year-old perished:** Lizzie Johnson, "A Fire's Unfathomable Toll," *San Francisco Chronicle,* April 26, 2019, projects.sfchronicle .com/2019/redwood-fire-victims/.

49   **seventeen of the twenty-one wildfires:** Ivan Penn and Peter Eavis, "PG&E Is Cleared in Deadly Tubbs Fire of 2017," *New York Times,* January 24, 2019, nytimes.com/2019/01/24/business/energy-environment/ pge-tubbs-fire.html.

49   **eight times the size of San Francisco:** Kimberly Veklerov, "Cal Fire Releases Details of Probe into Cause of Wine Country Fires," *San Francisco Chronicle,* June 9, 2018, sfchronicle.com/bayarea/article/Cal-Fire-releases -details-of-probe-into-cause-of-12981639.php.

49   **The culprit, Cal Fire investigators found:** California Department of

Forestry and Fire Prevention, *CAL FIRE Investigators Determine the Cause of the Tubbs Fire,* Sacramento: 2019, fire.ca.gov/media/5124/tubbscause1v .pdf.

49 **a ninety-one-year-old woman:** The woman was Ann Zink. She was at her home in Riverside County when the Tubbs Fire broke out. Kurtis Alexander, Evan Sernoffsky, and Lizzie Johnson, "Tubbs Fire: State Blames Private Electrical Equipment for Deadly Wine Country Blaze," *San Francisco Chronicle,* January 24, 2019, sfgate.com/california-wildfires/article/ Tubbs-Fire-State-blames-private-electrical-13559073.php.

50 **the utility cut electricity to distribution lines in seven counties:** Lizzie Johnson and Michael Cabanatuan, "PG&E Power Shutdown: No Coffee, No Gas. But Calistoga Takes Shutdown in Stride," *San Francisco Chronicle,* October 15, 2018, sfchronicle.com/california-wildfires/article/ PG-E-power-shutdown-No-coffee-no-gas-But-13309296.php.

50 **San Diego Gas & Electric had pioneered this approach:** "People didn't understand why we had to turn the power off for public safety," a SDG&E executive explained. "Nothing we did at the beginning was easy. . . . We owe it to our communities to do what is right and be a safe operator of the power grid." Caroline Winn, chief operating officer of San Diego Gas and Electric, "Lessons Learned in San Diego," panel presentation at the Wildfire Technology Innovation Summit, California State University, Sacramento, March 20, 2019. Recording available at youtube .com/playlist?list=PLsgixh8pRZUBuk0O7MeqpyfD1zhvjutCc.

50 **after sparking two huge blazes in 2007:** Cal Fire found that the Witch and Rice fires were caused by SDG&E. A third blaze, the Guejito Fire, ignited after a Cox Communications fiber optic cable came into contact with an SDG&E power line. The Guejito eventually merged with the Witch Fire. California Public Utilities Commission, Consumer Protection and Safety Division, Utilities Safety and Reliability Branch, *Investigation of the Witch Fire Near San Ysabel, California* and *Investigation of the Rice Fire Near Fallbrook, California,* by Mahmoud (Steve) Intably, I.08-11-006, October 2007, accessed May 27, 2020, docs.cpuc.ca.gov/word_pdf/ FINAL_DECISION/93739.pdf.

50 **nighttime temperatures had risen:** The occurrence of nighttime heat waves in particular has nearly doubled in the last thirty-year period in California, compared with the average from 1950 to 2016, from eleven days to twenty-one days per year. California Senate Office of Research, *Climate Change and Health: Understanding How Global Warming Could Impact Public Health in California,* November 2018; Kendra Pierre-Louis and Nadja Popovich, "Nights Are Warming Faster Than Days. Here's Why That's Dangerous," *New York Times,* July 11, 2018, nytimes.com/ interactive/2018/07/11/climate/summer-nights-warming-faster-than-days -dangerous.html.

50 **In the previous four years alone:** The California Public Utilities Com-

mission began requiring utilities to track wildfires in 2014. Between 2014 and 2017, PG&E reported igniting 1,554 wildfires. Some were small; others were large enough to make the news. Pacific Gas and Electric Company, *Fire Incident Data Collection Plan,* 2014–2017, accessed May 27, 2020, cpuc.ca.gov/uploadedFiles/CPUC_Website/Content/About_Us/Organization/Divisions/News_and_Outreach_Office/PGE_Fire%20Incident%20Data%202014-2017.pdf.

50 **nearly sixty thousand customers:** Johnson and Cabanatuan, "PG&E Power Shutdown."

50 **SDG&E, followed PG&E's lead:** Ibid.

50 **had used shutoffs as many as twelve times:** The first shutoff—known officially as a Public Safety Power Shutoff—by SDG&E occurred October 5–6, 2013. They also occurred January 14–15, 2014; May 14, 2014; November 24–25, 2014; September 21–22, 2017; October 20–21, 2017; October 23–25, 2017; December 4–12, 2017; December 14–15, 2017; January 27–29, 2018; October 15–16, 2018; and October 19–20, 2018. The thirteenth—and last—shutoff of 2018 was on November 16. SDG&E (@SDGE), "Hi Don, hopefully we can help clear up the confusion. Here is a chart of the public safety power shutoffs we've done since our first in 2013," October 10, 2019, 11:45 A.M., twitter.com/SDGE/status/1182366579679354880.

50 **parts of eight counties:** Pacific Gas and Electric Company, "PG&E Continues to Closely Monitor Weather Conditions Ahead of Possible Public Safety Power Shutoff in Parts of Eight Counties," November 7, 2018, pge.com/en/about/newsroom/newsdetails/index.page?title=2018 1107_pge_c ontinues_to_closely_monitor_weather_conditions_ahead_of _possible_public_safety_po wer_shutoff_in_parts_of_eight_counties.

50 **Lauren Gill sent out an email:** Lauren Gill, Paradise town manager, email to Town Council, November 7, 2018.

51 **more than 3,100 miles of lines:** The distance from San Francisco to Boston is about 3,096 miles. Pacific Gas and Electric Company, "PG&E Continues to Closely Monitor."

51 **More than seventy thousand customers:** Pacific Gas and Electric Company, *PG&E Public Safety Power Shutoff Report to the CPUC,* by Meredith E. Allen, November 27, 2018, accessed May 28, 2020, p. 1, pge.com/pge_global/common/pdfs/safety/emergency-preparedness/natural-disaster/wildfires/PSPS-Report-Letter-11.27.18.pdf.

51 **"We have just been informed by PG&E":** David Copp, email to Doug Teeter, Butte County supervisor, November 7, 2018.

52 **PG&E had plans to avoid such outages:** Pacific Gas and Electric Company, *PG&E Adds More Weather Stations and High-Definition Cameras to Monitor Wildfire Conditions,* December 11, 2019, pge.com/en/about/newsroom/newsdetails/index.page?title=20191211_pge_adds_more_weather_stations_and_highdefinition_cameras_to_monitor_wildfire_conditions.

52   **the utility had activated its new Wildfire Safety Operations Center:** Pacific Gas and Electric Company, *PG&E Public Safety Power Shutoff Report.*

52   **had only just opened eight months before:** Pacific Gas and Electric Company, "PG&E Opens New Wildfire Safety Operations Center, Marks California Wildfire Awareness Week with $2 Million for Fire Safe Councils," May 8, 2018, pge.com/en/about/newsroom/newsdetails/index .page?title=20180508_pge_o pens_new_wildfire_safety_operations_center _marks_california_wildfire_awareness_week_with_2_million_for_fire _safe_councils.

52   **A woman from the Bay Area had bought the ghost town:** Betsy Ann Cowley, owner of Pulga, in discussion with the author, November 4, 2019.

52   **The fifty-six mile Caribou–Palermo Line:** Katherine Blunt and Russell Gold, "PG&E Delayed Safety Work on Power Line That Is Prime Suspect in California Wildfire," *Wall Street Journal,* February 27, 2019, wsj .com/articles/pg-e-delayed-safety-work-on-power-line-that-is-prime -suspect-in-california-wildfire-11551292977.

52   **considered for inclusion on the National Register of Historic Places:** As part of PG&E's regulatory compliance obligations, the Caribou-Palermo Line was investigated for eligibility with the California Office of Historic Preservation in 2017. Federal permits for new or updated transmission lines trigger such reviews. While it is uncommon, other transmission lines have made it on the National Register, including the Palo Verde–Devers Transmission Line Corridor in Arizona and the Big Creek Hydroelectric System Historic District, which runs from Fresno County to Los Angeles. The Caribou-Palermo Line didn't qualify because it didn't possess "high or artistic value," wasn't associated with the lives of significant historical figures, and didn't illuminate "our understanding of the past." Its only historical associations were with the Caribou-Valona Line, which "lacks both significance and integrity as a cohesive representative of early twentieth century transmission development. Although the line, in its original configuration, was the first to run at 165 kV (15 kV higher than previously established lines), it held this record for only 6 months before being surpassed." Paul Lusignan, historian, National Register of Historic Places, in email to the author, May 29, 2020. Dudek Environmental Consulting Firm, "Draft Initial Study and Mitigated Negative Declaration for Pacific Gas and Electric Company South of Palermo 115 kV Power Line Reinforcement Project," Application No. 16-04-023, May 2017, p. 225, cpuc.ca.gov/environment/info/dudek/Palmero/SPRP _Draft_MND_Recirc_ May2017.pdf.

52   **average tower near Jarbo Gap was sixty-eight years old:** Pacific Gas and Electric Company, "Electric Overhead Steel Structure Strategy Overview," June 2017, s1.q4cdn.com/880135780/files/doc_downloads/wild fire_updates/1078.pdf.

52   **a contractor painting a tower:** According to an internal spreadsheet, PG&E safety specialist supervisor Eric Matthew Rubio considered the incident a "near miss." He said that the contractor had "reached to reposition himself grasping a piece of flat cross bracing when the J hook hardware used to secure the flat bracing to the tower leg failed and broke at the J part of the J hook hardware." The bolt had been corroded to about 20 percent of its original thickness. Rubio's advice: "Crews working on these towers need to use caution when working on or near towers."

52   **they assumed any damage would occur during a wet winter storm:** An internal PG&E email explains that the Caribou–Palermo Line scored under 300 for a pole replacement project. The rationale for its low score was that "there is no likely large environmental event (if structures fail, it will likely be due to heavy rain and no wildfires are possible then). Also no likely public safety issue with live wires down because it is in a remote area. Reliability score is not that high because although the likelihood of failed structures happening is high, the affected customers are likely in the order of >1K." Manho Yeung, email to Carlos Gonzalez et al., February 26, 2014.

53   **knocked down five steel towers:** The memo from PG&E reads: "Five towers collapsed and were removed on the powerline in December 2012, and the line is currently supported on wood poles that PG&E installed to temporarily restore service. After transferring the line to the new towers, PG&E will either remove the wood poles or leave them in place for distribution use." Replacing the towers triggered the Caribou–Palermo's candidacy for the National Register of Historic Places. Brian K. Cherry, PG&E vice president, regulation and rates, email to Edward F. Randolph, director, CPUC Energy Division, September 6, 2013.

53   **It had been designed in 1917:** The line was installed between 1919 and 1921 by the Great Western Power Company. Mike Ramsey, Butte County district attorney, in discussion with the author, November 7, 2019.

53   **put into service on May 6, 1921:** California Public Utilities Commission, Safety and Enforcement Division, Electric Safety and Reliability Branch, "Incident Investigation Report," E20181108-01, November 8, 2019, p. 8.

53   **shortly after World War I ended:** By way of perspective: World War I ended on November 11, 1918, after Germany formally surrendered. Warren G. Harding became president on March 4, 1921. The first Miss America Pageant was held on September 8, 1921. Prohibition began in 1920. The Nineteenth Amendment, giving women the right to vote, was ratified on August 18, 1920.

53   **posted fifteen tweets:** Pacific Gas and Electric Company, *PG&E Public Safety Power Shutoff Report*, p. 4.

53   **Gill dispatched a final email to her colleagues:** Lauren Gill, Paradise town manager, email to Town Council, November 7, 2018.

THE FIRE: THE CARIBOU-PALERMO LINE

Interviews: Butte County district attorney Mike Ramsey; fire scientist Michael J. Gollner.

57   **The 143-pound, 115-kilovolt braided aluminum wire:** California Public Utilities Commission, Safety and Enforcement Division, Electric Safety and Reliability Branch, "Incident Investigation Report," p. 11.

57   **reaching temperatures up to 10,000 degrees Fahrenheit:** Michael J. Gollner, associate professor at the University of Maryland Department of Fire Protection Engineering, in discussion with the author, August 29, 2019.

57   **Droplets of molten metal:** About 28,000 fires occur annually in the United States because of molten metal from power lines, railroads, and equipment. The Witch and Guejito fires (SDG&E) also ignited this way. The 2011 Bastrop County Complex Fire in Texas is another. The study of how hot metal sparks ignite fires is still relatively new—and fascinating. James Urban, Casey Zak, and A. Carlos Fernandez-Pello, "Cellulose Spot Fire Ignition Caused by Hot Metal Particles," *Proceedings of the Combustion Institute,* December 2015; Yudong Lui et al., "Temperature and Motion Tracking of Metal Spark Sprays," *Fire Technology,* April 2019.

THE FIRE: DESCENDING INTO CONCOW

Interviews: Professors Don Hankins, Scott Stephens, and Crystal Kolden.

65   **the water and sap stored in tree trunks began to boil:** Don Hankins, professor of geography and planning at California State University, Chico, in discussion with the author, August 7, 2019.

65   **Their shriveled roots carried fire laterally:** Scott Stephens, professor of fire science at the University of California, Berkeley, in discussion with the author, December 10, 2019.

66   **melted and hardened to slivers:** Don Hankins, 2019.

66   **now visible to two satellites:** Robert Sanders, "New Satellite View of Camp Fire as It Burned Through Paradise," *Berkeley News,* November 15, 2018, news.berkeley.edu/2018/11/15/new-satellite-view-of-camp-fire-as-it-burned-through-paradise/; Crystal Kolden, associate professor of fire science at the University of Idaho, in conversation with the author, August 13, 2019.

CHAPTER 4: CODE RED

Interviews: Cal Fire captain Matt McKenzie; Pulga owner Betsy Ann Cowley; Cal Fire battalion chief Curtis Lawrie; Paradise Fire chief David Hawks; Cal Fire Emergency Command Center staff Beth Bowersox, captain Marcus Ekdahl, Jennifer Burke, Shannon Delong, and captain Stacer Harshorn; Butte County analyst Mike Thompson; California Department of Water Resources guide Jana Frazier; Cal Fire captain Miguel Watson; Paradise Police dispatcher Carol

Ladrini; Cal Fire battalion chief Gus Boston; Cal Fire division chiefs John Messina and Garrett Sjolund; contract pilot David Kelly; CodeRed spokespeople Sue Holub and Troy Harper; Cal Fire director Ken Pimlott; Brad Meyer; Shem Hawkins; Todd Derum; Beth Bowersox; Terrie Prosper; Cindi Dunsmoor.

67   **a young man who had been inside the bar:** "This guy just came out of nowhere and came out with a gun and shot people in Thousand Oaks, California," Holden Harrah said. "And that's what's really blowing my mind, it's a really safe area." Faith Karimi and Joe Sutton, "At Least 12 Killed In Shooting at a Bar in Thousand Oaks; Gunman Also Dead," Action News Now, November 8, 2018, actionnewsnow.com/content/news/At-least-12-killed-in-shooting-at-a-bar-in-Thousand-Oaks-gunman-also-dead-500018092.html.

68   **six wildland firefighters:** The six firefighters killed were Cal Fire heavy equipment operator Braden Varney, Arrowhead Hotshots captain Brian Hughes, Cal Fire heavy equipment mechanic Andrew Brake, contract bulldozer operator Don Ray Smith, Redding Fire Department inspector Jeremy Stoke, and Draper City (Utah) fire chief Matthew Burchett. Lizzie Johnson, "Bringing Home Braden," *San Francisco Chronicle,* September 12, 2019, projects.sfchronicle.com/2019/bringing-home-braden/.

68   **his phone blinked with a notification:** The exact time was 6:29:55.

69   **The hook had fallen forty-seven feet:** Using the area beneath the J hook as a reference point, investigators found that before it broke, it had hung forty-seven feet above the ground. Tower 27/222 is on a steep incline, so the precise distance it fell varies depending on the reference point. *United States of America v. Pacific Gas and Electric Company,* Case 14-CR-00175-WHA, United States District Court, Northern District of California, 2019, p. 4.

70   **The town, which a geologist had built in 1904:** Pulga was known as Big Bar until 1916. Its name changed because of confusion with a neighboring town, which was also dubbed Big Bar. In 1994, Pulga was turned into a meditation and hypnotherapy institution called Mystic Valley. For more information on Pulga's history, visit pulgatown.com/town-history.

70   **a right-of-way consultant working for PG&E:** The consultant was Steve Hertstein. Matthias Gafni, "PG&E Email: Work on Pulga Transmission Tower Different than Power Line Linked to Camp Fire," *San Jose Mercury News,* November 14, 2018, mercurynews.com/2018/11/14/pge-email-work-on-pulga-transmission-tower-different-than-power-line-linked-to-camp-fire/.

71   **On a cruise in the Dominican Republic:** Betsy Ann Cowley, 2019.

71   **"Engine 2161 responding":** Karla Larsson, "Listen to 5 Hours of Camp Fire Scanner Traffic," YouTube video, 5:00:52, November 27, 2018, youtube.com/watch?v=NQRQHFmzegY.

72   **oversaw units in thirty-five of California's counties:** Some counties

are grouped into a single "unit." These units include Humboldt–Del Norte, Butte, Sonoma-Lake-Napa, Mendocino, San Mateo–Santa Cruz, Santa Clara, Lassen-Modoc, Nevada-Yuba-Placer, Amador–El Dorado, Shasta-Trinity, Siskiyou, and Tehama-Glenn in Northern California, and Riverside, San Bernardino, San Diego, Fresno-Kings, Madera-Mariposa-Merced, San Benito–Monterey, Tulare, and Tuolumne-Calaveras in Southern California. In total, this amounts to 802 fire stations, 42 conservation camps, 12 air attack bases, and 10 helitack bases. More information can be found at fire.ca.gov/media/4925/whatiscalfire.pdf.

72 **an email from a PG&E public safety specialist:** The public safety specialist was Rob Cone. In the email, Hawks said: "Obviously they are monitoring the weather and conditions may change, but Rob Cone told me this evening that as of now the models are showing that Butte County may receive the strongest of the forecasted winds." David Hawks, Paradise Fire division chief, email to Marc Mattox, Lauren Gill, and Curtis Lawrie, November 6, 2018.

72 **The email had nagged at Lawrie:** Curtis Lawrie, Cal Fire battalion chief, in discussion with the author, April 1, 2019.

73 **The community's population numbered 710:** Concow's population had declined considerably after years of devastating wildfires. In 1990, it was home to 1,392 residents. By 2000, the number had dropped to 1,095. And in 2010, the population stood at 710—little more than half of what it had been two decades earlier.

73 **The median household income was about $25,000:** U.S. Census Bureau, *Quick Facts,* 2010, data.census.gov/cedsci/table?q=2010%20 census%20data%20for%20Concow,%20CA&g=1600000US0616035& hidePreview=false&tid=ACSDP5Y2010.DP03&y=2010& vintage=2010 &layer=VT_2010_160_00_PY_D1&cid=DP02_0001E.

73 **There was one elementary school with six teachers:** Concow Elementary is part of the tiny Golden Feather Union Elementary School District. "Concow Elementary School," *Public School Review,* accessed June 1, 2020, publicschoolreview.com/concow-elementary-school-profile.

73 **Lawrie's radio crackled:** Larsson, "Listen to 5 Hours of Camp Fire Scanner Traffic."

74 **dispatchers monitored the footage:** The captain in charge of the Emergency Command Center is considered the "incident commander" on a wildfire until air or ground forces can arrive.

74 **more than twenty thousand 911 calls annually:** California Department of Forestry and Fire Protection, Butte Unit, *Butte County Cooperative Fire Protection Annual Report,* Oroville: 2017, p. 4.

74 **had already issued the first alert:** The CodeRed alert was dispatched at 7:13 A.M., sending 29 emails, 15 phone calls, and 6 text messages to residents of Pulga. Ten phone numbers were reached; five were not. Butte

County Sheriff's Office, *Code Red Alerts,* by Michael Thompson, Oroville: November 8, 2018.

74 **the one to suggest "Camp Fire" as the name:** Beth Bowersox, Cal Fire Emergency Command Center dispatcher, in discussion with the author, January 2, 2020.

75 **had developed it in 1974:** "NIMS and the Incident Command System," Federal Emergency Management Agency, accessed May 1, 2020, fema .gov/txt/nims/nims_ics_position_paper.txt.

76 **causing blackouts for four customers:** California Public Utilities Commission, Safety and Enforcement Division, Electric Safety and Reliability Branch, "Incident Investigation Report," p. 80.

77 **had arrived for her shift:** Carol Ladrini, Paradise Police dispatcher, in discussion with the author, September 12, 2019.

78 **wanted to know where the wildfire was burning:** Ladrini and other dispatchers stuck to scripted answers. This recording at 7:11 A.M. was released by Paradise Police. It, and other calls, can be accessed at drive.google .com/drive/folders/1JzYtze0SfGDrRxXcl1nJLwvxTTlH4Idz.

78 **A fellow battalion chief:** The Cal Fire battalion chief was Gus Boston, husband of Shelby Boston, Butte County's director of social services. Gus Boston, Cal Fire battalion chief, in discussion with the author, February 20, 2019.

79 **a Cal Fire captain was corkscrewing along a ridgeline:** The crew was led by Cal Fire captain Miguel Watson. As he led the engines down Rim Road, he remembered, "The wind was hitting the engine so hard it ripped open a compartment on top of the engine. If you ever stand up there, [you know] it takes a little bit of force to lift that compartment. It seals pretty well. It was crazy. It felt like the wind was going to knock the engine over." Miguel Watson, Cal Fire captain, in discussion with the author, March 14, 2019.

80 **from countries as far away as New Zealand and Australia:** Crews have been flying in from Australia to help the United States since 2000, when the two countries began collaborating. Firefighters came from Down Under in 2017 to help during the Wine Country wildfires. The United States sent personnel to Australia in 2009 and 2010. Firefighters are paid by their home country, and the terrain and tactics are similar enough, so the system works, officials have said. Lizzie Johnson, "Big Boost from Down Under: Australians and Kiwis Join California Firefight," *San Francisco Chronicle,* August 19, 2018, sfchronicle.com/california -wildfires/article/Big-Boost-from-Down-Under-Australians-and-Kiwis -13164685.php.

81 **911 calls continued to roll in:** Cal Fire released the 911 calls it received during the Camp Fire. They can be found here: calfire.app.box.com/s/ ioth04321tb09r0aqigxpyyxhza9ock5/folder/89509185902.

81   **fifty-one cell sites:** Terrie Prosper, director, news and outreach, California Public Utilities Commission, in discussion with the author, December 6, 2019.

81   **seventeen towers:** Cindi Dunsmoor, Butte County Emergency Services officer, in discussion with the author, December 13, 2019.

82   **Raised by small-town grocers:** John Messina, Cal Fire division chief, in discussion with the author, February 20, 2019.

82   **to order a regional incident management team:** "It was the middle of the afternoon; people had headlights on," remembered Todd Derum, a member of the team. "I was like, this is not going to be good—not that they are ever good—but this will be another one of those off-the-chart, heinous incidents. . . . When you see a fire that is moving like that through a community, calling it a wildfire is not even an accurate statement whatsoever." Todd Derum, Cal Fire operations chief, in discussion with the author, December 30, 2019.

83   **One woman—a mother of five:** The woman was Terra Hill, a thirty-six-year-old mother of five. She and her roommate, forty-eight-year-old Jesus Fernandez, sought shelter under the truck. Fernandez died there, waiting for help.

83   **asked the Butte County Sheriff's sergeant:** The Butte County Sheriff's sergeant was Brad Meyer.

84   **only one person . . . had shown up for work:** The one person to show up was Michael Thompson. Michael Thompson, Butte County Sheriff's Office information systems analyst, in discussion with the author, September 20, 2019.

84   **two people . . . calling in with identical requests:** The two people were Cal Fire dispatcher Jennifer Burke and Butte County Sheriff's sergeant Brad Meyer. Jennifer Burke, Cal Fire dispatcher, in discussion with the author, January 7, 2020.

84   **the California-based software company OnSolve:** Sue Holub, OnSolve chief marketing officer, and Troy Harper, OnSolve director of government strategy, in discussion with the author, January 8, 2020.

84   **only 11 percent of Butte County's population had registered:** Risa Johnson, "How to Sign Up for Emergency Alerts, Check Evacuation Routes," *Chico Enterprise-Record,* February 14, 2018, chicoer.com/2018/02/14/how-to-sign-up-for-emergency-alerts-check-evacuation-routes/.

84   **At 7:57 A.M., the analyst keyed Messina's first evacuation order:** The first evacuation alert in Paradise reached 1,699 numbers. Another 965 numbers weren't reached. Butte County Sheriff's Office, *Code Red Alerts.*

84   **he never received the order to evacuate Concow:** At 11:52 A.M., the Butte County Sheriff's Office dispatched an alert reading, "Due to the fire in the area, an evacuation order has been issued for the area of Highway 70 from Concow South including all of Yankee Hill on both sides of 70.

If you need assistance in evacuating, please call 911." The message reached 191 phone numbers; it failed to deliver to 207 numbers. And the alert came hours too late. Butte County Sheriff's Office, *Code Red Alerts.*

84   **the Integrated Public Alert and Warning System:** Plans for alert and warning protocols had last been established in Butte County in 2011—seven years before the Camp Fire. In a government report, Butte County said that it attempted to issue an alert through the IPAWS system, but it failed to go through. Despite guidance from FEMA saying that counties should test their software *before* a disaster, staff acknowledged they had never done so. Auditor of the State of California, *California Is Not Adequately Prepared to Protect Its Most Vulnerable Residents from Natural Disasters,* 2019-103, Sacramento: December 2019, auditor.ca.gov/reports/2019-103/chapters.html.

85   **took off at 7:48 A.M.:** David Kelly, Cal Fire air tanker contract pilot, in discussion with the author, May 31, 2019.

85   **Cal Fire operated twenty-three of these air tankers:** California Department of Forestry and Fire Prevention, "Aviation History," accessed June 1, 2020, fire.ca.gov/about-us/history/aviation-history/.

88   **Kelly's supervisor said over the radio:** The supervisor was Cal Fire battalion chief Shem Hawkins. He said: "What I really tried to impress upon the command staff [at the ECC] was the need for evacuations and how fast the fire was going to spread and that there was not going to be a way to stop the fire." Shem Hawkins, in discussion with the author, February 19, 2020.

88   **submerged themselves in the cold reservoir:** Kurtis Alexander, "Trapped by Camp Fire, More than a Dozen People—One 90—Survived in Chilly Lake," *San Francisco Chronicle,* November 13, 2018, sfchronicle.com/california-wildfires/article/Survival-in-the-lake-People-trapped-by-Camp-Fire-13389692.php.

89   **Messina called a fellow division chief:** The Cal Fire division chief was Garrett Sjolund. Garrett Sjolund in discussion with the author, February 19, 2019.

90   **had gathered at Skywalker Ranch:** Ken Pimlott, Cal Fire director, in discussion with the author, August 16, 2019.

90   **"If you stick around for that picture":** Garrett Sjolund, February 19, 2019.

CHAPTER 5: THE IRON MAIDEN

Interviews: Chris and Rachelle Sanders; retired fire chiefs Jim Broshears and George Morris; Paradise Police chief Eric Reinbold, lieutenant Anthony Borgman, and dispatcher Carol Ladrini; Paradise town manager Lauren Gill, assistant town manager Marc Mattox, assistant to the town manager Colette Curtis, and town clerk Dina Volenski; Paradise information technology manager Josh Marquis; Feather River hospital chief financial officer Ryan Ashlock, nurse Tammy Ferguson, patient Stephen Arrington, groundskeeper Tom Paxton, intensive care unit director Allyn Pierce, charge nurse Ed Beltran, and nurse supervisor Bev Roberson; Butte County district attorney Mike Ramsey; Paradise Unified School District superintendent Michelle John; *Paradise Post* editor Rick Silva.

93   **Gill was responsible for gathering intelligence:** Per the town's Emergency Operations Plan, Gill had "complete authority over all Town of Paradise department personnel and resources as directed in the Municipal Code" (section 2.40.060) during an emergency. This included the Police Department. Since the Camp Fire, the Town of Paradise and the Paradise Police force have met with Butte County officials and decided that the Sheriff's Office will be in charge of CodeRed alerts if a disaster is threatening any jurisdiction, like Paradise or Chico. Town of Paradise, *Emergency Operations Plan,* Paradise: 2011, pp. 43–44. townofparadise.com/index .php/forms-and-documents/town-clerk/1193-2011-town-of-paradise -eop-w-supporting-docs-final-1-12/file.

94   **Gill sent a text message:** Lauren Gill, text message to Paradise Town Council, November 8, 2018.

94   **Bolin had texted the chat group back:** Greg Bolin, text message to Paradise Town Council, November 8, 2018.

94   **this time with a message from councilman Mike Zuccolillo:** Mike Zuccolillo, text message to Paradise Town Council, November 8, 2018.

94   **asked her to send an internal CodeRed:** Dina Volenski, Paradise town clerk, in discussion with the author, December 30, 2019.

95   **had already posted an update:** The staffer was assistant town manager Marc Mattox.

96   **By 2013, she had been appointed to the most powerful position:** Anthony Siino, "Lauren Gill Begins First Day as Interim Town Manager," *Paradise Post,* January 2, 2013, paradisepost.com/2013/01/02/lauren-gill -begins-first-day-as-interim-town-manager/.

97   **thirteen monster blazes had barely missed the town limits:** Matthias Gafni, "Rebuild Paradise? Since 1999, 13 Large Wildfires Burned in the Footprint of the Camp Fire," *San Jose Mercury News,* December 2, 2018, mercurynews.com/2018/12/02/rebuild-paradise-since-1999-13 -large-wildfires-burned-in-the-footprint-of-the-camp-fire/.

97   **the Oakland-Berkeley Hills Fire killed twenty-five people:** Rachel Swan, "25 Years Later: Oakland Hills Ripe for Another Firestorm," *San Francisco Chronicle,* October 20, 2016, sfchronicle.com/bayarea/article/25 -years-later-Oakland-hills-ripe-for-another-9984731.php.

97   **as much as 5.5 to 19 million acres annually:** Robert E. Martin and Scott L. Stephens, "Prehistoric and Recent Fire Occurrence in California," State Resources Water Control Board (1997), accessed May 26,

2020, p. 4, waterboards.ca.gov/water_issues/programs/tmdl/records/region_1/2003/ref1 039.pdf.

98   **burned in 1923, 1970, and 1980:** U.S. Fire Administration, *The East Bay Hills Fire,* USFA-TR-060, Oakland-Berkeley: October 1991, accessed January 12, 2019, p. 11, usfa.fema.gov/downloads/pdf/publications/tr-060.pdf.

98   **"catastrophic life and property loss":** The full excerpt reads: "Heavy fuel loads, steep terrain, poor access and light flashy fuels create severe fire hazards. The increased population in this area creates a high potential for catastrophic life and property loss." California Department of Forestry and Fire Protection, Butte Unit, *Butte County Community Wildfire Protection Plan* (Sacramento, CA: January 5, 2006), p. 102. This information is available by public records request. See fire.ca.gov/programs/fire-protection/reports/.

98   **It blackened 8,333 acres:** "Butte County Community Wildfire Protection Plan 2015–2020," p. 7.

98   **and caused $6 million in damage:** Laura Smith, "County Officials 'Not Optimistic' Poe Fire Will Be Declared Disaster," *Chico News & Review,* October 11, 2001, newsreview.com/chico/county-officials-not-optimistic-poe-fire-will-be-declared-disaster/content?oid=6338.

99   **"The greatest fear is fire on the Ridge":** The thirty-seven-year-old woman was Shauna Robbins. Matthai Kuruvila, "74 Paradise homes destroyed by Humboldt Fire," *San Francisco Chronicle,* June 15, 2008, sfgate.com/bayarea/article/74-Paradise-homes-destroyed-by-Humboldt-Fire-3209635.php#.

99   **it carved Paradise and Magalia into fourteen zones:** Town of Paradise, *Town of Paradise Evacuation Traffic Control Plan,* Paradise: Town Hall, March 10, 2015. This information is available by public records request. See townofparadise.com/index.php/our-government/departments/town-clerk.

99   **a local Civil Grand Jury issued a report:** The full excerpt reads: "The unpredictability, intensity, and locations of the 2008 wildfires near the towns of Magalia, Paradise, Concow, and Forest Ranch emphasized the critical shortcomings of the area's readiness for extreme fire situations." Butte County Grand Jury, *Wildfire & Safety Considerations for Butte County General Plan 2030,* p. 46. Paradise: Town Hall, 2008.

99   **responded with a scathing public letter:** Mayor Frankie Rutledge to the Honorable James F. Reilley, August 11, 2009.

99   **thirteen people had been injured:** Eleven people were injured in 2011, 16 in 2012, 9 in 2013, 13 in 2014, and 17 in 2015. Two people died in 2013 and 2015. Ironically, after the Downtown Paradise Safety Project was completed on the Skyway, injuries actually went up. There were 29 incidents in the four-month period between November 2014 and March

2015. Town of Paradise Public Works Department, *Downtown Paradise Safety Project Collision Data Report,* Paradise: Town Hall, December 31, 2015.

99 **reduce pedestrian accidents by means of a "road diet":** After the Camp Fire, reporters at the *Los Angeles Times* were the first to break news about the Town Council's 2014 vote to narrow the Skyway. Paige St. John, Rong-Gong Lin II, and Joseph Serna, "Paradise Narrowed Its Main Road by Two Lanes Despite Warnings of Gridlock During a Major Wildfire," *Los Angeles Times,* November 20, 2018, latimes.com/local/california/la-me-ln-paradise-evacuation-road-20181120-story.html.

100 **Eselin called the project "insane":** Mildred Eselin stated that she had not seen more than one person crossing the Skyway since the last Town Council meeting she had attended. "All the people I ever talked to have never heard of [the town's survey]," she said. "And when I told them what the point of it was . . . the general response was, 'Insane.'" Eselin said that she didn't think the project would increase business in the downtown and said that she was very concerned about how the improvements would affect evacuation during wildland fires. Mildred Eselin, Paradise Town Council meeting, May 13, 2014, 01:47:27, livestream.com/townofparadise/events/2983185.

100 **The Bay Area engineering firm:** The firm, Whitlock & Weinberger Transportation, Inc., of Santa Rosa, said in its report, "Because of the emergency evacuation needs of the corridor, raised medians are not recommended in the corridor." Later, the report added: "Due to the potential for fires in the Paradise hillside areas, Skyway should be designed to accommodate two travel lanes in the downhill direction, which may be needed to serve vehicle evacuation during emergencies such as fire. . . . Because of this operation, the [road] cannot be designed with raised medians within the center lane's space." Butte County Association of Governments and the Town of Paradise, *Skyway Corridor Study,* Santa Rosa, CA: February 12, 2009, p. 35, bcag.org/Planning/Skyway-Corridor-Study/.

100 **up to twenty-four thousand vehicles traveled the road every day:** About 13,000 to 24,000 vehicles traveled on the Skyway every day, averaging speeds of 30 to 40 mph, according to a report published seven years before the wildland urban interface drill. By 2016, rush hour traffic fell in the upper limit of that range. Butte County Association of Governments and the Town of Paradise, *Skyway Corridor Study,* p. 1.

100 **simply avoided downtown:** Jim Broshears of the Paradise Ridge Fire Safe Council told the editor of the local newspaper that he thought "residents may have decided to avoid the drill," something that "officials had hoped would not happen." Rick Silva, "Evacuation, Firefighting Simulation Goes Well; Improvements Identified," *Paradise Post,* June 22, 2016, paradisepost.com/2016/06/22/officials-say-evacuation-drill-a-success/.

100   **firefighters staged their own drill:** The drill ran from 6:30 to 8:30 A.M. on June 22, 2016. Officials wanted residents to practice evacuating and make sure they knew their evacuation zone. Rick Silva, "Upper Ridge Fire Drill Set for June 22," *Paradise Post,* June 8, 2016, paradisepost.com/2016/06/08/upper-ridge-fire-drill-set-for-june-22/.

101   **a system for emptying the entire Ridge at once:** California Department of Homeland Security, Exercise and Evaluation Program, *2016 Wildland Urban Interface and Evacuation Exercise After Action Report,* by Jim Broshears, Butte County: June 22, 2016.

101   **the town's roster hadn't been updated in years:** While Butte County has a program that identifies vulnerable residents, a roster overseen by the Town of Paradise hadn't been updated in more than a decade. Jim Broshears, former Paradise Fire chief, in discussion with the author, January 15, 2019.

102   **was readied for rotator cuff surgery:** Patient Zero was Stephen Arrington. Stephen Arrington in discussion with the author, May 29, 2019.

103   **("Don't talk crap about anyone"):** The intensive care unit director was Allyn Pierce. Allyn Pierce in discussion with the author, September 2, 2019.

103   **She received training for this role four times a year:** Jim Broshears, former Paradise Fire chief, in discussion with the author, January 15, 2019.

105   **It was Reinbold's responsibility to call Gill:** Eric Reinbold, Paradise Police chief, in discussion with the author, December 23, 2019.

106   **A sergeant who would have been assigned:** Anthony Borgman, Paradise Police lieutenant, in discussion with the author, January 22, 2020.

108   **four staff members were trained to use CodeRed:** The other staff members were assistant to the town manager Colette Curtis, information technology manager Josh Marquis, town clerk Dina Volenski, and public information officer Ursula Smith.

108   **the state had been confronted with the grave consequences:** Lizzie Johnson, "Wildfires Emphasize Need to Improve Emergency Alert Systems," *San Francisco Chronicle,* December 4, 2017, sfchronicle.com/bayarea/article/Wildfires-emphasize-need-to-improve-emergency-12405556.php; Lizzie Johnson, "Unlike in North Bay, Ventura County Officials Issued Wide Alert," *San Francisco Chronicle,* December 5, 2017, sfchronicle.com/news/article/Unlike-in-North-Bay-Ventura-County-officials-12408111.php.

109   **The state criticized Sonoma County's decision:** Joaquin Palomino and Lizzie Johnson, "Sonoma County Warnings Fell Short in Wine Country Fires, State Report Says," *San Francisco Chronicle,* February 26, 2018, sfchronicle.com/news/article/Sonoma-County-emergency-readiness-warnings-fell-12709628.php.

109 **Marquis clicked on the zones:** Meaghan Mackey, "New Analytics Show Camp Fire CodeRed Alerts Failed to Reach More than Half the Residents," KRCR News, May 6, 2019, krcrtv.com/news/deprecated -camp-fire/new-analytics-show-camp-fire-code-red-alerts-failed-to-reach -more-than-half-the-residents.

109 **The alert went out to fewer than half:** The first CodeRed alert from the Town of Paradise reached 3,262 phone numbers. It failed to reach another 2,927 phone numbers. When an alert is sent, the system tries to reach each number up to six times before timing out.

110 **bottlenecks at the bulb-outs:** The bulb-outs were added on the Skyway from Pearson Road to Elliott Road.

110 **a live interview to a television journalist:** The interview was with KRCR, News Channel 7. The footage has since been taken down from KRCR's website, so information from the broadcast was from a transcription taken by the author at the time and by Colette Curtis's recollection.

111 **The notice was dispatched to 2,765 phone numbers:** Of the 2,765 numbers, 1,163 weren't reached, equaling 42 percent of the residents. Of the 6,189 alerts from the first alert, 2,927 numbers weren't reached, equaling 47 percent of the residents.

113 **her cellphone buzzed with a call from Rick Silva:** Rick Silva, "Missed Birthday Dinners," *Paradise Post,* November 8, 2019, paradisepost .com/2019/11/08/silva-missed-birthday-dinners/.

CHAPTER 6: ABANDONING THE HOSPITAL

Interviews: Chris and Rachelle Sanders; Keven Page; Feather River hospital chief financial officer Ryan Ashlock, nurse Tammy Ferguson, patient Stephen Arrington, groundskeeper Tom Paxton, intensive care unit director Allyn Pierce, medical assistant J. D. Rasmussen, respiratory supervisor Chris Lephart, charge nurse Ed Beltran, and nurse supervisors Bev Roberson and Sarah McCain; Cal Fire battalion chief Bill Lopez and captain Rick Manson; California Highway Patrol commander Brandon Straw.

125 **"boiling liquid expanding vapor explosion":** Between 1940 and 2005, about eighty major BLEVEs claimed more than a thousand lives and injured more than ten thousand people. Tasneem Abbasi and S. A. Abbasi, "The Boiling Liquid Expanding Vapour Explosion (BLEVE): Mechanism, Consequence Assessment, Management," *Journal of Hazardous Materials* 141 (2007), pp. 489–519.

126 **60 to 80 percent of the town's thoroughfares:** The main purpose of access roads is to reach homes in neighborhoods. Town of Paradise, *Municipal Service Review,* Paradise: August 2007, p. 3.4-1.

127 **increased travel times by as much as fifty-four seconds:** According to the presentation, northbound travel times on the Skyway increased by about twenty-nine seconds, while peak travel times (between 5 and 6 P.M.)

increased by fifty-four seconds. A video of the presentation can be found at youtu.be/TwryVczSDyk. Town of Paradise, *Downtown Paradise Safety Project,* Paradise: 2015, p. 65.

131   ***This might be a lost cause,* their captain thought:** He was Cal Fire captain Rick Manson.

131   **According to a 2007 report, a third of the town's public roadways:** Of the town's 98.5 center-line miles of public streets, 31.8 miles were considered to need repairs beyond routine maintenance, equaling 32 percent. Town of Paradise, *Municipal Service Review,* pp. 3.4–4.

131   **179 burning vehicles:** The 19,000 vehicles recovered included trailers, farming equipment, and boats—anything with some type of vehicle identification number. It took officers nine months to identify all of the burned vehicles. Of the 179 vehicles on evacuation routes, 48 were recovered from Bille Road, 40 from the Skyway, 42 from Pearson Road, and 7 from Clark Road. Brandon Straw (California Highway Patrol Commander), in discussion with the author, January 16, 2020.

OBSERVATION: A FATAL BREATH

Interviews: Feather River hospital cardiopulmonary manager Ben Mullin and respiratory supervisor Chris Lephart.

CHAPTER 7: A BLIZZARD OF EMBERS

Interviews: Paradise Fire chief David Hawks; Cal Fire division chief John Messina; Paradise Town Council candidate Julian Martinez and town clerk Dina Volenski; Feather River hospital nurses Tammy Ferguson and Crissy Foster.

137   **Honey Run Covered Bridge:** The bridge was the first of its kind to connect Paradise to Butte Creek Canyon. More information can be found on the website of the Honey Run Covered Bridge Association (HRCBA): hrcoveredbridge.org.

138   **(Records would later show):** John Messina, Cal Fire division chief, in discussion with the author, August 13, 2019.

138   **"rate of spread" card:** According to Hawks's card, "moderate" was any wildfire less than 1 mph, or 80 feet per minute. "Dangerous" was 1 to 3 mph, or 80 to 264 feet per minute. And "critical spread" was anything greater than 3 mph, or 264 feet per minute.

139   **which he presented at a Town Council meeting:** That meeting can be viewed on the Town of Paradise's website at livestream.com/townof paradise/events/8325375/videos/178945109.

139   **who had recently died in the Carr Fire:** Lizzie Johnson, "150 Minutes of Hell," *San Francisco Chronicle,* December 12, 2018, projects .sfchronicle.com/2018/carr-fire-tornado/.

139   **twice the land area of New York City:** New York City covers 205,000

acres, according to its Department of City Planning. The Mendocino Complex Fire burned 459,000 acres.

140   **$1 billion worth of new road projects:** "Jody Jones Retiring as Caltrans District 3 Director," California Department of Transportation, June 23, 2014, accessed May 26, 2020, eterritorialdispatch.blogspot.com/ 2014/06/jody-jones-retiring-as-caltrans.html.

140   **couldn't handle the crush of a mass exodus:** This included 1,348 mobile homes and 64 recreational vehicles.

140   **only 3,700 vehicles an hour could evacuate:** The memo continued: "An evacuation of this magnitude certainly will not happen on its own. It will require very careful planning, a good communication system, proper preparation, close cooperation between citizens and all the involved agencies, proper execution, and more than a little bit of luck." Dennis J. Schmidt, Public Works director/town engineer, memo to Charles L. Rough, Jr., town manager, June 11, 2002.

THE FIRE: LEAPING THE WEST BRANCH

Interviews: President of the Association for Fire Ecology Chris Dicus; University of Maryland Department of Fire Protection Engineering associate professor Michael J. Gollner.

145   **two dozen embers accumulated:** James Urban et al., "Temperature Measurement of Glowing Embers with Color Pyrometry," *Fire Technology,* February 2019.

145   **A small pile of firebrands:** Michael J. Gollner in discussion with the author, August 29, 2019.

CHAPTER 8: SAVING TEZZRAH

Interviews: Jamie and Tezzrah Mansanares; Heritage executive administrator Sonya Meyer, nurse Trudy Vaughn, and chef Jill Fassler; Butte County director of social services Shelby Boston; Butte County Emergency Services officer Cindi Dunsmoor.

147   **begged a nurse to run back inside:** The nurse was Trudy Vaughn.

147   **built to handle sixteen hundred vehicles an hour:** Schmidt memo to Rough, June 11, 2002.

147   **elderly adults were twice as likely to die:** "Older Adults," National Fire Protection Association, accessed June 10, 2020, nfpa.org/Public -Education/Fire-causes-and-risks/Specific-groups-at-risk/Older -adults.

148   **About 25 percent of residents were older than sixty-five:** U.S. Census Bureau, *Quick Facts,* 2010, census.gov/quickfacts/CA.

148   **compared to 14 percent statewide:** Ibid.

148   **the disability rate was nearly twice the state average:** In 2015, 10.6

percent of Californians had a disability, compared to 25 percent in Paradise. Laura Newberry, "Poor, Elderly and Too Frail to Escape: Paradise Fire Killed the Most Vulnerable Residents," *Los Angeles Times,* February 10, 2019, latimes.com/local/lanow/la-me-ln-camp-fire-seniors-mobile-home-deaths-20190209-story.html.

148 **three women were left behind:** The three women were Teresa Ammons, Dorothy Mack, and Helen Pace.

148 **enacted building regulations for mobile homes:** "Manufactured Housing and Standards," U.S. Department of Housing and Urban Development, accessed June 9, 2020, hud.gov/program_offices/housing/rmra/mhs/mhshome.

148 **The first was the In Home Support Services program:** The director of social services was Shelby Boston. You may remember that she is married to Cal Fire battalion chief Gus Boston.

148 **The Ridge had an enrollment of 960:** Shelby Boston in discussion with the author, April 1, 2019.

149 **SNAP was more of a voluntary educational program:** Though people who weren't in the county's In Home Support Services program could sign up for SNAP, most hadn't. In a 2019 report, the state auditor found that everyone from the Ridge in the SNAP program had come from IHSS. Auditor of the State of California, *California Is Not Adequately Prepared,* pp. 39–40.

149 **twenty-page document, drafted by:** San Diego County, "Emergency Preparedness: Taking Responsibility for Your Safety; Tips for People with Disabilities and Activity Limitations," 2019.

152 **the highway had been paved in 2013:** Steve Schoonover, "Upper Ridge Escape Route Handles First Test," *Chico Enterprise-Record,* November 12, 2018, chicoer.com/2018/11/12/upper-ridge-escape-route-handles-first-test/.

153 **was going to hitch a ride with them:** The colleague was Shay Hindry, a licensed vocational nurse.

OBSERVATION: PARADISE IRRIGATION DISTRICT

Interviews: Paradise Irrigation District manager Kevin Phillips and workers Ken Capra and Jeremy Gentry.

CHAPTER 9: THE LOST BUS

Interviews: Kevin McKay; Ponderosa Elementary School teachers Mary Ludwig and Abbie Davis; student Rowan Stovall and her mother, Nicole Alderman; Paradise Ridge Fire Safe Council chairman Phil John; Butte County Fire Safe Council director Calli-Jane Deanda.

157 **capable of accommodating nine hundred cars per hour:** Schmidt memo to Rough, June 11, 2002.

161 **The Butte County General Plan for 2030:** "Health and Safety Element," in *Butte County General Plan for 2030,* pp. 11–45, buttecounty.net/dds/Planning/Butte-County-General-Plan.

161 **comprised one full-time employee:** Calli-Jane Deanda, director, Butte County Fire Safe Council, in discussion with the author, March 19, 2019.

163 **Two of the school district's assistant superintendents:** The men were Tom Taylor and David McCready.

CHAPTER 10: THE BEST SPOT TO DIE

Interviews: Cal Fire captain Sean Norman.

168 **A thirty-seven-year-old city fire:** Johnson, "150 Minutes of Hell."

169 **killed twenty-three people:** Bettina Boxall, "The Same Elements That Made the Thomas Fire Such a Monster Also Created Deadly Debris Flows," *Los Angeles Times,* January 12, 2018, latimes.com/local/lanow/la-me-mudflows-science-montecito-20180112-story.html.

169 **The town was similar to Paradise:** Kurtis Alexander, "Fearful of Being the Next Paradise, Grass Valley Confronts Its Fire Vulnerability," *San Francisco Chronicle,* June 30, 2019, sfchronicle.com/california-wildfires/article/Fearful-of-being-the-next-Paradise-Grass-Valley-14061525.php.

170 **taken an aggressive stance toward wildfires:** Timothy Egan, "Teddy Roosevelt and the 'Burn' That Saved Forests," *Fresh Air,* National Public Radio, Philadelphia, September 10, 2010, npr.org/transcripts/129750575.

171 **The agency's "ranger" title was swapped for "chief":** Stephen Pyne, *California: A Fire Survey* (Tucson: University of Arizona Press, 2016), p. 38.

171 **fire season had increased by seventy-eight days:** Ken Pimlott, "Wildfire Resilience in California," California Adaptation Forum, 2020, californiaadaptationforum.org/2018/02/15/wildfire-resilience-in-california/.

171 **Ken Pimlott cautioned in a 2016 memo:** The memo was prepared to support Cal Fire's efforts to stop wildfires by conducting prescribed burning and defensible space inspections. They were a minor effort—20,000 acres of prescribed burning and 15,000 acres of treatment projects including chipping, fuel breaks, and mechanical thinning. Ken Pimlott to State of California Natural Resources Agency, September 27, 2016, documentcloud.org/documents/4176986-2016-Fuels-Reduction-Memo.html.

171 **Across the state, 189 communities:** Peter Fimrite, "California Neigh-

borhoods Prepare for Wildfires with Help from Federal Program," *San Francisco Chronicle,* July 7, 2019, sfchronicle.com/news/article/California -neighborhoods-prepare-for-wildfires-14077846.php.

171 **Cal Fire's expenditures had skyrocketed to $947 million:** Judy Lin, "Should California Buy Disaster Insurance?" KQED, February 17, 2019, kqed.org/news/11727057/should-california-buy-disaster-insurance.

171 **within the first two months of its fiscal year:** For the eighth time in ten years, Cal Fire was forced to dip into its budget reserves to put out wildfires in 2018. "It's a reasonable request," said Assemblyman Phil Ting, D–San Francisco, chair of the Assembly Budget Committee. "Unfortunately, it seems these requests are happening every year, earlier in the year, and on a larger scale each year." Melody Gutierrez, "Disastrous Fire Season Has Cal Fire Running Out of Money," *San Francisco Chronicle,* September 6, 2018, sfchronicle.com/politics/article/Disastrous-fire-season -has-Cal-Fire-running-out-13210912.php.

172 **blazes scorched 12.7 million acres:** Lizzie Johnson, "The Heart Is Still Pumping," *San Francisco Chronicle,* November 7, 2019, sfchronicle.com/ california-wildfires/article/Starting-over-A-year-after-the-Camp-Fire-14 811903.php.

172 **Jerry Brown told reporters:** Scott Neuman, "California's Gov. Brown: Wildfires Are Evidence of Changing Climate 'in Real Time,'" NPR, August 2, 2018, wamu.org/story/18/08/02/californias-gov-brown-wildfires -are-evidence-of-changing-climate-in-real-time/.

173 **three distinct phases in a human's response:** Amanda Ripley, *The Unthinkable: Who Survives When Disaster Strikes—and Why* (New York: Harmony, 2009), p. xviii.

173 **Psychologists have found:** Laurence Gonzales, *Deep Survival: Who Lives, Who Dies, and Why* (New York: Norton, 2003), p. 196.

173 **passengers in airplane crashes:** Ripley, *Unthinkable,* p. 132.

174 **known as the Lake Wobegon effect:** Ibid., p. 39.

174 **it took a specific moment:** Ibid., p. 138.

OBSERVATION: THE SHERIFF'S DAUGHTER

Interviews: Paradise Police officer Kassidy Honea; Sheriff Kory Honea; Paradise Town Council member Mike Zuccolillo.

THE FIRE: SIEGE IN SIMI VALLEY

178 **torched small ranches, movie sets:** Sopan Deb, "Set for 'Westworld' and Other Shows Destroyed in California Fire," *The New York Times,* November 10, 2018, nytimes.com/2018/11/10/arts/westworld-set-woolsey -fire-california.html.

CHAPTER 11: "THE SAFETY OF OUR COMMUNITIES"

Interviews: Travis Wright; Suzie Ernest; Shauna Jarocki; Mike Ranney.

180 **one woman strolled with her pets:** The woman was Shauna Jarocki.

181 **Farther down Edgewood:** Butte County District Attorney's Office, *A Summary of the Camp Fire Investigation,* Oroville: 2020.

184 **It had opened earlier that year:** David R. Baker, "Worried About Wildfires, PG&E Unveils New Fire Forecasting Center," *San Francisco Chronicle,* March 22, 2018, sfchronicle.com/business/article/Inside-PG-E -s-new-fire-forecasting-center-12897514.php.

185 **emailed four department heads:** They included Elizaveta Malashenko, Leslie Palmer, Danjel Bout, and Terrie Prosper. Meredith Allen, email to CPUC, November 8, 2018.

185 **analysts sitting at computer keyboards:** PG&E declined to talk with me for this book. Descriptions of the Wildfire Safety Operations Center and the Emergency Operations Center came from looking at photos published in newspaper articles. This piece in particular, was helpful: Jason Fagone, "Inside PG&E's 'War Room,' the Chaos of Shut-off Week Seems Far Away," *San Francisco Chronicle,* October 12, 2019, sfchronicle.com/ california-wildfires/article/Inside-PG-E-s-war-room-the -chaos-of-14516046.php.

186 **the company's Emergency Operations Center:** "State Operation Center Situation Status Report," California Office of Emergency Services, November 9, 2018.

186 **This room was white:** Fagone, "Inside PG&E's 'War Room.'"

186 **PG&E was required to submit:** These rules are required for any state utility. "Incident Investigations for Electric and Communication Facilities," California Public Utilities Commission, accessed July 29, 2020, cpuc .ca.gov/General.aspx?id=2090.

186 **an outage had registered on the Caribou-Palermo transmission line:** "PG&E Publicly Releases Supplemental Report on Electric Incidents Near the Camp Fire," Pacific Gas and Electric Company, December 11, 2018, accessed July 27, 2020, pge.com/en/about/newsroom/news details/index.page?title=20181211_pge_p ublicly_releases_supplemental _report_on_electric_incidents_near_the_camp_fire.

CHAPTER 12: THE LONGEST DRIVE

Interviews: Chris and Rachelle Sanders.

197 **Forest Service ranger Edward Pulaski:** Timothy Egan, *The Big Burn* (New York: Houghton Mifflin Harcourt, 2009), p. 167.

198 **Citizens were offered a choice:** Sarah McCaffrey, Robyn Wilson, and Avishek Konar, "Should I Stay or Should I Go Now? Or Should I Wait

and See? Influences on Wildfire Evacuation Decisions," *Risk Analysis* 38, no. 7 (2018), p. 1390.

198 **(An alert during the 2020 bush fire siege):** "Incidents and Warnings," VIC Emergency, accessed June 30, 2020, emergency.vic.gov.au/public/ event%2Fwarning%2F12031.html.

198 **After the 2009 Black Saturday bush fires killed 173 people:** Victorian Bush Fires Royal Commission, *Final Report,* Bernard Teague, Ronald McLeod, and Susan Pascoe, 978-0-9807408-1-3, Victoria: Parliament of Victoria, 2010, royalcommission.vic.gov.au/finaldocuments/summary/PF/ VBRC_Summary_PF.pdf.

198 **fifty-seven people had hidden in bathrooms:** Diana Leonard, "Too Late to Leave: Australia's Advice for Surviving Bush Fire When Surrounded," *Washington Post,* January 9, 2020, washingtonpost.com/weather/2020/01/ 09/too-late-leave-australias-advice-surviving-bush-fire-when-surrounded/.

199 **thousands of people were also assembled at ad hoc shelters:** According to the researcher Thomas Cova, the decision to shelter in place happens when exit roads are blocked, when evacuating is seen as being too risky, or when residents have prepared and made plans to shelter at home. Thomas Cova et al., "Protective Actions in Wildfires: Evacuate or Shelter-in-Place?" *Natural Hazards Review* 10, no. 4 (July 2009), pp. 151–62.

199 **People were polite:** The researcher Enrico Quarantelli has found that panic and chaotic behavior are rare during disasters, explaining that the "social order [does] not break down." As the author Rebecca Solnit explained, in studies of more than two thousand people in nine hundred fires, behavior was "mostly rational, sometimes altruistic, and never about the beast within when the thin veneer of civilization is peeled off. Except in the movies and the popular imagination. And in the media." Rebecca Solnit, *A Paradise Built in Hell* (New York: Penguin, 2009), p. 124.

199 **people panicked most when they were alone:** Ripley, *Unthinkable,* p. 152.

OBSERVATION: SARA'S CALL

Interviews: Cal Fire dispatcher Beth Bowersox; Paradise Police sergeant Steve Bertagna.

204 **Her husband, Marshall:** Michael Cabanatuan, "Sara Magnuson, 75, Lived Alone in Paradise," *San Francisco Chronicle,* December 10, 2018, sfchronicle.com/california-wildfires/article/Sara-Magnuson-75-lived-alone -in-Paradise-13456007.php.

204 **had run a jewelry business:** Jason Green, "Sara Magnuson Lived Alone, Struggled After Husband's Death," *Chico Enterprise-Record,* January 23, 2019, chicoer.com/2019/01/23/camp-fire-profile-sara-magnuson-lived -alone-struggled-after-husbands-death/.

CHAPTER 13: NO ATHEISTS IN FOXHOLES

Interviews: Nurses Tammy Ferguson and Crissy Foster; pediatrician David Russell; Paradise Fire chief David Hawks; Heritage staffers Jill Fassler and Trudy Vaughn; paramedics Mike Castro and Sean Abrams; EMT Shannon Molarius; Butte County Search and Rescue volunteer Joe Grecco; Cal Fire firefighter Tim Moore.

207 **Tammy called her twenty-four-year-old daughter:** *Los Angeles Times* reporter Corina Knoll also documented Tammy's last phone calls in a striking and widely read article published just before Thanksgiving 2018. It's a beautiful piece and well worth the read. Corina Knoll, "As Deadly Flames Approached, a Mother Called Her Daughters to Say Goodbye," *Los Angeles Times,* November 22, 2018, latimes.com/local/california/la -me-the-last-call-20181122-story.html.

209 **"I have never felt such genuine love":** The entire letter recommending Tammy for the Daisy Award can be found at daisyfoundation.org/ daisy-award/honorees/tammy-ferguson.

OBSERVATION: TEN THOUSAND FEET ABOVE THE FIRE

Interviews: California Highway Patrol officers Brent Sallis and Joe Airoso; California Highway Patrol chief Brent Newman.

THE FIRE: PARTICLES OF POISON

Interviews: San Mateo County public health officer Dr. Shruti Dhapodkar; San Francisco public health officer Dr. Jane Gurley; Bay Area Air Quality Management District member Lisa Fasano; Bay Area Air Quality Management District member Dr. Judith Cutino; San Francisco Department of Emergency Management employee Francis Zamora.

225 **More than 3.4 million metric tons of carbon dioxide:** Kurtis Alexander, "Camp Fire's Climate Toll: Greenhouse Gases Equal About a Week of California Auto Emissions," *San Francisco Chronicle,* November 30, 2018, sfchronicle.com/california-wildfires/article/California-wildfires -Staggering-toll-on-forests-13432888.php.

225 **The fine particles measured less than 2.5 microns:** The makeup of smoke varies widely depending on what is burning. Smoke particles are characterized under the umbrella of "particulate matter" and are about sixty times smaller than a thread of human hair. While research is still being done, it's believed that exposure to smoke increases the risk of cancer and other long-term health issues. "Wildfire Smoke: A Guide for Public Health Officials," Environmental Protection Agency, 2001, www3 .epa.gov/ttnamti1/files/ambient/smoke/wildgd.pdf.

226 **cardiac arrest for humans:** Caitlin G. Jones et al., "Out-of-Hospital Cardiac Arrests and Wildfire-Related Particulate Matter During 2015–

2017 California Wildfires," *Journal of the American Heart Association* 9, no. 8 (April 2020), ahajournals.org/doi/epub/10.1161/JAHA.119.014125.

226 **as much as 70 percent more likely to occur on smoky days:** Yohannes Tesfaigzi, Lisa Ann Miller, and Jed Bassein, "Wildfire Smoke Exposure and Human Health: Significant Gaps in Research for a Growing Public Health Issue," *Environmental Toxicology and Pharmacology* 55 (August 2017), researchgate.net/publication/319385514_Wildfire_Smoke _Exposure_and_H uman_Health_Significant_Gaps_in_Research_for_a _Growing_Public_Health_Issue.

CHAPTER 14: PARADISE ABLAZE

Interviews: Bus driver Kevin McKay; teachers Abbie Davis and Mary Ludwig; student Rowan Stovall and her mother, Nicole Alderman; Cal Fire captain Sean Norman and firefighter Calin Moldovan; Paradise Police officer Rob Nichols.

230 **Paradise Memorial Trail:** Mary Nugent, "Paradise Celebrates Its Yellowstone Kelly Heritage Trail," *Chico Enterprise-Record,* August 20, 2018, chicoer.com/2018/08/20/paradise-celebrates-its-yellowstone-kelly-heritage -trail/.

231 **Minimum-security inmates**: Lizzie Johnson, "Fewer Prison Inmates Signing Up to Fight California Wildfires," *San Francisco Chronicle,* September 1, 2017, sfchronicle.com/bayarea/article/Fewer-prison-inmates -signing-up-to-fight-12165598.php.

234 **going into diabetic shock:** Lizzie Johnson, "Survival in Paradise: Remarkable Escapes as Firestorm Swept Through Town," *San Francisco Chronicle,* November 18, 2018, sfchronicle.com/california-wildfires/article/ A-fire-a-newborn-baby-and-a-pact-Tales-of-13402034.php.

235 **California had become the first state:** The updated building codes can be found on the state's website at hcd.ca.gov/building-standards/state -housing-law/wildland-urban-interface.shtml, chapter 7A.

235 **nine out of ten houses:** Paige St. John, Joseph Serna, and Ron-Gong Lin II, "Here's How Paradise Ignored Warnings and Became a Deathtrap," *Los Angeles Times,* December 30, 2018, latimes.com/local/california/la -me-camp-fire-deathtrap-20181230-story.html.

235 **inspected 29,776 properties:** This data was obtained from Cal Fire through a public records request. When asked why the one citation was issued, the records keeper couldn't provide an answer—that information had been deleted at some point over the years.

235 **new nuisance abatement ordinance:** John Messina, who took David Hawks's job to become the fire chief in Paradise, explained that the new ordinance, 508, encompassed many things, including nuisance abatement and California's fire code. It was adopted after the old ordinance, 461,

was rescinded. "It just went away, unbeknownst to a lot of people," Messina said. "The fire department had been enforcing the old ordinance illegally."

OBSERVATION: THE MAN IN THE TRASH TRUCK

Interviews: Dane Ray Cummings and Margaret Newsum.

CHAPTER 15: PROMISE

Interviews: Travis Wright; Suzie Ernest; Mike Ranney; Jamie Niedermeyer; Feather River hospital surgical unit manager Jeff Roach and nurse Tammy Ferguson; Rachelle and Chris Sanders; Jamie, Erin, and Tezzrah Mansanares; Heritage Paradise staffer Jill Fassler; bus driver Kevin McKay; teachers Abbie Davis and Mary Ludwig; student Rowan Stovall and her mother, Nicole Alderman; Ridgeview High principal Mike Lerch; Cal Fire captains Sean Norman and Matt McKenzie, battalion chief Joe Tapia, firefighters Calin Moldovan and Andrew Goose, and engineer Sam Layton.

242 **he crossed paths with a man:** The man was Greg Woodcox. He filmed a video of what remained of his friends—to much controversy. Some fire survivors felt that the world needed to know how bad a blaze could be. Others, including the families of the dead, felt that showing their loved ones' skeletons was disrespectful. Evan Sernoffsky, "He Couldn't Save His Friends. Now Camp Fire Survivor's Video Is Drawing Anger," *San Francisco Chronicle,* November 11, 2018, sfchronicle.com/california-wildfires/article/He-couldn-t-save-his-friends-Now-Camp-Fire-13382 947.php.

251 **two firefighters from Nevada County:** They were engineer Sam Layton and firefighter Andrew Goose, both based in Nevada City.

OBSERVATION: NIGHTFALL

Interviews: Butte County Sheriff Kory Honea and spokeswoman Megan McCann.

254 **gathered reporters for a press conference:** The press conference can also be watched online at the Butte County Sheriff's Office Facebook page, facebook.com/watch/live/?v=509369919541941&ref=watch_perma link.

256 **PG&E dispatched a tweet:** Pacific Gas & Electric Company (@PGE4Me), "PG&E has determined that it will not proceed with plans today for a Public Safety Power Shutoff in portions of 8 Northern CA counties, as weather conditions did not warrant this safety measure. We want to thank our customers for their understanding. Bit.ly/2SVpRtw," November 8, 2018, 3:14 P.M., twitter.com/pge4me/status/10606720009 29267713?lang=en.

CHAPTER 16: UNCONFIRMED DEATHS

Interviews: Butte County Sheriff's Office chaplains Jeremy Carr and Dan Wysong and investigator Tiffany Larson; Sheriff Kory Honea; Alameda County Sheriff's Office sergeant Howard Baron.

263   **Jeremy Carr noticed them:** Parts of this chapter first appeared in one of my earlier articles for the *San Francisco Chronicle.* Lizzie Johnson, "Camp Fire Chaplain's Challenge: Making Sense of So Much Loss," *San Francisco Chronicle,* December 16, 2018, sfchronicle.com/california-wildfires/article/Camp-Fire-chaplain-s-challenge-making-sense-of-13469245.php.

263   **More than five thousand firefighters:** Priyanka Boghani, "Camp Fire: By the Numbers," *Frontline,* October 29, 2019, pbs.org/wgbh/frontline/article/camp-fire-by-the-numbers/.

264   **Air Quality Index registered:** On Friday, November 9, 2018, Oroville's air quality was rated at 535 AQI, particularly troublesome for vulnerable populations including children, the elderly, and those with respiratory illnesses like asthma. Tiffany Jeung, "Animation Shows California Air Quality Is Worst in the World from Wildfires," *Inverse,* November 16, 2018, inverse.com/article/50925-california-wildfire-air-quality-worst-in-world.

264   **the equivalent of smoking eight cigarettes:** Zoë Schlanger, "In Parts of California, Breathing Is Like Smoking Half a Pack of Cigarettes a Day," *Quartz,* November 9, 2018, qz.com/1458615/the-camp-fire-is-making-california-air-quality-as-bad-as-smoking-half-a-pack-of-cigarettes-a-day/.

264   **a race director explained:** The director was Lauri Abrahamsen, who explained that the race wasn't possible because "October and November is the new fire season." Nearly eight thousand runners had registered for the annual race. Alexandra Casey, " 'You Can't Run in the Smoke': Berkeley Half Marathon Canceled," *The Daily Californian,* November 14, 2018, dailycal.org/2018/11/14/you-cant-run-in-the-smoke-berkeley-half-marathon-canceled/.

264   **More than 180 public school districts:** Ricardo Cano, "School Closures from California Wildfires This Week Have Kept More than a Million Kids Home," *Cal Matters,* November 15, 2018, calmatters.org/environment/2018/11/school-closures-california-wildfires-1-million-students/.

265   **a survivor had written:** The account, written by Sylvan Creecy for the University of Southern California's student newspaper, the *Daily Trojan,* was reprinted in *The Washington Post.* Theresa Vargas, " 'Oh God! Help Me!': In California's Deadliest Fire, Survivors Watched Co-Workers Die," *The Washington Post,* December 15, 2017, washingtonpost.com/news/retropolis/wp/2017/12/15/oh-god-help-me-in-californias-deadliest-fire-survivors-watched-co-workers-die/.

265   **a clothing company would eventually sell:** Stephanie Schmieding,

"'Honea Is My Homie' T-Shirts Raise $120,000 for Camp Fire Relief Efforts," Action News Now, December 27, 2018, actionnewsnow.com/content/video/503570322.html.

267 **$280 million "Beleza Emerald":** Mike Murphy, "Was a $280 Million Emerald Destroyed in California Wildfire? PG&E Is Dubious," *Market-Watch,* November 20, 2019, marketwatch.com/story/was-a-280-million -emerald-destroyed-in-california-wildfire-pge-is-dubious-2019-11-19.

267 **Sheriff's deputy Tiffany Larson:** I first spoke with Larson for a story in the *San Francisco Chronicle.* Lizzie Johnson, "Camp Fire: In the Ruins of Paradise, a Grim Search for Signs of Life and Death," *San Francisco Chron-icle,* November 13, 2018, sfchronicle.com/california-wildfires/article/ Camp-Fire-In-the-ruins-of-Paradise-a-grim-13386493.php.

268 **a chaplain whose house had burned down twice before:** The chaplain was Dan Wysong.

269 **a deputy from the Alameda County Sheriff's Office:** The deputy was Sergeant Howard Baron.

270 **John Digby, seventy-eight, was a retired mail carrier:** His son, Roman Digby, told the *San Francisco Chronicle:* "It's a pretty desperate situation. I've been calling the sheriff a couple times a day for updates. I just want to know either way if he's been killed or alive, so I can try to move on and deal with what comes next." Kurtis Alexander, Evan Sernoffsky, and Megan Cassidy, "Camp Fire: Death Toll Rises to 42 as Coroner's Recovery Crews Find More Bodies," *San Francisco Chronicle,* November 12, 2018, sfgate.com/california-wildfires/article/camp-fire-butte -county-death-toll-victims-missing-13384222.php.

OBSERVATION: IDENTIFICATION AND HOT SAUCE

Interviews: Sacramento County coroner Kim Gin; ANDE Corp. CEO Richard Selden; Placer County Sheriff's Office forensic pathologist Greg Reiber; California assemblyman Jim Wood.

273 **the bodies—or more often just fragments of them:** This observation is based on a story that I published in the *San Francisco Chronicle.* Lizzie Johnson, "A Name for 'Doe D': Wedding Rings, Hot Sauce and 'Rapid DNA': Inside the Intricate Effort to Identify Every Victim of the Historic Camp Fire," *San Francisco Chronicle,* May 8, 2019, sfchronicle .com/california-wildfires/article/The-Camp-Fire-coroner-Inside-the-effort -to-13826177.php.

274 **Of the eighteen dentists living in Paradise:** These numbers were provided by Assemblyman Jim Wood, a former family dentist of twenty-seven years and forensic odontologist. Interestingly enough, he's helped with identification efforts in fire zones within his own legislative territory, including the Tubbs Fire of 2017, as well as far-off places like Manhattan after 9/11. Teeth are the hardest substance in the body and much more

resistant to burning, hence their role in identification efforts. "The more you do this over the years, the cumulative effect is it just wears on you," Wood told me. "For me, personally, it is hard to see over the years, the people who have died, having been part of so many different examinations. It really is about knowing I am bringing the beginnings of closure. A process of closure of the loss of a family member." Jim Wood in discussion with the author, April 2019.

CHAPTER 17: MAYOR OF NOWHERE

Interviews: Paradise mayor Jody Jones, town manager Lauren Gill, and assistant town manager Marc Mattox; Town Council members Michael Zuccolillo, Melissa Schuster, Greg Bolin, and Steve Crowder; Paradise Ridge Fire Safe Council chairman Phil John; Paradise Fire chief David Hawks, director Ken Pimlott, and captain Sean Norman; Pulga owner Betsy Ann Cowley; director of California Office of Emergency Services Mark Ghilarducci; Kevin McKay; Rachelle and Chris Sanders; Jamie and Erin Mansanares; Travis and Carole Wright; Irma Enriquez; Paradise High School teacher Virginia Partain; Butte County director of social services Shelby Boston and manager of animal control Ryan Soulsby; Todd Kelman; Mel (Melissa) Contant.

276   **the Town Council convened on a chilly Tuesday evening:** I was present at the November 13, 2018, Town Council meeting in Chico. The meeting can be viewed online at livestream.com/townofparadise/events/ 8446221.

276   **The meeting agenda for November 13:** An earlier version of this chapter first appeared as a story in the *San Francisco Chronicle*. Lizzie Johnson, "Paradise Town Council Meets, Shattered by Fire's Devastation but Doing Its Duty," *San Francisco Chronicle*, November 13, 2018, sfchronicle .com/california-wildfires/article/Paradise-Town-Council-s-agenda-road -projects-13390335.php.

277   **"It's emotionally hard":** Phil John, Paradise Ridge Fire Safe Council chairman, in discussion with the author, November 13, 2018.

277   **having recently won reelection with 4,417 votes:** Amanda Hovik, "Jody Jones Appointed as Mayor," *Paradise Post*, December 14, 2018, paradisepost.com/2018/12/14/jody-jones-appointed-as-mayor/.

281   **Betsy Ann Cowley arrived home:** Betsy Ann Cowley in discussion with the author, November 4, 2019.

282   **President Donald Trump landed:** I was taking my first time off in ten days when Trump visited the Camp Fire burn zone. These scenes are based on pool reports released by the White House as well as reporting by my colleagues at the *San Francisco Chronicle* and interviews with those who were with Trump, including Paradise mayor Jody Jones and Cal Fire director Ken Pimlott. Kurtis Alexander, "Trump Views Paradise Fire Devastation, Promises 'to Take Care of the People,'" *San Francisco Chronicle*,

November 17, 2018, sfchronicle.com/california-wildfires/article/President
-travels-to-Paradise-to-view-wildfire-13401646.php.

283 **Governor Brown trained his eyes on the ground:** Governor Brown
and President Trump had long had a contentious relationship. Earlier
that year, Trump had derisively called him "Moonbeam" in a tweet attacking one of Brown's recent policy decisions. Donald J. Trump
(@realDonaldTrump), "Governor Jerry 'Moonbeam' Brown pardoned 5
criminal illegal aliens whose crimes include (1) Kidnapping and Robbery
(2) Badly beating wife and threatening a crime with intent to terrorize
(3) Dealing drugs. Is this really what the great people of California want?"
@FoxNews, March 31, 2018, 5:53 A.M., twitter.com/realDonaldTrump/
status/980065427375128576. At the time, I also wrote about this tweet
for the *San Francisco Chronicle*. Lizzie Johnson, "Pardoned by Brown,
Blasted by Trump: Bay Area Man Grateful for 2nd Chance," *San Francisco
Chronicle*, April 7, 2018, sfchronicle.com/bayarea/article/Pardoned-by
-Brown-blasted-by-Trump-Bay-Area-man-12810642.php.

284 **Trump had tweeted about the catastrophe:** Donald J. Trump
(@realDonaldTrump), "There is no reason for these massive, deadly and
costly forest fires in California except that forest management is so poor.
Billions of dollars are given each year, with so many lives lost, all because
of gross mismanagement of the forests. Remedy now, or no more Fed
payments!" November, 2018, 12:08 A.M., twitter.com/realDonaldTrump/
status/1061168803218948096.

285 **issued a blistering response:** The rest of the statement reads: "The
president's message attacking California and threatening to withhold aid
to the victims of the cataclysmic fires is ill-informed, ill-timed and demeaning to those who are suffering as well as the men and women on the
front lines. At a time when our every effort should be focused on vanquishing the destructive fires and helping the victims, the president has
chosen instead to issue an uninformed political threat aimed squarely at
the innocent victims of these cataclysmic fires. At this moment, thousands
of our brother and sister firefighters are putting their lives on the line to
protect the lives and property of thousands. Some of them are doing so
even as their own homes [lie] in ruins. In my view, this shameful attack on
California is an attack on all our courageous men and women on the front
lines. The president's assertion that California's forest management policies
are to blame for catastrophic wildfire is dangerously wrong. Wildfires are
sparked and spread not only in forested areas but in populated areas and
open fields fueled by parched vegetation, high winds, low humidity and
geography. Moreover, nearly 60 percent of California forests are under
federal management, and another one-third under private control. It is the
federal government that has chosen to divert resources away from forest
management, not California. Natural disasters are not 'red' or 'blue'—they
destroy regardless of party. Right now, families are in mourning, thousands
have lost homes, and a quarter-million Americans have been forced to

flee. At this desperate time, we would encourage the president to offer support in word and deed, instead of recrimination and blame." Brian K. Rice, "CPF President Brian Rice Responds to President Attack on CA Fire Response," *California Professional Firefighters,* cpf.org/go/cpf/news -and-events/news/cpf-president-brian-rice-responds-to-president-attack -on-ca-fire-response/.

285 **Newsom had added his bit:** Gavin Newsom (@GavinNewsom), "Lives have been lost. Entire towns have been burned to the ground. Cars abandoned on the side of the road. People are being forced to flee their homes. This is not a time for partisanship. This is a time for coordinating relief and response and lifting those in need up," November 10, 2018, 10:27 A.M., twitter.com/GavinNewsom/status/1061324385628221440.

287 **Five days later, it was Thanksgiving:** This part of the chapter also originated in a *San Francisco Chronicle* article. Like many previous holidays, I spent it with fire evacuees who were homeless. Lizzie Johnson, "Butte County Fire Survivors Share Strange, Sad Thanksgiving," *San Francisco Chronicle,* November 23, 2018, sfchronicle.com/california-wildfires/ article/Butte-County-pulls-together-but-Thanksgiving-13415243.php.

287 **the director of the Office of Emergency Services:** Mark Ghilarducci in discussion with the author, February 10, 2020.

291 **One evacuee mused about past holidays:** She was Irma Enriquez.

292 **fourteen hundred people were crammed into twelve temporary shelters:** Nanette Asimov and Kevin Fagan, "Thousands of Camp Fire Evacuees in Shelters, Tents Face Long Wait for Normalcy," *San Francisco Chronicle,* November 17, 2018, sfchronicle.com/california-wildfires/ article/Thousands-of-Camp-Fire-evacuees-in-shelters-13397067.php.

292 **An outbreak of the highly contagious norovirus:** "Outbreak of No- rovirus Illness Among Wildfire Evacuation Shelter Populations—Butte and Glenn Counties, California, November 2018," *Centers for Disease Con- trol and Prevention: Morbidity and Mortality Weekly Report* 69, no. 20 (2020): 613–17.

292 **"That's who you see in the shelters":** Shelby Boston, April 1, 2019.

292 **They housed more than thirty species:** Ryan Soulsby, who shared details about animal operations, is used to handling domestic animal issues in unincorporated areas. In the months before the Camp Fire, Soulsby— the program manager of Butte County's Animal Control Division—dealt with a Concow woman who owned 130 malamutes, one of which turned out to be a wolf. Ryan Soulsby (Butte County Manager of Animal Con- trol) in discussion with the author, October 9, 2019.

293 **the chickens continued laying eggs:** Soulsby added, "Lots and lots of chickens—it's unbelievable how many chickens we [had]." One of his fa- vorite reunification stories, he said involved "the chickens again." Accord-

ing to Soulsby, chickens are hard "because a lot of them are the same color and it's difficult to identify one red rooster versus the other. We had a family—two kids, a wife, and a father—come in looking for their four chickens. We took them to the 'chicken tent,' as we called it, [which is] full of runs with unidentified-owner chickens. This family went in with the two little girls. The chickens literally ran to the little girls. It was very cool to see, even the chickens knew who the family was."

293 **which volunteers drove by the hundreds to UC Davis:** The study was run by veterinarian Todd Kelman, who collected 372 eggs from Butte County. "There were some studies that found over two thousand different chemicals that they detect[ed] from ash in rural wildfires," he said. "What do chickens do all day? They're eating off the ground for hours at a time. It doesn't take much mental math to come up with that logical conjecture—I wonder if they're getting exposed to things. Not a lot of people had studied that." Because of financial constraints, the chemicals in the eggs were never analyzed in a lab. Todd Kelman in discussion with the author, May 8, 2020.

293 **dubbed Wallietown's "mayor":** I spent a day driving around Butte County with Mel Contant, whose son lived in Chico at the time of the Camp Fire. Less than a year later, she had given up her home in the Bay Area to stay in Butte to help out. Her most recent nickname is the FEMA Ninja for how many people she's helped cut through bureaucratic tape. "Have I slept in my truck? Yeah. Have I bounced around not knowing where I'll be the next night? Yeah. Do I know what it's like to be a survivor? Not even close," Contant told me in November 2018, when we first met. "At the end of the day, I have a home to go home to. I can't even enjoy that—how could I possibly go home and sleep in my nice warm bed with a roof over my head and a blanket on me knowing that there are these thousands of people out there that don't have that?" Melissa Contant in discussion with the author, February 11, 2020.

293 **had opened a Disaster Recovery Center:** More information on the center can be found at fema.gov/news-release/20200220/disaster -recovery-center-opens-paradise.

295 **destroyed 14 percent of the county's housing:** Anna M. Phillips, "'Where do people go?': Camp fire makes California's housing crisis worse," *San Francisco Chronicle,* November 24, 2018, latimes.com/local/ lanow/la-me-paradise-housing-shortage-20181123-story.html.

295 **approved permits for no more than thirty-five new homes:** By some measures, only eight new homes actually hit the housing stock in Paradise annually. Michelle Wiley, Sonja Hutson, and Lisa Pickoff-White, "A Year After the Camp Fire, Locals Are Rebuilding Paradise," KQED, November 7, 2019, kqed.org/news/11785247/a-year-after-the-camp -fire-locals-are-rebuilding-paradise.

295   **243 homes had been listed:** Phillips, "'Where do people go?'"

296   **knew she had to stop:** When I talked with Virginia Partain on Thanksgiving, she said: "Most of all, I miss my routine. My drive through Juice
and Java: 'iced tea, an asiago bagel, super-toasted with a little bit of cream
cheese.' My trek to Paradise High, where I had papers to grade, lessons to
write, my agenda to be placed on the board, my students to greet me with
jokes that made me chuckle, and the sound of the bell starting my day,
which gave my life meaning and significance for twenty-five years. Most
of all, I miss my purpose, my reason for life was so alive and full: teaching."
Virginia Partain in discussion with the author, November 22, 2018.

CHAPTER 18: SECONDARY BURNS

Interviews: Butte County sheriff Kory Honea; former Paradise Town Council
member Dona Gavagan Dausey and former town manager Donna Mattheis;
Butte County district attorney Mike Ramsey; PG&E workers Bob Dean, Luke
Bellefeuille, and Tom Dalzell.

297   **Though controversial:** The near collapse of the Oroville Dam spillway
was almost one of the biggest water disasters in state history. At the time,
I covered the emergency for the *San Francisco Chronicle.* Later, journalists
from the Associated Press and *The Sacramento Bee* uncovered how fraught
the decision to evacuate was. Butte County sheriff Kory Honea told Ryan
Sabalow and Dale Kasler of the *Bee:* "It sounded to me that thousands of
lives are at risk, so in a loud and a rather authoritative tone, I yelled for
everybody to be quiet and listen to me. . . . I said, 'It sounds to me that I
need to order the evacuation of the southern part of Butte County. If
there is anybody in this room who thinks that's the wrong move or has a
better idea then you need to speak up now. Tell me now.' The room fell
quiet, and everybody stayed quiet. So I said, 'I've got to do this.'" In notes,
he's described as calling the decision to evacuate pulling "the big red
handle." Ryan Sabalow and Dale Kasler, "Frustration, friction flashed
behind the scenes as Oroville Dam emergency grew," *The Sacramento Bee,*
September 8, 2017, sacbee.com/news/california/water-and-drought/
article172103682.html.

297   **what would amount to 27,784 insurance claims:** Kurtis Alexander,
"Camp Fire: Paradise in Recovery, Six Months Later," *San Francisco Chron-
icle,* May 3, 2019, sfchronicle.com/california-wildfires/article/Camp-Fire
-Paradise-in-recovery-six-months-later-13815152.php.

297   **nearly six thousand homeowners had been denied coverage:** The
data on who is insured came from the California Department of Insurance. Figures for all of the state's counties can be found at insurance.ca
.gov/0400-news/0100-pressreleases/2020/upload/nr104Charts-New
RenewedNon-RenewedData-2015-2019-101920.pdf.

297   **customers in zip codes affected by past wildfires:** Gireesh Shrimali,
"In California, more than 340,000 lose wildfire insurance: Residents are

left with little to no options in the state's fire-prone areas," *High Country News,* October 22, 2019, hcn.org/articles/wildfire-in-california-more -than-340000-lose-wildfire-insurance.

298 **the California legislature would propose a law:** Data released in August 2019 by the California Department of Insurance showed that insurance is becoming harder to find for communities across the state, with six counties from the Sierra to San Diego seeing a greater than 10 percent increase in nonrenewals in 2018 alone. This data does not account for the full impact of insurance companies' nonrenewal response to the Camp Fire and Woolsey and Hill fires—catastrophic wildfires that killed 89 people, destroyed 13,000 homes and businesses, and cost more than $11.4 billion in damages—in addition to all fires in 2019 and 2020. Assembly Bill 2367, called "Renew California," would require insurance companies to write or renew policies for existing homes in communities that meet a new statewide standard for fire hardening.

298 **twelve had been participants:** Shelby Boston, April 1, 2019.

298 **(Wrongful death suits against PG&E):** Camille von Kaenel, a reporter at *The Chico Enterprise-Record,* was the first to write in depth about what it means to die from a wildfire but not be tallied among its fatalities. She writes: "People who died as an indirect result of the disaster aren't memorialized with crosses on the Skyway. They don't get profiles in the newspaper. They aren't counted by any local, state or national agency. That has consequences: Most often, they had preexisting vulnerabilities and passed away because their care was interrupted. But the expanded death toll isn't being factored into future emergency preparedness or health system capacity planning." Camille von Kaenel, "Families Mourn Indirect, 'Forgotten' Deaths from Camp Fire: At Least 50 Indirect Deaths Appear in Vetted Claims Against PG&E," *Chico Enterprise-Record,* February 11, 2020, chicoer.com/2020/02/11/families-mourn-indirect-forgotten-deaths-from -camp-fire/.

298 **the company submitted two electric incident reports:** The two reports can be found at s1.q4cdn.com/880135780/files/doc_downloads/ 2018/wildfire/12/12-11-18.pdf.

299 **the only structure on his 160-acre homestead:** Ruby J. Swartzlow, "Indians in Paradise," *Tales of the Paradise Ridge,* vol. 1, no. 1, June 1960.

300 **"We just kept growing and growing":** Dona Gavagan Dausey in discussion with the author, August 1, 2019.

300 **"The future of Paradise":** The California Digital Newspaper Collection is an excellent source of past articles. Another edition of the *Chico Record* from April 1909 boasts: "So many transfers in Paradise real-estate have been made and so many contracts of sale entered into the past ten months that it would be impossible to give a complete list of all those who have seen the great opportunities at Paradise and have seized the present moment to secure a little land before the prices rise any higher. . . . As one

buyer expressed it, 'Paradise is bound to grow. It simply cannot help it.' After all, the chief reason for the development of the Ridge is that it is inevitable." "Paradise: The Town with a Future," *Chico Record,* July 3, 1908.

301 **Newsom would later tell reporters:** Kathleen Ronayne, "California Governor Won't Block Building in High-Fire Areas," Associated Press, April 15, 2019, apnews.com/b17b5c9200a64466b49f3f605f9202fe.

302 **to speak during public comment:** A video recording of the meeting on November 29, 2018, can be found at adminmonitor.com/ca/cpuc/voting_meeting/20181129/.

302 **wasn't intimidated:** Mike Ramsey in discussion with the author, November 7, 2019.

304 **PG&E's tainted track record:** My colleague J. D. Morris and I first worked on this story at the *San Francisco Chronicle.* Lizzie Johnson and J. D. Morris, "At PG&E, a Workforce on Edge—and Under Attack—as Fire Season Arrives," *San Francisco Chronicle,* June 9, 2019, sfchronicle.com/california-wildfires/article/At-PG-E-a-workforce-on-edge-and-under-attack-13962723.php.

CHAPTER 19: REBIRTH

Interviews: FEMA administrator Bob Fenton; Paradise Irrigation District manager Kevin Phillips; Butte County district attorney Mike Ramsey; Concow residents Peggy and Pete Moak; Olivia Carmin of the Heffern family; Cal Fire captain Stacer Harshorn; former Paradise Town Council member Steve "Woody" Culleton; Paradise schools superintendent Michelle John and student Faith Brown; Butte County Sheriff's Office investigator Tiffany Larson; Paradise Police sergeant Steve Bertagna.

309 **"The amount of metals in that ash":** Bob Fenton in discussion with the author, October 15, 2019.

310 **the $2 billion clean-up:** Kurtis Alexander, "Reclaiming Paradise: Six Months After the Camp Fire, a Devastated Community Hopes to Rebuild—If It Can," *San Francisco Chronicle,* May 3, 2019, projects.sfchronicle.com/2019/rebuilding-paradise/.

310 **had been contaminated with cancer-causing benzene:** After the Tubbs Fire of 2017, the Santa Rosa community of Fountaingrove faced a similar issue. Five miles of water pipes, which served 350 homes, were destroyed. As sections of the polyvinyl chloride pipe melted during the blaze, carcinogens leached into the system. And when the water pressure dropped, ash and chemicals from burnt-out homes were sucked back into the main lines, permanently contaminating them. The polluted zone covered 184 acres there. "To be blunt, a lot of what we are dealing with, no city has ever had to deal with before," Chris Rogers, Santa Rosa's vice mayor, told me about the contaminated system. "Even during our re-

search and talking with experts, there is not a large body of evidence or plans that have been developed to address something of this nature. We are building the plane as we fly it." As wildfires burn hotter, more communities are likely to face similar issues. Lizzie Johnson, "Another Gut Punch for Santa Rosa: Fire Destroys Neighborhood Water System," *San Francisco Chronicle,* April 15, 2018, sfchronicle.com/bayarea/article/Another-gut -punch-for-Santa-Rosa-Fire-destroys-12834914.php.

310 **struggled to absorb the overflow:** Chico was impacted in other ways, too. Enloe Hospital saw a 40 percent uptick in baby deliveries. Emergency room visits went from 180 per day to 260. After the fire, car crashes increased by 23 percent. Tony Bizjak, "Chico's Post Camp Fire World: Car Crashes, Frayed Nerves and Bare-Knuckle Politics," *The Sacramento Bee,* May 23, 2019, sacbee.com/news/california/article230015334.html.

310 **organized a recall effort:** Residents were unhappy with Mayor Randall Stone and councilman Karl Ory. The effort ended in November 2019.

311 **settled in places like Hawaii:** Chico State University did a fascinating project tracking where, and why, Camp Fire survivors have ended up in the places that they did. Researchers found that half of the over-65 demographic had left—more than in any other age group. Meanwhile, people with children tended to end up in Chico or within thirty miles of their old home. The study can be found at today.csuchico.edu/mapping-a -displaced-population/.

311 **of the roughly three thousand houses:** About 75 percent of the homes destroyed by wildfires in California were located in the wildland-urban interface. Christopher Flavelle, "Why Is California Rebuilding in Fire Country? Because You're Paying for It," *Bloomberg,* March 1, 2018, bloomberg.com/news/features/2018-03-01/why-is-california-rebuilding -in-fire-country-because-you-re-paying-for-it.

311 **legislators had met at the state capitol:** A video recording of the meeting on May 8, 2019, can be found at assembly.ca.gov/media/assembly -joint-hearing-local-government-governmental-organization-housing -community-development-20190508/video.

312 **the Paradise Town Council held a special meeting:** A video recording of the meeting can be found on the town's Facebook page: facebook .com/watch/live/?v=289528595325881&ref=watch_permalink.

318 **dozens of people had sought refuge in a lake:** Alexander, "Trapped by Camp Fire."

319 **Faith Brown, seventeen, could hardly wait:** I originally wrote about Faith for the *San Francisco Chronicle.* Lizzie Johnson, "She Couldn't Wait to Leave Paradise. Now Faith, 17, Aches for What She Lost," *San Francisco Chronicle,* December 30, 2018, sfchronicle.com/california-wildfires/ article/She-couldn-t-wait-to-leave-Paradise-Now-Faith-13497737.php.

322 **(her husband, Phil John):** This book is dedicated to Phil John because

he was the bedrock of Paradise, its unofficial ambassador. His death came as a shock to the community. "It is just the highest of highs and lowest of lows," former fire chief Jim Broshears told me. "There was this beautiful graduation in this spectacular setting that no one initially thought would happen. A few days later, Phil has a heart attack. I'm still trying to wrap my brain around it." Lizzie Johnson, "'He Tried to Make Paradise All Its Name Implied': Long After the Camp Fire, Resident's Death Opens Wounds," *San Francisco Chronicle,* June 21, 2019, sfchronicle.com/california-wildfires/article/He-tried-to-make-Paradise-all-its-name-14026656.php.

322 **Principal Loren Lighthall, forty-six, paced:** Lizzie Johnson, "The Paradise Principal: Loren Lighthall Guided Paradise High Through the Devastation of the Camp Fire. But in the Wake of the Disaster, the Beloved School Leader and His Family Had to Confront What Was Lost, and Where They Stood," *San Francisco Chronicle,* June 28, 2019, sfchronicle.com/california-wildfires/article/The-Paradise-principal-In-the-wake-of-disaster-14056148.php.

324 **after reading about the Camp Fire in his local newspaper:** That article appeared in the *Los Angeles Times.* Hailey Branson-Potts and Louis Sahagun, "The Epic Undertaking to Start School Again Amid the Devastation of the Camp Fire," *Los Angeles Times,* December 2, 2018, latimes.com/local/lanow/la-me-ln-camp-fire-schools-restarting-20181203-story.html.

OBSERVATION: ANNIVERSARY

325 **Only 2,034 residents had returned:** Lizzie Johnson, "The Sky Was Blue. The Day Was Perfect. But in Paradise on Friday, Nothing Felt Right," *San Francisco Chronicle,* November 8, 2019, sfchronicle.com/california-wildfires/article/The-sky-was-blue-The-day-was-perfect-But-in-14821551.php.

327 **all the things they had lost:** The list of beloved things that were lost came from a Google survey of twenty-five residents, along with interviews with sources and conversations I had with residents while living in Paradise, waiting in line at the Holiday Market, eating at Sophia's Thai, getting coffee at Starbucks.

CHAPTER 20: RECKONING

Interviews: Butte County district attorney Mike Ramsey; Arielle Funk and Jessee Ernest.

328 **to be only indirectly related:** Butte County District Attorney's Office, *Summary of the Camp Fire Investigation.*

328 **the petroleum company BP had made news:** J. D. Morris and Lizzie Johnson, "PG&E to Plead Guilty to Involuntary Manslaughter, Fire-

Starting in Camp Fire," *San Francisco Chronicle,* March 23, 2020, sfchronicle
.com/california-wildfires/article/PG-E-pleads-guilty-to-arson-manslaughter
-in-Camp-15150340.php.

329 **if PG&E had been an actual person:** Lizzie Johnson and J. D. Morris,
"PG&E Legal Saga Reaches Climax with Camp Fire Sentencing, Bankruptcy Approval," *San Francisco Chronicle,* June 18, 2020, sfchronicle.com/
business/article/PG-E-legal-saga-reaches-climax-with-Camp-Fire-1535
0339.php.

329 **Ramsey could only seek $10,000:** J. D. Morris and Lizzie Johnson,
"'Guilty, Your Honor': PG&E Enters Pleas for 85 Camp Fire Felonies,"
San Francisco Chronicle, June 16, 2020, sfchronicle.com/california-wild
fires/article/PG-E-pleads-guilty-to-84-counts-of-involuntary-15344269
.php.

330 **A 696-page report released by the California Public Utilities
Commission:** J. D. Morris, "Camp Fire Failure Part of PG&E's 'Pattern'
of Poor Maintenance, Regulators Say," *San Francisco Chronicle,* December
3, 2019, sfchronicle.com/california-wildfires/article/Regulators-PG-E
-could-have-prevented-Camp-Fire-14877131.php.

331 **listed her luxury Marin County home for sale:** Williams, the CEO,
received an 8.1 percent raise in 2018, according to corporate filings, despite the Camp Fire. As part of her compensation while working at PG&E,
she was given a car and driver, health club benefits, financial services, and
a $51,000 security system for her home. PG&E has said these offerings
were tied to "corporate performance." Claudine Zap, "Former Pacific Gas
and Electric CEO Geisha Williams Selling $4.7M Tiburon Mansion,"
Realtor.com, March 19, 2020, realtor.com/news/celebrity-real-estate/
former-pge-ceo-geisha-williams-selling-tiburon-mansion/.

331 **the ninety-two-page report:** The report can be found on the Butte
County District Attorney's website: buttecounty.net/districtattorney/
CampFire.

331 **the Caribou-Palermo Line had been erected:** J. D. Morris and Lizzie
Johnson, "'Guilty, Your Honor.'"

332 **less than one eighth of an inch:** Butte County District Attorney's Office, *Summary of the Camp Fire Investigation,* p. 22.

332 **bought for 22 cents in 1919:** Mike Ramsey in discussion with the author, November 7, 2019.

332 **tasked with climbing 5 percent of the towers annually:** Butte
County District Attorney's Office, *Summary of the Camp Fire Investigation,*
p. 23.

332 **"This inability to determine who made decisions":** Butte County
District Attorney's Office, *Summary of the Camp Fire Investigation,* p. 43.

333 **PG&E's $57.65 billion bankruptcy restructuring plan:** More infor-
mation can be found on PG&E's website: pge.com/en/about/newsroom/

newsdetails/index.page?title=20200620_pge_achieves_bankruptcy_court
_confirmation_of_its_plan_of_reorganization.

333   **a move to nearby Oakland:** J. D. Morris and Roland Li, "PG&E Will
Relocate to Oakland After More than 100 Years in San Francisco," *San
Francisco Chronicle,* June 8, 2020, sfchronicle.com/business/article/PG-E
-will-relocate-to-Oakland-after-more-than-100-15325849.php#:~:text=
In%2DDepth-,PG%26E%20will%20relocate%20to%20Oakland%20after
%20more%20than%20100%20years,8%2C%202020%209%3A17%20p.m.
&text=The%20company%20announced%20Monday%20that,by%20Lake
%20Merritt%20in%202022.

IN MEMORY OF THOSE WHO DIED

Joyce Acheson, 78
Herbert Alderman, 79
Teresa Ammons, 82
Rafaela Andrade, 84
Carol Arrington, 88
Julian Binstock, 88
David Bradburd, 70
Cheryl Brown, 75
Larry Brown, 72
Richard Clayton Brown, 74
Andrew Burt, 36
Joanne Caddy, 75
Barbara Carlson, 71
Vincent Mario Carota, 65
Dennis Clark, 49
Evelyn Cline, 83
John Arthur Digby, 78
Gordon Dise, 66
Paula Dodge, 70
Randall Dodge, 67
Andrew Downer, 54
Robert Duvall, 76
Paul Ernest, 72
Rose Farrell, 99
Jesus Fernandez, 48

Jean Forsman, 83
Ernest Foss, 63
Elizabeth Gaal, 80
Sally Gamboa, 69
James Garner, 63
Richard Jay Garrett, 58
Bill Godbout, 79
Shirley Haley, 67
Dennis Hanko, 56
Anna (Toni) Hastings, 67
Jennifer Hayes, 53
Christina Heffern, 40
Ishka Heffern, 20
Matilde Heffern, 68
Dorothy Herrera, 93
Lou Herrera, 86
Evva Holt, 85
TK Huff, 71
Gary Hunter, 67
James Kinner, 84
Warren Lessard, 68
Dorothy Lee Mack, 88
Sara Magnuson, 75
Joanne Malarkey, 90
John Malarkey, 89

Chris Maltby, 69

David Marbury, 66

Deborah Morningstar, 66

Helen Pace, 84

Joy Porter, 72

Beverly Powers, 64

Robert Quinn, 74

Joseph Rabetoy, 39

Forrest Rea, 89

Venice Regan, 95

Ethel Riggs, 96

Lolene Rios, 56

Jerry Rodrigues, 73

Christopher Salazar, 72

Phyllis Salazar, 76

Sheila Santos, 64

Ronald Schenk, 75

Berniece Schmidt, 93

John Sedwick, 82

Don Shores, 70

Kathy Shores, 65

Judith Sipher, 68

Larry Smith, 80

Russell Stewart, 63

Victoria Taft, 67

Shirlee Teays, 90

Joan Tracy, 80

Ellen Walker, 72

Donna Ware, 86

Isabel Webb, 68

Marie Wehe, 78

Kimber Wehr, 53

Carl Wiley, 77

David Young, 69

Unknown

INDEX

ABOUT THE AUTHOR

LIZZIE JOHNSON is a staff writer at *The Washington Post*. Previously, she worked at the *San Francisco Chronicle,* where she reported on fifteen of the deadliest, largest, and most destructive blazes in modern California history and covered over thirty communities impacted by wildfires. Originally from Nebraska, she lived part-time in Paradise while reporting this book and currently lives in Washington, D.C.

lizziejohnson.net
Facebook.com/lizziejohnsonmedia
Twitter: @LizzieJohnsonnn
Instagram: @Lizziee.J